Lecture Notes in Computer Science 8819

Commenced Publication in 1973
Founding and Former Series Editors:
Gerhard Goos, Juris Hartmanis, and Jan van Leeuwen

T0212600

Hendrik Blockeel   Matthijs van Leeuwen
Veronica Vinciotti (Eds.)

Advances in Intelligent Data Analysis XIII

13th International Symposium, IDA 2014
Leuven, Belgium, October 30 – November 1, 2014
Proceedings

 Springer

Volume Editors

Hendrik Blockeel
KU Leuven, Department of Computer Science
Heverlee, Belgium
E-mail: hendrik.blockeel@cs.kuleuven.be

Matthijs van Leeuwen
KU Leuven, Department of Computer Science
Heverlee, Belgium
E-mail: matthijs.vanleeuwen@cs.kuleuven.be

Veronica Vinciotti
Brunel University, Uxbridge, UK
E-mail: veronika.vinciotti@brunel.ac.uk

ISSN 0302-9743 e-ISSN 1611-3349
ISBN 978-3-319-12570-1 e-ISBN 978-3-319-12571-8
DOI 10.1007/978-3-319-12571-8
Springer Cham Heidelberg New York Dordrecht London

Library of Congress Control Number: 2014951873

LNCS Sublibrary: SL 3 – Information Systems and Application,
incl. Internet/Web and HCI

Typesetting: Camera-ready by author, data conversion by Scientific Publishing Services, Chennai, India

Printed on acid-free paper

Springer is part of Springer Science+Business Media (www.springer.com)

Preface

We are proud to present the proceedings of IDA 2014: the 13th International Symposium on Intelligent Data Analysis, which was held from October 30 to November 1, 2014, in Leuven, Belgium.

The series started in 1995 and was held biennially until 2009. In 2010, the symposium re-focused to support papers that go beyond established technology and offer genuinely novel and game-changing ideas, while not always being as fully realized as papers submitted to other conferences. IDA 2014 continued this approach and sought first-look papers that might elsewhere be considered preliminary, but contain potentially high-impact research.

The IDA symposium is open to all kinds of modeling and analysis methods, irrespective of discipline. It is an interdisciplinary meeting that seeks abstractions that cut across domains. IDA solicits papers on all aspects of intelligent data analysis, including papers on intelligent support for modeling and analyzing data from complex, dynamical systems. Intelligent support for data analysis goes beyond the usual algorithmic offerings in the literature.

Papers about established technology were only accepted if the technology was embedded in intelligent data analysis systems, or was applied in novel ways to analyzing and/or modeling complex systems. The conventional reviewing process, which tends to favor incremental advances on established work, can discourage the kinds of papers that IDA 2014 has published.

The reviewing process addressed this issue explicitly: referees evaluated papers against the stated goals of the symposium, and an informed, thoughtful, positive review written by a program chair advisor could outweigh other, negative reviews and result toward acceptance of the paper. Indeed, it was noted that this had notable impact on some of the papers included in the program. In addition, IDA 2014 introduced the "First Look Track", which allowed researchers to present their ground breaking research at the symposium without publishing it in the proceedings. This resulted in the presentation of ideas and visions that were not yet mature enough for publication.

We were pleased to have a very strong program. We received 76 submissions in total, from 215 different authors from 30 different countries on six continents. In all, 70 papers were submitted to the regular proceedings track, of which 33 were accepted for inclusion in this volume. Six papers were submitted to the First Look Track, of which three were accepted for presentation at the symposium. The IDA Frontier Prize was awarded to the most visionary contribution. As in previous years, we included a poster and video track for PhD students to promote their work. The best 2-minute video, as decided by the participants of the symposium, was awarded the Video Prize.

We were honored to have three distinguished invited speakers at IDA 2014:

- Arnoldo Frigessi from the University of Oslo, Norway, talked about the renewed interest in the analysis of ranked data due to novel applications in the era of big data. In particular, he presented a Bayesian approach to rank estimation.
- Jan Van den Bussche from Hasselt University, Belgium, presented DNAQL, a language for DNA programming, which was developed to better understand, from a theoretical perspective, the database aspects of DNA computing.
- Chris Lintott from Oxford University, UK, talked about citizen science, the involvement of many volunteers in the scientific process, and reflected on the lessons learned when data analysis needs millions of people.

The conference was held at the Faculty Club in Leuven, situated in the historical Grand Béguinage, a UNESCO World Heritage Site. We wish to express our gratitude to all authors of submitted papers for their intellectual contributions; to the Program Committee members and the additional reviewers for their effort in reviewing and commenting on the submitted papers; to the members of the IDA Steering Committee for their ongoing guidance and support; and to the Program Committee advisors for their active involvement. Special thanks go to the poster and video chair, Elisa Fromont; the local chair, Tias Guns; the publicity chair, Márcia Oliveira; the sponsorship chair, David Martens; the Frontier Prize chairs, Arno Siebes and Allan Tucker; and the webmaster, Vladimir Dzyuba. We gratefully acknowledge those who were involved in the local organization of the symposium: Behrouz Babaki, Thanh Le Van, Ashraf Masood Kibriya, Benjamin Negrevergne, and Vincent Nys. Finally, we are grateful to our sponsors and supporters: KNIME, which funded the IDA Frontier Prize for the most visionary contribution presenting a novel and surprising approach to data analysis; SAS, which funded the IDA Video Prize for the best video presented in the PhD poster and video track; the Research Foundation – Flanders (FWO); the *Artificial Intelligence* journal; the City of Leuven; and Springer.

August 2014   Hendrik Blockeel
                                                                   Matthijs van Leeuwen
                                                                   Veronica Vinciotti

Organization

General Chair

Hendrik Blockeel   KU Leuven, Belgium

Program Chairs

Matthijs van Leeuwen   KU Leuven, Belgium
Veronica Vinciotti   Brunel University, UK

Poster and Video Chair

Elisa Fromont   Jean Monnet University, France

Local Chair

Tias Guns   KU Leuven, Belgium

Publicity Chair

Márcia Oliveira   University of Porto, Portugal

Sponsorship Chair

David Martens   Universiteit Antwerpen, Belgium

Frontier Prize Chairs

Arno Siebes   Universiteit Utrecht, The Netherlands
Allan Tucker   Brunel University, UK

Advisory Chairs

Jaakko Hollmén   Aalto University, Finland
Frank Höppner   Ostfalia University of Applied Sciences,
                 Germany

Webmaster

Vladimir Dzyuba   KU Leuven, Belgium

Local Organizing Committee

Behrouz Babaki	KU Leuven, Belgium
Thanh Le Van	KU Leuven, Belgium
Ashraf Masood Kibriya	KU Leuven, Belgium
Benjamin Negrevergne	KU Leuven, Belgium
Vincent Nys	KU Leuven, Belgium

Program Committee Advisors

Niall Adams	Imperial College London, UK
Michael Berthold	University of Konstanz, Germany
Hendrik Blockeel	KU Leuven, Belgium
Liz Bradley	University of Colorado, USA
João Gama	University of Porto, Portugal
Jaakko Hollmén	Aalto University School of Science, Finland
Frank Klawonn	Ostfalia University of Applied Sciences, Germany
Joost Kok	Leiden University, The Netherlands
Xiaohui Liu	Brunel University, UK
Arno Siebes	Universiteit Utrecht, The Netherlands
Hannu Toivonen	University of Helsinki, Finland
Allan Tucker	Brunel University, UK
David Weston	Imperial College, UK

Program Committee

Fabrizio Angiulli	University of Calabria, Italy
Alexandre Aussem	Université Lyon 1, France
Barbro Back	Åbo Akademi University, Finland
Jose Balcazar	Universitat Politècnica de Catalunya, Spain
Christian Borgelt	European Centre for Soft Computing, Spain
Henrik Bostrom	Stockholm University, Sweden
Jean-Francois Boulicaut	INSA Lyon, France
Andre Carvalho	University of São Paulo, Brazil
Bruno Cremilleux	Université de Caen, France
Bernard De Baets	Ghent University, Belgium
Tijl De Bie	University of Bristol, UK
José Del Campo-Ávila	University of Málaga, Spain
Wouter Duivesteijn	Technische Universität Dortmund, Germany
Saso Dzeroski	Jožef Stefan Institute, Slovenia
Fazel Famili	IIT - National Research Council Canada, Canada
Ad Feelders	Universiteit Utrecht, The Netherlands

Additional Reviewers

Aksehirli, Emin
Alvares Cherman, Everton
Batista, Gustavo
Braune, Christian
Béchet, Nicolas
Chakeri, Alireza
Doell, Christoph
Faria de Carvalho, Alexandre
Fassetti, Fabio
Gossen, Tatiana
Held, Pascal
Ienco, Dino
Jeudy, Baptiste

Kotzyba, Michael
Low, Thomas
Maslov, Alexandr
Palopoli, Luigi
Rönnqvist, Samuel
Souza, Vinícius
Spolaôr, Newton
Tanevski, Jovan
Trajanov, Aneta
Van Brussel, Thomas
Zhou, Mu
Zou, Lei

Invited Talks

Bayesian Inference for Ranks

Arnoldo Frigessi

University of Oslo

Abstract. Analysis of rank data has received renewed interest, due to novel applications in the era of big data. Examples include the ranking and comparison of products, services, films or books, by thousands of volunteer users over the internet. The basic purpose is to estimate the latent ranks of n items on the basis of N samples, each of which is the ranking of the same n items by an independent assessor. If the N assessors do not represent a homogeneous population sharing the same latent ranking of the n items, one also needs to partition the N assessors in homogeneous clusters, each sharing their own unknown latent rank. In many situations N is very large. In some cases the number of items n is small, like when a panel of voters rank a few alternative political candidates in an election. In other cases, n is so large that assessors are not able to produce a full ranking of all items at all; instead one resorts on either a partial ranking (of the top 5 items, say) or on a series of pairwise comparisons. In these cases (and many other with a similar flavour) the data are highly incomplete. There are also situations, where observations are quantitative but each assessor uses different scales, and in these cases passing to ranks is clearly a possibility.

Expressing preferences is in general a very useful way of collecting information, and the analysis of such data is receiving increasing attention. There is a very large literature on statistical and machine learning methods to estimate such latent ranks, which goes under the title preference learning, see for example the excellent book [1]. Much inference for rank data is frequentist in nature, based on exponential models in a distance between ranks [2]. Such a classical family of distance based models are Mallows models [3, 4]. Here one starts with a distance between the observed and the latent ranks, and assumes that the observed ranks of the n items are exponentially distributed in such a distance. The normalising constant of such a distribution is a sum over all permutations of n items, and therefore prohibitive in general. For some special (and useful) distances, including Kendall's correlation, this partition function can be computed analytically, but for other distances (also very useful ones!) it cannot. The footrule distance [3], which is the l_1 norm of the difference between the observed and latent ranks, and Spearman's distance (l_2 norm) are not tractable.

We take a Bayesian approach, where the latent ranks are assumed to be random and have a prior distribution over the set of all permutations. We use a uniform distribution in our examples. There is a further parameter in Mallows' distribution, say α, which acts as inverse variance in the exponential model. On this parameter we assume an exponential prior, and argue for a certain form of the hyperparameter. We develop an MCMC algorithm for sampling the latent ranks, α, and other design parameters, from the posterior distribution. Here we need to specify a proposal distribution which allows moving easily in the space of permutations. For the partition function, we used a simple importance sampling scheme, which appeared to be satisfactory for n in the order of the tens, while other approaches are needed for larger n. The algorithm is then tested on simulated data, on some simple preference experiments that we collected and on benchmark data sets. We also show an application on the ranking of football teams, where each game in the season is a comparison. Finally we discuss limitations of our approach, in particular how we expect the scaling in n to behave. This is joint work with Øystein Sørensen, Valeria Vitelli and Elja Arjas [5].

References

1. Fürnkranz, J., Hüllermeier, E.: Preference learning. Springer, US (2010)
2. Marden, J.I.: Analyzing and modeling rank data. CRC Press (1996)
3. Mallows, C.: Non-null ranking models. Biometrika 44(1/2), 114–130 (1957)
4. Diaconis, P.: Group representations in probability and statistics. Lecture Notes-Monograph Series, p. i-192 (1988)
5. Sørensen, Ø., Vitelli, V., Frigessi, A., Arjas Bayesian, E.: inference from rank data. arXiv preprint, 1405.7945 (2014)

The DNA Query Language DNAQL

Jan Van den Bussche

Hasselt University & Transnational University of Limburg

Abstract. This invited talk presents an overview of our work on databases in DNA performed over the past four years, joint with my student
Joris Gillis and postdoc Robert Brijder [1–5]. Our goal is to better understand, at a theoretical level, the database aspects of DNA computing. The talk will be self-contained and will begin with an introduction
to DNA computing. We then introduce a graph-based data model of so-
called sticker DNA complexes, suitable for the representation and manipulation of structured data in DNA. We also define DNAQL, a restricted
programming language over sticker DNA complexes. DNAQL stands to
general DNA computing as the standard relational algebra for relational
databases stands to general-purpose conventional computing. We show
how DNA program can be statically typechecked. Thus, nonterminating
reactions, as well as other things that could go wrong during DNA manipulation, can be avoided. We also investigate the expressive power of
DNAQL and show how it compares to the relational algebra.

References

1. Gillis, J.J.M., Van den Bussche, J.: A formal model for databases in DNA. In:
   Horimoto, K., Nakatsui, M., Popov, N. (eds.) ANB 2010. LNCS, vol. 6479, pp.
   18–37. Springer, Heidelberg (2012)
2. Brijder, R., Gillis, J.J.M., Van den Bussche, J.: A comparison of graph-theoretic
   DNA hybridization models. Theoretical Computer Science 429, 46–53 (2012)
3. Brijder, R., Gillis, J.J.M., Van den Bussche, J.: A type system for DNAQL. In: Ste-
   fanovic, D., Turberfield, A. (eds.) DNA 2012. LNCS, vol. 7433, pp. 12–24. Springer,
   Heidelberg (2012)
4. Brijder, R., Gillis, J.J.M., Van den Bussche, J.: Graph-theoretic formalization of
   hybridization in DNA sticker complexes. Natural Computing 12(2), 223–234 (2013)
5. Brijder, R., Gillis, J.J.M., Van den Bussche, J.: The DNA query language DNAQL.
   In: Proceedings 16th International Conference on Database Theory. ACM Press
   (2013)

The DNA Query Language DNAQL

Jan Van den Bussche

Hasselt University & Transnational University of Limburg

Abstract. This invited talk presents an overview of our work on query languages for DNA computing.

References

1.
2.
3.
4.

Table of Contents

Selected Contributions

Malware Phylogenetics Based on the Multiview Graphical Lasso 1
  Blake Anderson, Terran Lane, and Curtis Hash

Modeling Stationary Data by a Class of Generalized Ornstein-Uhlenbeck
Processes: The Gaussian Case 13
  Argimiro Arratia, Alejandra Cabaña, and Enrique M. Cabaña

An Approach to Controlling the Runtime for Search Based
Modularisation of Sequential Source Code Check-ins................. 25
  Mahir Arzoky, Stephen Swift, Steve Counsell, and James Cain

Simple Pattern Spectrum Estimation for Fast Pattern Filtering with
CoCoNAD 37
  Christian Borgelt and David Picado-Muiño

From Sensor Readings to Predictions: On the Process of Developing
Practical Soft Sensors 49
  *Marcin Budka, Mark Eastwood, Bogdan Gabrys, Petr Kadlec,
  Manuel Martin Salvador, Stephanie Schwan, Athanasios Tsakonas,
  and Indrė Žliobaitė*

Comparing Pre-defined Software Engineering Metrics with Free-Text
for the Prediction of Code 'Ripples' 61
  *Steve Counsell, Allan Tucker, Stephen Swift, Guy Fitzgerald,
  and Jason Peters*

ApiNATOMY: Towards Multiscale Views of Human Anatomy 72
  *Bernard de Bono, Pierre Grenon, Michiel Helvensteijn, Joost Kok,
  and Natallia Kokash*

Granularity of Co-evolution Patterns in Dynamic Attributed Graphs ... 84
  *Élise Desmier, Marc Plantevit, Céline Robardet,
  and Jean-François Boulicaut*

Multi-user Diverse Recommendations through Greedy Vertex-Angle
Maximization 96
  Pedro Dias and João Magalhães

ERMiner: Sequential Rule Mining Using Equivalence Classes 108
  *Philippe Fournier-Viger, Ted Gueniche, Souleymane Zida,
  and Vincent S. Tseng*

Mining Longitudinal Epidemiological Data to Understand a Reversible
Disorder   120
    Tommy Hielscher, Myra Spiliopoulou, Henry Völzke,
    and Jens-Peter Kühn

The BioKET Biodiversity Data Warehouse: Data and Knowledge
Integration and Extraction   131
    Somsack Inthasone, Nicolas Pasquier, Andrea G.B. Tettamanzi,
    and Célia da Costa Pereira

Using Time-Sensitive Rooted PageRank to Detect Hierarchical Social
Relationships   143
    Mohammad Jaber, Panagiotis Papapetrou, Sven Helmer,
    and Peter T. Wood

Modeling Daily Profiles of Solar Global Radiation Using Statistical and
Data Mining Techniques   155
    Pedro F. Jiménez-Pérez and Llanos Mora-López

Identification of Bilingual Segments for Translation Generation   167
    Kavitha Karimbi Mahesh, Luís Gomes, and José Gabriel P. Lopes

Model-Based Time Series Classification   179
    Alexios Kotsifakos and Panagiotis Papapetrou

Fast Simultaneous Clustering and Feature Selection for Binary Data   192
    Charlotte Laclau and Mohamed Nadif

Instant Exceptional Model Mining Using Weighted Controlled Pattern
Sampling   203
    Sandy Moens and Mario Boley

Resampling Approaches to Improve News Importance Prediction   215
    Nuno Moniz, Luís Torgo, and Fátima Rodrigues

An Incremental Probabilistic Model to Predict Bus Bunching in
Real-Time   227
    Luis Moreira-Matias, João Gama, João Mendes-Moreira,
    and Jorge Freire de Sousa

Mining Representative Frequent Patterns in a Hierarchy of Contexts . . .   239
    Julien Rabatel, Sandra Bringay, and Pascal Poncelet

A Deep Interpretation of Classifier Chains   251
    Jesse Read and Jaakko Hollmén

A Nonparametric Mixture Model for Personalizing Web Search   263
    El Mehdi Rochd and Mohamed Quafafou

Widened KRIMP: Better Performance through Diverse Parallelism   276
  Oliver Sampson and Michael R. Berthold

Finding the Intrinsic Patterns in a Collection of Time Series   286
  Anke Schweier and Frank Höppner

A Spatio-temporal Bayesian Network Approach for Revealing
Functional Ecological Networks in Fisheries   298
  Neda Trifonova, Daniel Duplisea, Andrew Kenny, and Allan Tucker

Extracting Predictive Models from Marked-Up Free-Text Documents
at the Royal Botanic Gardens, Kew, London   309
  Allan Tucker and Don Kirkup

Detecting Localised Anomalous Behaviour in a Computer Network   321
  Melissa Turcotte, Nicholas Heard, and Joshua Neil

Indirect Estimation of Shortest Path Distributions with Small-World
Experiments   333
  Antti Ukkonen

Parametric Nonlinear Regression Models for Dike Monitoring
Systems   345
  Harm de Vries, George Azzopardi, André Koelewijn,
  and Arno Knobbe

Exploiting Novel Properties of Space-Filling Curves for Data
Analysis   356
  David J. Weston

RealKrimp — Finding Hyperintervals that Compress with MDL for
Real-Valued Data   368
  Jouke Witteveen, Wouter Duivesteijn, Arno Knobbe,
  and Peter Grünwald

Real-Time Adaptive Residual Calculation for Detecting Trend
Deviations in Systems with Natural Variability   380
  Steven P.D. Woudenberg, Linda C. van der Gaag, Ad Feelders, and
  Armin R.W. Elbers

Author Index   393

Malware Phylogenetics Based on the Multiview Graphical Lasso

Blake Anderson[1], Terran Lane[2], and Curtis Hash[1]

[1] Los Alamos National Laboratory
banderson@lanl.gov
[2] Google, Inc.

Abstract. Malware phylogenetics has gained a lot of traction over the past several years. More recently, researchers have begun looking at directed acyclic graphs (DAG) to model the evolutionary relationships between samples of malware. Phylogenetic graphs offer analysts a better understanding of how malware has evolved by clearly illustrating the lineage of a given family. In this paper, we present a novel algorithm based on graphical lasso. We extend graphical lasso to incorporate multiple views, both static and dynamic, of malware. For each program family, a convex combination of the views is found such that the objective function of graphical lasso is maximized. Learning the weights of each view on a per-family basis, as opposed to treating all views as an extended feature vector, is essential in the malware domain because different families employ different obfuscation strategies which limits the information of different views. We demonstrate results on three malicious families and two benign families where the ground truth is known.

Keywords: Gaussian Graphical Models, Malware, Multiview Learning.

1 Introduction

In addition to malware aimed at a more general audience, advanced persistent threats (APT) are becoming a serious problem for many corporations and government agencies. APT typically involves malware that has been tailored to accomplish a specific goal against a specific target. During the life cycle of APT malware, the authors generally follow software engineering principles. This naturally leads to a phylogenetic graph, a graph demonstrating the evolutionary relationships between the different software versions.

When beginning the process of understanding a new, previously unseen sample of malware, it is advantageous to leverage the information gained from reverse engineering previously seen members of that instance's *family* because techniques learned from related instances can often be applied to a new program instance. A malware family is a group of related malware instances which share a common codebase and exhibit similar functionality (e.g. different branches in a software repository). In this paper, we focus on extracting a more detailed picture about the relationships between the malware instances within a given family.

H. Blockeel et al. (Eds.): IDA 2014, LNCS 8819, pp. 1–12, 2014.

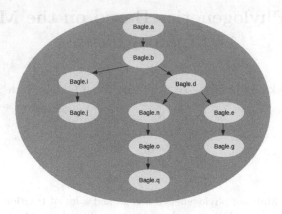

Fig. 1. An example of creating a phylogenetic graph for the bagle worm

Figure 1 exhibits the proposed output of the ideal phylogenetic algorithm. This figure clearly shows the evolution of the bagle virus [15] as a directed graph. The information presented in Figure 1 is invaluable for a reverse engineer tasked with understanding specific instances within a malware family as well as the general evolution of the family.

We describe a method based on the graphical lasso that finds a Gaussian graphical model where the malware instances are represented as nodes, and the evolutionary relationships are represented as edges. We present a novel extension to a standard algorithm that incorporates multiple views of the data. In our algorithm, a convex combination of views, both static and dynamic, is learned in order to maximize a multiview graphical lasso optimization problem. This is an important distinction to make from previous work [13] because it allows our algorithm to learn the importance of views while it is learning the phylogenetic graph for a given family. The obfuscation strategies vary between different malicious families, meaning an adaptive strategy would have an advantage because the more reliable views are weighted more heavily.

2   Data

We begin this section by describing the six views of a program used in this paper. We go on to explain how we represent these views, and finally introduce how we measure similarity between different samples in our datasets. It is important to note that the views described are not an exhaustive list, and the framework presented in this paper can be easily extended to incorporate other views of a program. The only restriction that must be met is that a positive-semidefinite similarity (kernel) matrix must be defined on the view, a restriction that is easily met in practice.

2.1   Data Views

We take advantage of six different types of data with the aim of covering the most popular data views that have been used for malware analysis in the literature. We use three static, meaning the executable is not run, data views: the binary file [17], the disassembled binary [4], and the control flow graph of the disassembled binary [18]. We use two dynamic, meaning the data is collected while the program is being executed, data views: the dynamic instruction trace [1] and the dynamic system call trace [12]. Finally, we use a file information data view which contains seven statistics that provide a summary of the previous data views. A more in-depth description of the views used in this paper can be found in [2].

2.2   Markov Chain Data Representation

As an illustrative example, we focus on the dynamic trace data, although this representation is suitable for any sequence-based data view. The dynamic trace data are the instructions the program executes, typically in a virtual machine to reduce the risk of contamination. Given an instruction trace $\mathcal{P}$, we are interested in finding a new representation, $\mathcal{P}'$, such that we can make unified comparisons in graph space while still capturing the sequential nature of the data. We achieved this by transforming the dynamic trace data into a Markov chain which is represented as a weighted, directed graph. A graph, $G = \langle V, E \rangle$, is composed of two sets, V and E. The elements of V are called vertices and the elements of E are called edges. In this representation, the edge weight, e_{ij}, between vertices i and j corresponds to the transition probability from state i to state j in the Markov chain, hence, the edge weights for edges originating at v_i are required to sum to 1, $\sum_{i \rightsquigarrow j} e_{ij} = 1$. We use an $n \times n$ ($n = |V|$) adjacency matrix to represent the graph, where each entry in the matrix, $a_{ij} = e_{ij}$ [16].

The nodes of the graph are the instructions the program executes. To find the edges of the graph, we first scan the instruction trace, keeping counts for each pair of successive instructions. After filling in the adjacency matrix with these values, we normalize the matrix such that all of the non-zero rows sum to one. This process of estimating the transition probabilities ensures a well-formed Markov chain. The constructed graphs approximate the pathways of execution of the program, and, by using graph kernels (Section 2.3), the structure of these pathways can be exploited.

The Markov chain graph can be summarized as $G = \langle V, E \rangle$, where

- V is the vertex set composed of unique instructions,
- E is the weighted edge set where the weights correspond to the transition probabilities and are estimated from the data.

2.3   Measure of Similarity

Kernels [5,14] are used to make comparisons between the different views of a program. A kernel, $K(\mathbf{x}, \mathbf{x}')$, is a generalized inner product and can be thought of as a measure of similarity between two objects [21]. A well-defined kernel must satisfy two properties: it must be symmetric (for all $\mathbf{x}$ and $\mathbf{y} \in X$: $K(\mathbf{x}, \mathbf{y}) = K(\mathbf{y}, \mathbf{x})$) and positive-semidefinite (for any $x_1, \ldots, x_n \in X$ and $\mathbf{c} \in \mathbb{R}^n$: $\sum_{i=1}^{n} \sum_{j=1}^{n} c_i c_j K(x_i, x_j) \geq 0$).

For the Markov chain representations and the file information feature vector, we use a standard squared exponential kernel:

$$K_{SE}(\mathbf{x}, \mathbf{x}') = \sigma^2 e^{-\frac{1}{2\lambda^2} \sum_i (x_i - x'_i)^2} \tag{1}$$

where $\mathbf{x}_i$ represents one of the seven features for the file information data view, or a transition probability for the Markov chain representations. σ and λ are the hyperparameters of the kernel function.

For the control flow graph data view, we attempted to find a kernel that closely matched previous work in the literature [18]. Although our approach does not take the instruction information of the basic blocks into account, we selected the graphlet kernel due to its computational efficiency. A k-graphlet is defined as a subgraph of a graph G, with the number of nodes of the subgraph equal to k. If $\boldsymbol{f}_G$ is a a feature vector, where each feature is the number of times a unique graphlet of size k occurs in G, the normalized probability vector is:

$$D_G = \frac{\boldsymbol{f}_G}{\# \text{ of all graphlets of size } k \text{ in } G} \tag{2}$$

and the graphlet kernel is defined as:

$$K_g(G, G') = \boldsymbol{D}_G^T \boldsymbol{D}_{G'} \tag{3}$$

3   Phylogenetic Graphs

Most malware phylogeny techniques in the literature rely on bifurcating tree-based approaches [8,25]. These methods do not give ancestral relationships to the data. We are interested in the problem of phylogenetic graph reconstruction where the goal is to find a graph, $\mathcal{G} = \langle V, E \rangle$. The vertices of the graph, V, are the instances of malware and the edges of the graph, E, represent phylogenetic relationships between the data such as "child of" or "parent of".

The main advantage of this approach is that the graph explicitly states the phylogenetic relationships between instances of malware. For a reverse engineer, knowing that an instance of malware is similar to known malware, from the same family for example, rapidly speeds up the reverse engineering process as the analyst will have a general idea of the mechanisms used by the sample. However, it is more informative to know the set of parents from which a given sample's functionality is derived. Having malware which is a composite of several other samples is becoming more widespread as malware authors are beginning to use standard software engineering practices, thus making the reuse of code more prevalent.

3.1 Overview of Graphical Lasso

The solution we propose is unsupervised and based on the graphical lasso (glasso) [9], which estimates the phylogenetic graph by finding a sparse precision matrix based on the combined kernel matrix. Glasso maximizes the Gaussian log-likelihood with respect to the precision matrix, $\Theta = \Sigma^{-1}$, of the true covariance (kernel) matrix, Σ:

$$\max_{\Theta} \{\log(\det(\Theta)) - \mathrm{tr}(K\Theta) - ||\Theta \circ P||_1\} \tag{4}$$

where K is the sample covariance matrix, which in this case means the kernel matrix for a given view of malware. $||\cdot||_1$ is the classic L_1 norm used in standard lasso, P is a matrix penalizing specific edges of the precision matrix, and $\circ$ is the Hadamard product. There have been several efficient algorithms developed to solve Equation 4 that we take advantage of in this work [9].

By using the L_1 penalty, a sparse graph that captures the conditional independencies of the true covariance matrix is found. For instance, if there exists three examples of malware with a direct lineage (x_c is derived from x_b, x_b is derived from x_a), then the naïve approach of creating links between similar examples would create a completely connected graph between these samples. This would not be unreasonable as they are all similar. The strength of glasso is that it leverages the precision matrix to discover that x_c and x_a are conditionally independent given x_b, which would omit the edge between x_c and x_a.

3.2 Modifying Graphical Lasso for Multiple Views

Equation 4 solves the problem of finding a Gaussian graphical model for a single view. The obfuscation techniques of malware make this an insufficient solution as individual views can often be unreliable. Instead, Equation 4 is modified to accommodate multiple views of the data (i.e., a convex combination of multiple kernel matrices) with the following multiview problem:

$$\max_{\Theta,\beta} \ \{\log(\det(\Theta)) - \mathrm{tr}(\sum_{i=1}^{M}(\beta_i K_i)\,\Theta) - ||\Theta \circ P||_1 - \lambda||\beta||_2\} \tag{5}$$

where β is the mixing weights and λ is the regularization penalty on β.

Using the linearity of the **trace** function and rearranging terms:

$$\min_{\Theta,\beta} \ \{\sum_{i=1}^{M}\beta_i \mathrm{tr}(K_i\Theta) - \log(\det(\Theta)) + ||\Theta \circ P||_1 + \lambda||\beta||_2\} \tag{6}$$

subject to $\sum_i \beta_i = 1$ and $\forall_i \beta_i \geq 0$.

The algorithm we employ is based on alternating projections, first finding the optimal Θ while holding β fixed, and then finding the optimal β while holding Θ fixed. As we mentioned previously, there are many efficient algorithms to solve

---

Algorithm 1. Multiple View Graphical Lasso, iteratively finds optimal β and Θ, return Θ

Require: initial β

$\quad$ score$_0 \leftarrow \log(\det(\Theta)) - \text{tr}\left(\sum_{i=1}^{M} (\beta_i K_i)\Theta\right) - ||\Theta \circ P||_1$

$\quad$ **while** score$_t <$ score$_{t-1}$ **do**

$\quad\quad a_i = \text{tr}(K_i\Theta)$

$\quad\quad \beta \leftarrow \min \; a^T\beta + \frac{1}{2}\beta^T C\beta$

$\quad\quad\quad$ s.t. $\;\; G\beta \preceq h$

$\quad\quad K \leftarrow \sum_{i=1}^{M} \beta_i K_i$

$\quad\quad \Theta \leftarrow \max_\Theta \{\log(\det(\Theta)) - \text{tr}(K\Theta) - ||\Theta \circ P||_1\}$

$\quad\quad$ score$_t \leftarrow \log(\det(\Theta)) - \text{tr}\left(\sum_{i=1}^{M} (\beta_i K_i)\Theta\right) - ||\Theta \circ P||_1$

$\quad$ **end while**

$\quad$ **return** precision matrix, Θ

---

for the optimal Θ [9]. To solve for the optimal β assuming a fixed Θ, we first note that $-\log(\det(\Theta)) + ||\Theta \circ P||_1$ is independent of β and can therefore be ignored in the optimization of β leaving us with:

$$\min_{\beta} \; \{\sum_{i=1}^{M} \beta_i\text{tr}(K_i\Theta) + \lambda||\beta||_2\} \tag{7}$$

If we let $a_i = \text{tr}(K_i\Theta)$, this problem can be stated as a quadratic program [6] allowing for the use of many efficient algorithms:

$$\min_{\beta} \; a^T\beta + \frac{1}{2}\beta^T C\beta \tag{8}$$

$$\text{s.t.} \;\; G\beta \preceq h$$

where

$$G = \begin{bmatrix} -1 & 0 & \cdots \\ 0 & \ddots & 0 \\ \vdots & 0 & -1 \\ \cdots & 1 & \cdots \\ \cdots & -1 & \cdots \end{bmatrix} \tag{9}$$

and

$$h = [\mathbf{0} \; 1 \; -1]^T \tag{10}$$

The top negative identity matrix of G and $\mathbf{0}$ vector in h enforce non-negative β's while the last two constraints force the β's to sum to one. $\preceq$ is the componentwise inequality. $\frac{1}{2}\beta^T C\beta$ removes the incentive of having $\beta_i = 1, \beta_{j\neq i} = 0$. In the case of the degenerate solution, $\beta_i = 1, \beta_{j\neq i} = 0$, this procedure reduces to a feature selector basing all further optimizations on the best view. Intuitively, a more robust solution would make use of all available information. We defend this intuitive claim in Section 4.2. Algorithm 1 outlines the procedure for finding the optimal precision matrix, Θ, when given multiple views of the data.

Algorithm 2. Multiple View Graphical Lasso, iteratively finds optimal β and Θ, return Θ. Leverages cluster information to have different penalization throughout the precision matrix.

Require: initial β
    clusters, $C \leftarrow$ multiview clustering algorithm [3]
    for $c_k \in C$ **do**
        $\rho \leftarrow (\sum_{i,j}^{|c_k|} K_{i,j})/|c_k|^2$
        $\Theta_k \leftarrow$ perform Algorithm 1 with penalization ρ on cluster c_k
    end for
    $K' \leftarrow$ compute inter-cluster similarity matrix
    $\Theta_C \leftarrow$ perform Algorithm 1 with K'
    return precision matrices, Θ_C, Θ_k

3.3 Leveraging Clusters in Graphical Lasso

While Algorithm 1 solves Equation 6, we found that the uniform penalization of the L_1-norm led to an interesting phenomenon, namely, the resulting precision matrix was globally sparse with a small subset of the nodes being highly connected. This suggests that a uniform penalty is not appropriate for this problem. Instead, a clustering pre-processing step is first used. Once similar versions of the program are clustered together, penalizing the L_1-norm uniformly within that cluster becomes more appropriate.

Leveraging clusters in the multiview graphical lasso is straightforward. First, the clusters, $C = c_1, \ldots, c_n$, are found using a multiview clustering algorithm [3]. Then, the multiview graphical lasso (Algorithm 1) is applied to each cluster, adjusting the penalty term by using the heuristic, $(\sum_{i,j}^{|c_k|} K_{i,j})/|c_k|^2$, which effectively penalizes self-similar clusters more heavily, allowing those clusters to be more sparse. Finally, we create a new set of similarity matrices, but instead of measuring the similarity between different instances in the dataset, this similarity matrix measures similarity between the different clusters. This matrix is created by taking the average similarity between every $x \in c_i$ and $y \in c_j$. Performing the multiview glasso on these matrices finds the conditional independences between the different clusters. Algorithm 2 details the procedure.

4 Evaluation

4.1 Datasets

To be able to accurately quantify our results, we must have access to ground truth, i.e., the true phylogenetic graph of the program. As one can imagine, determining the true phylogeny is a very difficult, time-consuming process. Fortunately, it is made easier for some benign programs as their subversion or github repositories can be used to gather this information. Unfortunately, malware authors do not typically use these tools in an open setting, making the phylogenetic graphs of malware much more difficult to obtain. To understand the evolution of a malware family, we elicited the help of several experts. These experts used several sources of information to come up with an informed graph depicting the evolution of malware, such as the time the malware was first seen in the wild, the compile-time timestamp, the functionality of the sample, and the obfuscation methods employed by the sample. While the phylogenetic graphs the domain experts have manually found are by no means 100% accurate, they do provide a reasonable baseline in a setting where ground truth is not available. We use three malicious programs: Mytob, Koobface, and Bagle. Mytob is a mass-mailing worm that seeks a user's address book to then send itself to all of that user's contacts [24]. The Koobface worm spreads through social networking sites with the intent of installing software for a botnet [23]. Bagle is another mass-mailing worm that creates a botnet [22].

In addition to the malicious programs, we validate our methods on two benign programs, NetworkMiner and Mineserver, due to the ground truth for these graphs being more readily available. NetworkMiner is a network forensics anaylsis tool specializing in packet sniffing and parsing PCAP files [20]. Mineserver is a way to host worlds in the popular Minecraft game [19].

4.2   Results

Table 1 lists all of the results for the 5 datasets previously described with respect to precision, recall, and F-norm. The Frobenius norm, or F-norm, is defined as $||A - B||_F = \sqrt{\sum_i \sum_j (A_{ij} - B_{ij})^2}$. Precision is defined to be the number of true edges in the graph found divided by the number of total edges in the graph found, and recall is defined to be the number of true edges in the graph found divided by the number of true edges in the ground truth graph. We compare our multiview glasso + clustering approach to our multiview glasso approach and regular glasso for the best, single view and a uniform combination of views. We also compare our approach to the Gupta algorithm [11] and a naïve baseline, the minimum spanning tree.

As Table 1 demonstrates, the proposed method performs well on a variety of datasets, both malicious and benign. The Gupta algorithm performs well with respect to precision, and even out-performs our algorithm on the mineserver dataset, but this is mainly because it finds sparser graphs, where precision will naturally be higher. The mineserver dataset is interesting for two reasons: the ground truth is known with absolute certainty, and merges and branches are present. Both of these cases are present in most real-word software engineering projects including malware. Figure 2 shows the ground truth as well as the graph acquired with the multiview glasso + clustering method. As the figure demonstrates, we were able to recover the majority of the branches and merges, and can recover most of the evolutionary flow of the program.

4.3   Computational Complexity

The multiview graphical lasso algorithm is an iterative algorithm with two main components. First, graphical lasso is solved given some fixed β vector. Graphical

Table 1. Phylogenetic graph reconstruction results in terms of F-norm, Precision, and Recall

Dataset	Method	F-norm	Precision	Recall
NetworkMiner	MKLGlasso+Clust	**4.5826**	**.4857**	**.85**
	MKLGlasso	5.5678	.3514	.65
	Glasso-Best View	6.0	.2895	.55
	Glasso-Best View+Clust	5.3852	.3902	.80
	Glasso-Uniform Comb	6.1644	.3043	.70
	Glasso-Uniform Comb+Clust	5.0	.4360	**.85**
	Gupta	5.0	.3810	.40
	Minimum Spanning Tree	5.6569	.35	.70
MineServer	MKLGlasso+Clust	**4.0**	0.7222	**0.8125**
	MKLGlasso	5.4772	0.5833	0.2188
	Glasso-Best View	5.8242	0.4118	0.1935
	Glasso-Best View+Clust	4.8134	0.4510	0.3871
	Glasso-Uniform Comb	5.6711	0.4314	0.1875
	Glasso-Uniform Comb+Clust	4.4655	0.4902	0.4194
	Gupta	4.7958	**0.8462**	0.3438
	Minimum Spanning Tree	7.4833	0.0	0.0
Bagle	MKLGlasso+Clust	5.7446	**0.20**	**0.3333**
	MKLGlasso	9.5394	0.0964	0.125
	Glasso-Best View	10.7731	0.0704	0.1176
	Glasso-Best View+Clust	**5.5813**	0.1480	.0909
	Glasso-Uniform Comb	10.2921	0.0812	0.0980
	Glasso-Uniform Comb+Clust	9.6476	.1351	.1220
	Gupta	6.5574	0.12	0.125
	Minimum Spanning Tree	8.3667	0.0208	0.0417
Mytob	MKLGlasso+Clust	7.9373	**0.1563**	**0.5263**
	MKLGlasso	8.5348	0.0988	0.2258
	Glasso-Best View	8.7388	0.0864	0.1935
	Glasso-Best View+Clust	8.2184	0.1282	0.2903
	Glasso-Uniform Comb	10.2766	0.0617	0.1951
	Glasso-Uniform Comb+Clust	8.8117	0.1081	0.2683
	Gupta	**6.0828**	0.05	0.0526
	Minimum Spanning Tree	7.2801	0.0526	0.1053
Koobface	MKLGlasso+Clust	**5.2915**	**0.5812**	**0.5**
	MKLGlasso	5.3852	0.2917	0.3889
	Glasso-Best View	6.8551	0.2391	0.3171
	Glasso-Best View+Clust	6.0427	0.2821	0.3235
	Glasso-Uniform Comb	6.6043	0.2195	0.2927
	Glasso-Uniform Comb+Clust	5.9486	0.3023	0.3636
	Gupta	5.9161	0.3158	0.3333
	Minimum Spanning Tree	7.2111	0.0278	0.0556

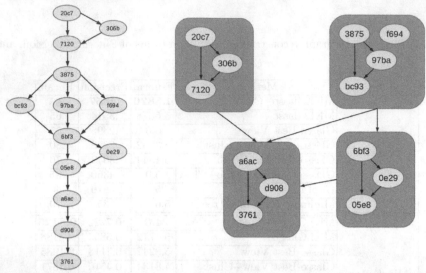

Fig. 2. Comparison between mineserver ground truth and the phylogenetic graph found

lasso has a computational complexity of $\mathcal{O}(n^3)$ where n is the number of samples. Next, β is updated with a quadratic program, which has a computational complexity $\mathcal{O}(n^3)$ where n is again the number of samples. This algorithm converged within 15 iterations for all datasets.

The multiview clustering preprocessing step is an iterative algorithm. The algorithm is composed of two main parts: computing the spectral clustering objective function and computing a semidefinite program to find β. Finding the new feature space can be done in $\mathcal{O}(\lfloor \log n \rfloor n^2)$ time where n is the number of samples. Solving the semidefinite program has a computational complexity of $\mathcal{O}(n^6)$ in the worst case but has been shown to be $\mathcal{O}(n^3)$ in the average case. More details of this algorithm can be found in [3].

For a more concrete view, we looked at five different dataset sizes, [10, 50, 100, 500, 1,000]. The average times Algorithm 2 took, averaged over 10 runs, were [0.29s, 1.06s, 1.16s, 31.75s, 238.33s]. These results were run on a machine with a Intel Core i7-2640M CPU @ 2.80GHz with 8 GBs of memory.

5   Related Work

Within the malware literature, most of the work done has centered around creating phylogenetic trees. For instance, in [25], Wagener et al. create a similarity matrix based on the system calls performed by the samples, and then use an agglomerative hierarchical clustering method to find the phylogenetic tree. Graph-pruning techniques have also been used [11] to find a tree. Gupta et al. begin with a fully connected similarity graph and incoming edges that are below a certain threshold are pruned for each node. All the remaining incoming edges are then pruned if their combined weight is less than some other predefined threshold.

In [13], Jang et al. infer directed acyclic graph lineage by using a minimum spanning tree algorithm with a post-processing step to allow nodes to have multiple parents. Jang et al. use a feature vector which combines both dynamic and static views. Unlike the method presented in this paper, these features are treated as a uniform vector, whereas we treat the feature sets as different views and learn the weights of each view to maximize our objective function. Graphical lasso also does not need a post-processing step to find multiple parents.

While there has not been any work done on a multiview graphical lasso that finds a single precision matrix, there has been some work that uses transfer learning [7,10]. In the problem posed by Danaher et al., the goal is to find several Gaussian graphical models where the underlying data is drawn from related distributions. The canonical example for this class of methods is learning gene regulation networks for cancer and normal tissue. Both types of tissue share many edges in the network so one would want to leverage all the data possible, but these networks also have significant differences. This goal can be accomplished by adding an additional L_1 penalty to penalize the difference between the two learned Gaussian graphical models.

Our proposed approach is different in that we find a single Gaussian graphical model conditioned on multiple views of the data. The joint graphical lasso will find multiple Gaussian graphical models, where a hyperparameter controls the amount of transfer learning between the different models.

6   Conclusions

In this paper, we have presented a novel extension to graphical lasso, which finds a weighted combination of views, both static and dynamic, to infer a phylogenetic graph for a family of programs. Finding a weighted combination of views that optimizes phylogenetic reconstruction is advantageous because malware families use different obfuscation strategies rendering certain views more reliable for different families. Our results show that we can efficiently find phylogenetic graphs, and that combining multiple views to maximize Equation 6 significantly increases performance compared to any single view as well as several baselines such as minimum spanning trees.

References

1. Anderson, B., Quist, D., Neil, J., Storlie, C., Lane, T.: Graph-Based Malware Detection using Dynamic Analysis, pp. 1–12. Springer, Paris (2011)
2. Anderson, B., Storlie, C., Lane, T.: Improving Malware Classification: Bridging the Static/Dynamic Gap. In: Proceedings of the Fifth ACM Workshop on Security and Artificial Intelligence, pp. 3–14. ACM (2012)
3. Anderson, B., Storlie, C., Lane, T.: Multiple Kernel Learning Clustering with an Application to Malware. In: IEEE Twelfth International Conference on Data Mining, pp. 804–809. IEEE (2012)

4. Bilar, D.: Opcodes as Predictor for Malware. International Journal of Electronic Security and Digital Forensics 1, 156–168 (2007)
5. Bishop, C.M.: Pattern Recognition and Machine Learning (Information Science and Statistics). Springer-Verlag New York, Inc., Secaucus (2006)
6. Boyd, S., Vandenberghe, L.: Convex Optimization. Cambridge University Press, New York (2004)
7. Danaher, P., Wang, P., Witten, D.M.: The Joint Graphical Lasso for Inverse Co-variance Estimation Across Multiple Classes. ArXiv e-prints (Nov 2011)
8. Darmetko, C., Jilcott, S., Everett, J.: Inferring Accurate Histories of Malware Evolution from Structural Evidence. In: The Twenty-Sixth International FLAIRS Conference (2013)
9. Friedman, J., Hastie, T., Tibshirani, R.: Sparse Inverse Covariance Estimation with the Graphical Lasso. Biostatistics 9(3), 432–441 (2008)
10. Guo, J., Levina, E., Michailidis, G., Zhu, J.: Joint Estimation of Multiple Graphical Models. Biometrika (2011)
11. Gupta, A., Kuppili, P., Akella, A., Barford, P.: An Empirical Study of Malware Evolution. In: First International Communication Systems and Networks and Workshops, pp. 1–10. IEEE (2009)
12. Hofmeyr, S.A., Forrest, S., Somayaji, A.: Intrusion Detection Using Sequences of System Calls. Journal of Computer Security 6(3), 151–180 (1998)
13. Jang, J., Woo, M., Brumley, D.: Towards Automatic Software Lineage Inference. In: Proceedings of the Twenty-Second USENIX Conference on Security, pp. 81–96. USENIX Association (2013)
14. Kashima, H., Tsuda, K., Inokuchi, A.: Kernels for Graphs. MIT Press (2004)
15. Kaspersky Lab Report: The Bagle Botnet, http://www.securelist.com/en/analysis/162656090/The_Bagle_botnet (accessed September 17, 2013)
16. Kolbitsch, C., Comparetti, P.M., Kruegel, C., Kirda, E., Zhou, X.Y., Wang, X.: Effective and Efficient Malware Detection at the End Host. In: Proceedings of the Eighteeth USENIX Security Symposium, pp. 351–366 (2009)
17. Kolter, J.Z., Maloof, M.A.: Learning to Detect and Classify Malicious Executables in the Wild. The Journal of Machine Learning Research 7, 2721–2744 (2006), http://dl.acm.org/citation.cfm?id=1248547.1248646
18. Kruegel, C., Kirda, E., Mutz, D., Robertson, W., Vigna, G.: Polymorphic Worm Detection Using Structural Information of Executables. In: Valdes, A., Zamboni, D. (eds.) RAID 2005. LNCS, vol. 3858, pp. 207–226. Springer, Heidelberg (2006)
19. Mineserver, https://github.com/fador/mineserver (accessed September 17, 2013)
20. NetworkMiner, http://sourceforge.net/projects/networkminer/ (accessed September 17, 2013)
21. Schölkopf, B., Smola, A.J.: Learning with Kernels. MIT Press (2002)
22. Symantec Bagle Security Report, http://www.symantec.com/security_response/writeup.jsp?docid=2004-011815-3332-99 (accessed September 17, 2013)
23. Symantec Koobface Security Report, http://www.symantec.com/security_response/writeup.jsp?docid=2008-080315-0217-99 (accessed September 17, 2013)
24. Symantec Mytob Security Report, http://www.symantec.com/security_response/writeup.jsp?docid=2005-022614-4627-99 (accessed September 17, 2013)
25. Wagener, G., State, R., Dulaunoy, A.: Malware Behaviour Analysis. Journal in Computer Virology 4(4), 279–287 (2008)

Modeling Stationary Data by a Class of Generalized Ornstein-Uhlenbeck Processes: The Gaussian Case

Argimiro Arratia[1,*], Alejandra Cabaña[2,**], and Enrique M. Cabaña[3]

[1] Dept. Computer Science, Universitat Politècnica de Catalunya, Barcelona, Spain
[2] Dept. Matemàtiques, Universitat Autònoma de Barcelona, Spain
[3] Depto. de Métodos Matemático-Cuantitativos, Universidad de la República, Montevideo, Uruguay

Abstract. The Ornstein-Uhlenbeck (OU) process is a well known continuous–time interpolation of the discrete–time autoregressive process of order one, the AR(1). We propose a generalization of the OU process that resembles the construction of autoregressive processes of higher order $p > 1$ from the AR(1). The higher order OU processes thus obtained are called *Ornstein-Uhlenbeck processes of order p* (denoted OU(p)), and constitute a family of parsimonious models able to adjust slowly decaying covariances. We show that the OU(p) processes are contained in the family of autoregressive moving averages of order $(p, p - 1)$, the ARMA($p, p - 1$), and that their parameters and covariances can be computed efficiently. Experiments on real data show that the empirical autocorrelation for large lags can be well modeled with OU(p) processes with approximately half the number of parameters than ARMA processes.

1 Introduction

The Ornstein-Uhlenbeck process (from now on OU) was introduced by L. S. Ornstein and E. G. Uhlenbeck [10] as a model for the velocities of a particle subject to the collisions with surrounding molecules. It is a well studied and accepted model for thermodynamics, chemical and other various stochastic processes found in physics and the natural sciences. Moreover, the OU process is the unique non-trivial stochastic process that is stationary, Markovian and Gaussian; it is also mean-reverting, and for all these properties it has found its way into financial engineering, first as a model for the term structure of interest rates in a form due to [11], and then under other variants or generalizations (e.g. where the underlying random noise is a Lévy process) as a model of financial time series with applications to option pricing, portfolio optimization and risk theory, among others (see [8] and references therein).

* Supported by MICINN project TIN2011-27479-C04-03 (BASMATI), Gen. Cat. 2009-SRG-1428 (LARCA) and MTM2012-36917-C03-03 (SINGACOM).
** Supported by MICINN projects MTM2012-31118 and TIN2011-27479-C04-03 (BASMATI).

H. Blockeel et al. (Eds.): IDA 2014, LNCS 8819, pp. 13–24, 2014.

The OU process can be thought of as continuous time interpolation of an autoregressive process of order one (i.e. an AR(1) process). We make this point clear in §2 and define OU in §3. Departing from this analogy, one can seek to define and analyze the result of iterating the application of the operator that maps a Wiener process onto an OU process, just as one iterates an AR process, in order to obtain a higher order OU process. This operator is defined in §4 and denoted $\mathcal{OU}$, with subscripts denoting the parameters involved. The p iterations of $\mathcal{OU}$, for each positive integer p, give rise to a new family of processes, the *Ornstein-Uhlenbeck processes of order p*, denoted OU(p), proposed as models for either stationary continuous time processes or the series obtained by observing these continuous processes at discrete instants, equally spaced or not. We show in §5 that this higher order OU process can be expressed as a linear combination of ordinary OU processes, and this allow us to derive a closed formula for its covariance. This has important practical implications, as shown in §7, since it allows to easily estimate the parameters of a OU(p) process by maximum likelihood or, as an alternative, by matching correlations, the latter being a procedure resembling the method of moments.

We give in §6 a state space model representation for the discrete version of a OU(p), from which we can show that for $p > 1$, a OU(p) turns out to be an ARMA(p,q), with $q \leq p - 1$. Notwithstanding this structural similarity, the family of discretized OU(p) processes is more parsimonious than the family of ARMA($p, p - 1$) processes, and we shall see empirically in §7 and §8 that it is able to fit well the auto covariances for large lags. Hence, OU processes of higher order appear as a new continuos model, competitive in a discrete time setting with higher order autoregressive processes (AR or ARMA).

Related Work. The construction and estimation of continuous–time analogues of discrete–time processes have been of interest for many years. More recently there has been an upsurge in interest for continuous–time representations of ARMA processes, due to their many financial applications. These models, known as CARMA, have been developed by Brockwell and others (see [3], the survey[4], and more recently [5]), from a state–space representation of a formal high order stochastic differential equation, which is not physically realizable. We provide an alternative constructive method to obtain continuous–time AR processes (CAR), a particular case of the CARMA models, from repeated applications of an integral operator (the OU operator). This construction can be extended to apply to a moving average of the noise as well, and hence provides a different approach to the CARMA models, that can lead to further generalizations. Regarding the estimation of our continuous AR model (the OU(p) process) from discretely observed data, we use the state–space representation method as in [5], and derive analogous results on their representability as discrete ARMA($p, p-1$) processes.

All omitted details and proofs in this presentation can be found in an extended version at [1].

2   From AR(1) to Ornstein Uhlenbeck Processes

The simplest ARMA model is an AR(1): $X_t = \phi X_{t-1} + \sigma \epsilon_t$ that can be written as $(1 - \phi B)X_t = \sigma \epsilon_t$, where ϵ_t, $t \in \mathbf{Z}$ is a white noise, and B is the back-shift operator that maps X_t onto $BX_t = X_{t-1}$. If $|\phi| < 1$, the process X_t is stationary. Equivalently, X_t can be written as $X_t = \sigma \mathcal{MA}(1/\rho)\epsilon_t$, where $\mathcal{MA}(1/\rho)$ is the moving average that maps ϵ_t onto $\mathcal{MA}(1/\rho)\epsilon_t = \sum_{j=0}^{\infty} \frac{1}{\rho^j} \epsilon_{t-j}$. The covariances of X_t are $\gamma_h = \mathbf{E}X_t X_{t+h} = \gamma_0/\rho^h$, where $\gamma_0 = \sigma^2/(1 - 1/\rho^2)$.

There are many possible ways of defining a continuous time analogue x_t, $t \in \mathbf{R}$, of AR(1) processes, for instance,

– by establishing that $\gamma(h) = \mathbf{E}x(t)x(t+h)$ be $\gamma_0 e^{-\kappa|h|}$
– by replacing the measure W concentrated on the integers and defined by
  $W(A) = \sum_{t \in A} \epsilon_t$, that allows writing $X_t = \int_{-\infty}^{t^+} \frac{1}{\rho^{t-s}} dW(s)$, by a measure
  w on $\mathbf{R}$, with stationary, i.i.d. increments and defining (with $\rho = e^{\kappa}$):
$$x(t) = \int_{-\infty}^{t} e^{-\kappa(t-s)} dw(s) \quad \Re(\kappa) > 0$$

Both ways lead to the same result: Ornstein-Uhlenbeck processes.

3   Ornstein-Uhlenbeck Processes

Let us denote by w a standard Wiener process, that is, a Gaussian, centered process with independent increments with variance $\mathbf{E}(w(t) - w(s))^2 = |t - s|$. We impose further (as usual) that $w(0) = 0$, but shall not limit the domain of the parameter to $\mathbf{R}^+$ and assume that $w(t)$ is defined for t in $\mathbf{R}$. Then, an Ornstein-Uhlenbeck process with parameters $\lambda > 0, \sigma > 0$ can be written as

$$\xi_{\lambda,\sigma}(t) = \sigma \int_{-\infty}^{t} e^{-\lambda(t-s)} dw(s) \tag{1}$$

or, in differential form,

$$d\xi_{\lambda,\sigma}(t) = -\lambda \xi_{\lambda,\sigma} \, dt + \sigma dw(t) \tag{2}$$

We may think of $\xi_{\lambda,\sigma}$ as the result of accumulating a random noise, with reversion to the mean (that we assume to be 0) of exponential decay with rate λ. The magnitude of the noise is given by σ.

4   Ornstein-Uhlenbeck Processes of Higher Order

We propose a construction of OU processes of order $p > 1$, obtained by a procedure that resembles the one that allows to build an autoregressive process of order p, AR(p), from an AR(1). We will see that the resulting higher order process is a parsimonious model, with few parameters, that is able to adjust slowly decaying covariances. The AR(p) process

$$X_t = \sum_{j=1}^{p} \phi_j X_{t-j} + \sigma\epsilon_t \quad \text{or} \quad \phi(B)X_t = \sigma\epsilon_t,$$

where $\phi(z) = 1 - \sum_{j=1}^{p} \phi_j z^j = \prod_{j=1}^{p}(1 - z/\rho_j)$ has roots $\rho_j = e^{\kappa_j}$, can be obtained by applying the composition of the moving averages $\mathcal{MA}(1/\rho_j)$ to the noise. Thus,

$$X_t = \sigma \prod_{j=1}^{p} \mathcal{MA}(1/\rho_j)\epsilon_t$$

Let us denote $\mathcal{MA}_\kappa = \mathcal{MA}(e^{-\kappa})$. A continuous version of the operator $\mathcal{MA}_\kappa$, that maps ϵ_t onto $\mathcal{MA}_\kappa \epsilon_t = \sum_{l \le t, \text{integer}} e^{-\kappa(t-l)}\epsilon_l$, is the Ornstein–Uhlenbeck operator $\mathcal{OU}_\kappa$ that maps $y(t)$ onto

$$\mathcal{OU}_\kappa y(t) = \int_{-\infty}^{t} e^{-\kappa(t-s)} dy(s)$$

and this suggests the use of the model OU(p), *Ornstein–Uhlenbeck process*:

$$x_{\kappa,\sigma}(t) = \sigma \prod_{j=1}^{p} \mathcal{OU}_{\kappa_j} w(t), \tag{3}$$

with parameters $\kappa = (\kappa_1, \ldots, \kappa_p)$, and σ.

5   OU(p) as a Superposition of OU(1)

Theorem 1. *The* Ornstein-Uhlenbeck process with parameters $\kappa = (\kappa_1, \ldots, \kappa_p)$, σ, $x_{\kappa,\sigma} = \prod_{j=1}^{p} \mathcal{OU}_{\kappa_j}(\sigma w)$, can be written as a linear combination of p processes of order 1.

(i) When the components of κ are pairwise different, the linear expression has the form:

$$x_{\kappa,\sigma} = \sum_{j=1}^{p} K_j(\kappa)\xi_{\kappa_j}, \quad \xi_{\kappa_j}(t) = \sigma \int_{-\infty}^{t} e^{-\kappa_j(t-s)} dw(s). \tag{4}$$

and the coefficients $K_j(\kappa) = \frac{1}{\prod_{\kappa_l \ne \kappa_j}(1 - \kappa_l/\kappa_j)}$.

(ii) When κ has components κ_h repeated p_h times ($h = 1, \ldots, q$, $\sum_{h=1}^{q} p_h = p$) the linear combination is:

$$x_{\kappa,\sigma} = \sum_{h=1}^{q} K_h(\kappa) \sum_{j=0}^{p_h-1} \binom{p_h-1}{j} \xi_{\kappa_h}^{(j)}$$

$$\xi_{\kappa_h}^{(j)}(t) = \sigma \int_{-\infty}^{t} e^{-\kappa_h(t-s)} \frac{(-\kappa_h(t-s))^j}{j!} dw(s)$$

The autocovariances of $x_{\kappa,\sigma}$ can be directly computed from this linear expression. In the case of κ having repeated components, the autocovariances are

$$\gamma_{\kappa,\sigma}(t)=\sum_{h'=1}^{q}\sum_{i'=0}^{p_{h'}-1}\sum_{h''=1}^{q}\sum_{i''=0}^{p_{h''}-1}K_{h'}(\kappa)\bar{K}_{h''}(\kappa)\binom{p_{h'}-1}{i'}\binom{p_{h''}-1}{i''}\gamma_{\kappa_{h'},\kappa_{h''},\sigma}^{(i',i'')}(t)$$

$$\gamma_{\kappa_1,\kappa_2,\sigma}^{(i_1,i_2)}(t) = \mathbf{E}\xi_{\kappa_1}^{(i_1)}(t)\overline{\xi_{\kappa_2}^{(i_2)}(0)}$$
$$= \sigma^2(-\kappa_1)^{i_1}(-\bar{\kappa}_2)^{i_2}\int_{-\infty}^{0}e^{-\kappa_1(t-s)}\frac{(t-s)^{i_1}}{i_1!}e^{-\bar{\kappa}_2(-s)}\frac{(-s)^{i_2}}{i_2!}ds,$$

and when the components of κ are pairwise different, the covariances can be written as $\gamma_{\kappa,\sigma}(t)=\sum_{h'=1}^{p}\sum_{h''=1}^{p}K_{h'}(\kappa)\bar{K}_{h''}(\kappa)\gamma_{\kappa_{h'},\kappa_{h''},\sigma}^{(0,0)}(t)$.

These formulas allow the computation of the covariances of the series obtained by sampling the OU(p) process at discrete times (equally spaced or not).

6   A State Space Representation of the OU(p) Process

The decomposition of the OU(p) process $x_{\kappa,\sigma}(t)$ as a linear combination of simpler processes of order 1 (Thm. 1), leads to an expression of the process by means of a state space model. This provides a unified approach for computing the likelihood of $x_{\kappa,\sigma}(t)$ through a Kalman filter. Moreover, it can be used to show that $x_{\kappa,\sigma}(t)$ is an ARMA($p,p-1$) whose coefficients can be computed from κ. In order to ease notation, we consider that the components of κ are all different.

The decomposition of $x_{\kappa,\sigma}(t)$ in (4) as a linear combination of the OU(1) processes

$$\xi_{\kappa_j}(t) = \sigma\int_{-\infty}^{t}e^{-\kappa_j(t-s)}dw(s) = \sigma e^{-\kappa_j}\xi_{\kappa_j}(t-1) + \sigma\int_{t-1}^{t}e^{-\kappa_j(t-s)}dw(s)$$

with innovations $\boldsymbol{\eta}_\kappa$ with components $\eta_{\kappa_j}(t) = \sigma\int_{t-1}^{t}e^{-\kappa_j(t-s)}dw(s)$ provides a representation of the OU(p) process in the space of states $\boldsymbol{\xi}_\kappa = (\xi_{\kappa_1},\ldots,\xi_{\kappa_p})^{\mathrm{tr}}$, as a VARMA model (this follows from Corollary 11.1.2 of [7]).

The transitions in the state space are

$$\boldsymbol{\xi}_\kappa(t) = \mathrm{diag}(e^{-\kappa_1},\ldots,e^{-\kappa_p})\boldsymbol{\xi}_\kappa(t-1) + \boldsymbol{\eta}_\kappa(t),$$

and

$$\boldsymbol{x}(t) = \boldsymbol{K}^{tr}(\kappa)\boldsymbol{\xi}(t)$$

The innovations have variance matrix $\mathrm{Var}(\boldsymbol{\eta}_{\kappa,\tau}(t)) = ((v_{j,l}))$, where $v_{j,l} = \sigma^2\mathbf{E}\int_{t-1}^{t}e^{-(\kappa_j+\bar{\kappa}_l)(t-s)}ds = \frac{1-e^{-(\kappa_j+\bar{\kappa}_l)}}{\kappa_j+\bar{\kappa}_l}$.

Now apply the AR operator $\prod_{j=1}^{p}(1-e^{-\kappa_j}B)$ to x_κ and obtain

$$\prod_{j=1}^{p}(1-e^{-\kappa_j}B)x_\kappa(t) = \sigma\sum_{j=1}^{p}K_jG_j(B)\eta_{\kappa_j}(t) =: \zeta(t),$$

with $G_j(z) = \prod_{l\neq j}(1-e^{-\kappa_j}z) := 1-\sum_{l=1}^{p-1}g_{j,l}z^l$.

This process has the same second-order moments as the ARMA$(p, p-1)$, $\prod_{j=1}^{p}(1 - e^{-\kappa_j}B)x_\kappa(t) = \sum_{j=0}^{p-1}\theta_j\epsilon(t-j) =: \zeta'(t)$ (ϵ is a white noise), when the covariances $c_j = \mathbf{E}\zeta(t)\bar{\zeta}(t-j)$ and $c'_j = \mathbf{E}\zeta'(t)\bar{\zeta}'(t-j)$ coincide.

The covariances c_j and c'_j are given respectively by the generating functions $\left(\sum_{h=0}^{p-1}\theta_h z^h\right)\left(\sum_{k=0}^{p-1}\bar{\theta}_k z^{-h}\right) = \sum_{l=-p+1}^{p-1}c_l z^l$ and

$$J(z) := \sum_{j=1}^{p}\sum_{l=1}^{p}K_j\bar{K}_l G_j(z)\bar{G}_l(1/z)v_{j,l} = \sum_{l=-p+1}^{p-1}c'_l z^l.$$

Since $J(z)$ can be computed once κ is known, the coefficients $\boldsymbol{\theta} = (\theta_0, \theta_1, \ldots, \theta_{p-1})$ are obtained by identifying the coefficients of the polynomials $z^{p-1}J(z)$ and $z^{p-1}\left(\sum_{h=0}^{p-1}\theta_h z^h\right)\left(\sum_{k=0}^{p-1}\bar{\theta}_k z^{-h}\right)$.

A state space representation

$$\boldsymbol{\xi}(t) = A\boldsymbol{\xi}(t-1) + \boldsymbol{\eta}(t)$$
$$x(t) = \boldsymbol{K}^{\mathrm{tr}}\boldsymbol{\xi}(t)$$

and its implications on the covariances of the OU process in the general case are slightly more complicated. When $\kappa_1, \ldots, \kappa_q$ are all different, $p_1, \ldots, p_q$ are positive integers, $\sum_{h=1}^{q}p_h = p$ and κ is a p-vector with p_h repeated components equal to κ_h, the OU(p) process x_κ is a linear function of the state space vector

$$(\xi_{\kappa_1}^{(0)}, \xi_{\kappa_1}^{(1)}, \ldots, \xi_{\kappa_1}^{(p_1-1)}, \ldots, \xi_{\kappa_q}^{(0)}, \xi_{\kappa_q}^{(1)}, \ldots, \xi_{\kappa_q}^{(p_q-1)})^{tr}$$

and the transition equation is no longer expressed by a diagonal matrix. We omit the details in this presentation, and note again that these are found in [1].

7   Estimation of the Parameters κ and σ of OU(p) Process

Though $\gamma(t)$ depends continuously on κ, the same does not happen with each term in the expression for the covariance, because of the lack of boundedness of the coefficients of the linear combination when two different values of the components of κ approach each other.

Since we wish to consider real processes x and the process itself and its covariance $\gamma(t)$ depend only of the unordered set of the components of κ, we shall reparameterise the process. With the notation $K_{j,i} = \frac{1}{(-\kappa_j)^i\prod_{l\neq j}(1-\kappa_l/\kappa_j)}$ (in particular, $K_{j,0}$ is the same as K_j), the processes $x_i(t) = \sum_{j=1}^{p}K_{j,i}\xi_j(t)$ and the coefficients $\boldsymbol{\phi} = (\phi_1, \ldots, \phi_p)$ of the polynomial $g(z) = \prod_{j=1}^{p}(1+\kappa_j z) = 1 - \sum_{j=1}^{p}\phi_j z^j$ satisfy $\sum_{i=1}^{p}\phi_i x_i(t) = x(t)$. Therefore, the new parameter $\boldsymbol{\phi} = (\phi_1, \ldots, \phi_p) \in \mathbf{R}^p$ shall be adopted.

7.1   Matching Correlations

From the closed formula for the covariance γ and the relationship between κ and ϕ, we have a mapping $(\phi, \sigma^2) \mapsto \gamma(t)$, for each t. Since $\rho^{(T)} := (\rho(1), \ldots, \rho(T))^{\mathrm{tr}}$ $= (\gamma(1), \ldots, \gamma(T))^{\mathrm{tr}}/\gamma(0)$ does not depend on σ^2, these equations determine a map $\mathcal{C} : (\phi, T) \mapsto \rho^{(T)} = \mathcal{C}(\phi, T)$, for each T. After choosing a value of T and obtaining an estimate $\rho_e^{(T)}$ of $\rho^{(T)}$ based on x, we propose as a first estimate of ϕ, the vector $\check{\phi}_T$ such that all the components of the corresponding κ have positive real parts, and such that the euclidean norm $\|\rho_e^{(T)} - \mathcal{C}(\check{\phi}_T, T)\|$ reaches its minimum, that is, a procedure that resembles the *method of moments*.

In our experiments we arbitrarily set the value of T to be the integral part of $0.9 \times n$, where n is the number of observations. This is in fact an upper bound value since, as we have observed, the graphs of $\check{\phi}_T$ for several values of T show in each case, that after T exceeds a moderate threshold, the estimates remain practically constant.

7.2   Maximum Likelihood Estimation of the Parameters of OU(p)

From the observations $\{\mu + x(i) : i = 1, \ldots, n\}$, obtain the likelihood L of the vector $x = (x(1)), \ldots, x(n))$:
$$\log L(\boldsymbol{x}; \phi, \sigma) = -\tfrac{n}{2}\log(2\pi) - \tfrac{1}{2}\log(\det(V(\phi, \sigma)) - \tfrac{1}{2}\boldsymbol{x}^{\mathrm{tr}}(V(\phi, \sigma))^{-1}\boldsymbol{x}$$
with $V(\phi, \sigma)$ equal to the $n \times n$ matrix with components $V_{h,i} = \gamma(|h - i|)$ $(h, i = 0, \ldots, n)$, that reduce to $\gamma(0)$ at the diagonal, $\gamma(1)$ at the 1st sub and super diagonals, etc. Obtain via numerical optimization the MLE $\hat{\phi}$ of ϕ and $\hat{\sigma}^2$ of σ^2. The estimator obtained from Matching Correlations can be used as an initial iterate. The estimations $\hat{\kappa}$ follow by solving $\prod_{j=1}^{p}(1 + \hat{\kappa}_j z) = 1 - \sum_{j=1}^{p} \hat{\phi}_j z^j$.

7.3   Some Simulations

We have simulated various series from OU(p) for different values of p and combinations of κ (real and complex, with repeated roots, etc.). These experiments have shown that the correlations of the series with the estimated parameters, either applying MC or ML, are fairly adapted to each other and to the empirical correlations. Here is one example.

Example 1. A series $(x_i)_{i=0,1,\ldots,n}$ of $n = 300$ observations of the OU_κ process x ($p = 3$, $\kappa = (0.9, 0.2 + 0.4\imath, 0.2 - 0.4\imath)$, $\sigma^2 = 1$) was simulated, and the parameters $\beta = (-1.30, -0.56, -0.18)$ and $\sigma^2 = 1$ were estimated by means of $\check{\beta}_T = (-1.9245, -0.6678, -0.3221)$, $T = 270$, $\hat{\beta} = (-1.3546, -0.6707, -0.2355)$ and $\hat{\sigma}^2 = 0.8958$. The corresponding estimators for κ are $\check{\kappa} = (1.6368, 0.1439 +0.4196\imath, 0.14389 -0.4196\imath)$ and $\hat{\kappa} = (0.9001, 0.2273 + 0.4582\imath, 0.2273 - 0.4582\imath)$. Figure 1 describes the theoretical, empirical and estimated covariances of x for different lags under the assumption $p = 3$, the actual order of x. The results obtained when the estimation is performed for $p = 2$ and $p = 4$ are also shown.

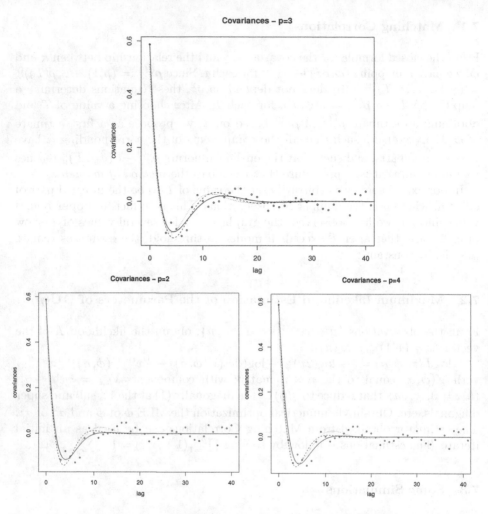

Fig. 1. Empirical covariances (○) and covariances of the MC (—) and ML (- - -) fitted OU models, for $p = 3$, 2 and 4, corresponding to Example 1. The covariances of OU_κ are indicated with a dotted line.

8   Applications to Real Data

We present two experimental results on sets of real data. The first data set is "Series A" from [2], and correspond to equally spaced observations of continuous time processes that can be assumed to be stationary. The second one is a series obtained by choosing one in every 100 terms of a high frequency recording of oxygen saturation in blood of a newborn child[1].

---

[1] The data were obtained by a team of researchers of Pereira Rossell Children Hospital in Montevideo, Uruguay, integrated by L. Chiapella, A. Criado and C. Scavone. Their permission to analyze the data is gratefully acknowledged by the authors.

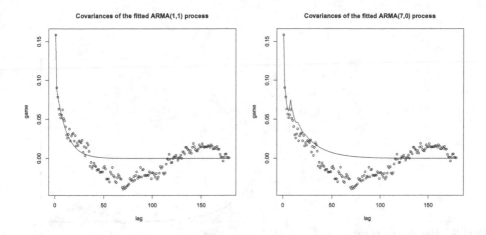

Fig. 2. Empirical covariances (∘) and covariances of the ML (—) fitted models ARMA(1,1) and AR(7) for Series A

8.1   Box, Jenkins and Reinsel "Series A"

The Series A is a record of $n = 197$ chemical process concentration readings, taken every two hours, introduced with that name and analyzed in [2, Ch. 4]. The authors suggest an ARMA(1,1) as a model for this data, and subsets of AR(7) are proposed in [6] and [9]. Figure 2 shows that these models fit fairly well the autocovariances for small lags, but fail to capture the structure of autocorrelations for large lags present in the series. On the other hand, the approximations obtained with the OU(3) process reflects both the short and long dependences, as shown in Figure 3.

The parameters of the OU(3) fitted by maximum likelihood for Series A are

$$\hat{\boldsymbol{\kappa}} = (0.8293, 0.0018 + 0.0330i, 0.0018 - 0.0330i) \quad \text{and} \quad \hat{\sigma} = 0.4401$$

The corresponding ARMA(3,2) is

$$(1 - 2.4316B + 1.8670B^2 - 0.4348B^3)x = 0.4401(1 - 1.9675B + 0.9685B^2)\epsilon$$

On the other hand, the ARMA(3,2) fitted by maximum likelihood is

$$(1 - 0.7945B - 0.3145B^2 + 0.1553B^3)x = 0.3101(1 - 0.4269B - 0.2959B^2)\epsilon.$$

The Akaike Information Criterion (AIC) of the parsimonious OU model is $8 - 2\ell'' = 109.90$, slightly better than the AIC of the unrestricted ARMA model, equal to $12 - 2\ell' = 110.46$. Finally we show in Figure 4 the predicted values of the continuous parameter process $x(t)$, for t between $n - 7$ and $n + 4$ (190-201), obtained as the best linear predictions based on the last 90 observed values, and on the correlations given by the fitted OU(3) model. The upper and lower lines are 2σ-confidence limits for each value of the process.

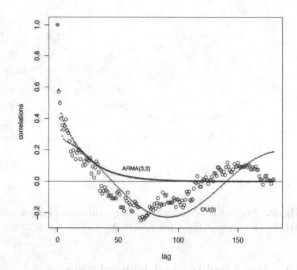

Fig. 3. Empirical covariances, and covariances of the ARMA(3,2) and OU(3) fitted by maximum likelihood

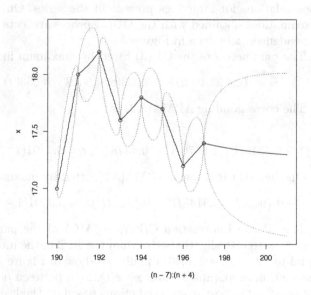

Fig. 4. Confidence bands for interpolated and extrapolated values of Series A for continuous domain

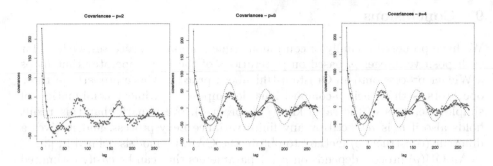

Fig. 5. Empirical covariances (○) and covariances of the MC (—) and ML (- - -) fitted OU(p) models for $p = 2, 3, 4$ corresponding to the series of O_2 saturation in blood

8.2   Oxygen Saturation in Blood

The oxygen saturation in blood of a newborn child has been monitored during seventeen hours, and measures taken every two seconds. We assume that a series $x_0, x_1, \ldots, x_{304}$ of measures taken at intervals of 200 seconds is observed, and fit OU processes of orders $p = 2, 3, 4$ to that series. Again the empirical covariances of the series and the covariances of the fitted OU(p) models for $p = 2$, $p = 3$ and $p = 4$ are plotted (see Figure 5) and the estimated interpolation and extrapolation, by using the estimated OU(3), are shown in Figure 6. In the present case, the actual values of the series for integer multiples of $1/100$ of the unit measure of 200 seconds are known, and plotted in the same figure.

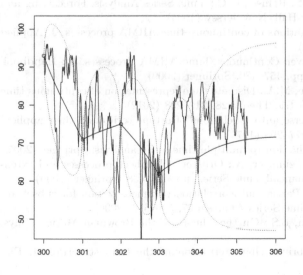

Fig. 6. Partial graph showing the five last values of the series of O_2 saturation in blood at integer multiples of the 200 seconds unit of time (○), interpolated and extrapolated predictions (—), 2σ confidence bands (- - -), and actual values of the series

9   Conclusions

We have proposed a family of continuous time stationary processes, OU(p), for each positive integer p, based on p iterations of the linear operator that maps a Wiener process onto an Ornstein-Uhlenbeck process. A nice property of these operators is that their p-compositions decompose as a linear combination of simple operators of the same kind (Theorem 1). We remark that this result holds also if w is replaced by any finite variance Lévy process, and it gives a different (constructive) method to obtain the CARMA models defined in [3].

An OU(p) process depends on $p + 1$ parameters that can be easily estimated by either maximum likelihood (ML) or matching correlations (MC) procedures. Matching correlation estimators provide a fair estimation of the covariances of the data, even if the model is not well specified. When sampled on equally spaced instants, the OU(p) family can be written as a discrete time state space model, and as it turns out, the families of OU(p) models constitute a parsimonious subfamily of the ARMA($p, p - 1$) processes. Furthermore, the coefficients of the ARMA can be deduced from those of the corresponding OU(p).

We have shown examples for which the ML-estimated OU model is able to capture a long term dependence that the ML-estimated ARMA model does not show. This leads to recommend the inclusion of OU models as candidates to represent stationary series to the users interested in such kind of dependence.

References

1. Arratia, A., Cabaña, A., Cabaña, E.M.: Modeling stationary data by a class of generalized Ornstein-Uhlenbeck processes (2014), Extended version available at http://www.lsi.upc.edu/$\sim$argimiro/mypapers/Journals/ACCOUp2014.pdf
2. Box, G.E.P., Jenkins, G.M., Reinsel, G.C.: Time Series Analysis. Forecasting and Control, 3rd edn. Prentice-Hall, New Jersey (1994)
3. Brockwell, P.J.: Representations of continuous–time ARMA processes. J. Applied Prob. 41A, 375–382 (2004)
4. Brockwell, P.J.: Lévy–Driven Continuous–Time ARMA Processes. In: Handbook of Financial Time Series, pp. 457–480. Springer (2009)
5. Chambers, M.J., Thornton, M.A.: Discrete time representation of continuous time ARMA processes. Econometric Theory 28, 219–238 (2012)
6. Cleveland, W.S.: The inverse autocorrelations of a time series and their applications. Technometrics 14, 277–298 (1971)
7. Lütkepohl, H.: New introduction to multiple time series analysis. Springer (2005)
8. Maller, R.A., Müller, G., Szimayer, A.: Ornstein-Uhlenbeck processes and extensions. In: Handbook of Financial Time Series, pp. 421–438. Springer (2009)
9. McLeod, A.I., Zhang, Y.: Partial autocorrelation parameterization for subset autoregression. Journal of Time Series Analysis 27, 599–612 (2006)
10. Uhlenbeck, G.E., Ornstein, L.S.: On the Theory of the Brownian Motion. Phys. Rev. 36, 823–841 (1930)
11. Vasicek, O.A.: An equilibrium characterisation of the term structure. J. Fin. Econ. 5, 177–188 (1977)

An Approach to Controlling the Runtime for Search Based Modularisation of Sequential Source Code Check-ins

Mahir Arzoky[1], Stephen Swift[1], Steve Counsell[1], and James Cain[2]

[1]Brunel University, Middlesex, UK
{mahir.arzoky,stephen.swift,steve.counsell}@brunel.ac.uk
[2]Quantel Limited, Newbury, UK
james.cain@quantel.com

Abstract. Software module clustering is the problem of automatically partitioning the structure of a software system using low-level dependencies in the source code to understand and improve the system's architecture. Munch, a clustering tool based on search-based software engineering techniques, was used to modularise a unique dataset of sequential source code software versions. This paper employs a seeding technique, based on results from previous modularisations, to improve the effectiveness and efficiency of the procedure. In order to reduce the running time further, a statistic for controlling the number of iterations of the modularisation based on the similarities between time adjacent graphs is introduced. We examine the convergence of the heuristic search technique and estimate and evaluate a number of stopping criterion. The paper reports the results of extensive experiments conducted on our comprehensive time-series dataset and provides evidence to support our proposed techniques.

Keywords: Software module clustering, modularisation, SBSE, seeding, time-series, fitness function.

1 Introduction

Large software systems tend to have complex structures that are often difficult to comprehend due to the large number of modules and inter-relationships that exist between them. As the modular structure of a software system tends to decay over time, it is important to modularise. Modularisation can facilitate program understanding and makes the problem at hand easier to understand, as it reduces the amount of data needed by developers [7]. Modularisation is the process of partitioning the structure of software system into subsystems. Subsystems group together related source-level components and can be organised hierarchically to allow developers to navigate through the system at various levels of details; they include resources such as modules, classes and other subsystems [7].

Directed graphs can be used to make the software structure of complex systems more comprehensible [16]. They can be described as language-independent, whereby components such as classes or subroutines of a system are represented as nodes and the inter-relationships between the components represented as edges. Such graphs are

H. Blockeel et al. (Eds.): IDA 2014, LNCS 8819, pp. 25–36, 2014.

referred to as Module Dependency Graph (MDG). Creating an MDG of the system does not make it easy to understand the system's structure; graphs could be partitioned to make them more accessible and easier to comprehend. Dependence information from system source code is used as input information. A file is considered as a module and the reference relationship between files is considered to be a relationship. Mancoridis et al. [12] were the first to use MDG as a representation of the software module clustering problem.

For various search algorithms [15], search-based software engineering has been shown to be highly robust. There have been a large number of studies [8] [9] [10] [13] [17] using the search-based software engineering approach to solve the software module-clustering problem. In previous studies, techniques that treat clustering as an optimisation problem were introduced. A number of various heuristic search techniques, including Hill Climbing were used to explore the large solution space of all possible partitions of an MDG.

In addition, there are a number of other alternative approaches for improving the efficiency and convergence of relatively sparse matrices [5]. However, some of these methods are specifically adapted to very sparse matrices and it is not guaranteed that the MDG will be of a particular sparseness. On the contrary, we are using a general purpose method that is more adaptable.

This paper introduces strategies to modularise source code check-ins, taking advantage of the fact that the dataset is time-series based. The nearer the source code in time, the more similar it is expected to be, and also the more similar the modularisation is expected to be. This paper extends [2] and [3] that introduced the seeding technique to improve the effectiveness and efficiency of the modularisation procedure. For this paper, we introduce a statistic for controlling the number of iterations of the modularisation based on similarities between time adjacent graphs. We aim to reduce the running time of the process by estimating and evaluating a number of stopping criterion.

The paper is organised as follows: Section 2 and 3 describe the experimental methods and highlight the creation and pre-processing of the source data. Section 4 explains the move operator and its implications. Section 5 and 6, respectively describes the experimental procedure and discusses the results. Finally, Section 7 draws conclusions and outlines future work of the project.

2   Experimental Methods

2.1   Clustering Algorithm

This work extends that of Arzoky et al. [2] and [3] and, follows Mancoridis et al., and Mitchell [12] [16], who first introduced search-based approach to software modularisation. The clustering algorithm was re-implemented from available literature on Bunch's clustering algorithm [12] to form a tool called Munch. Munch is a prototype implemented to carry out experimentations of different heuristic search approaches and fitness functions. Munch's uses an MDG as an input and produces a hierarchical decomposition of the system structure as an output. Closely related modules are grouped into clusters which are loosely connected to other clusters. A cluster is a set of the modules in each partition of the clustering.

The aim is to produce a graph partition which minimises coupling between clusters and maximises cohesion within each cluster. Coupling is defined as the degree of dependence between different modules or classes in a system, whereas cohesion is the internal strength of a module or class [18].

The clustering algorithm uses a simple random mutation Hill Climbing approach [15] to guide the search; refer to Algorithm 1 for the pseudo code. It is a simple, easy to implement technique that has proven to be useful and robust in terms of modularisation [16].

```
Algorithm 1. MUNCH(ITER,M)
Input: ITER- the number of iterations (runs),
M - An MDG
1) Let C be a random (or specified - for seeded)
   clustering arrangement
2) Let F = Fitness Function
3) For i = 1 to ITER (number of iterations)
4)    Choose two random clusters X and Y (X≠Y)
5)    Move a random variable from cluster X to Y
6)    Let F'= Fitness Function
7)    If F' is worse than F Then
8)       Undo move
9)    Else
10)      Let F = F'
11)   End If
12) End For
Output: C - a modularisation of M
```

2.2 Fitness Function

A fitness function is used to measure the relative quality of the decomposed structure of system into subsystems (clusters). In our previous work, we experimented with two fitness functions: the Modularisation Quality (MQ) metric of Mancoridis et al [12], and the EValuation Metric (EVM) of Tucker et al [19]. We also introduced EValuation Metric Difference (EVMD), a faster version of EVM. EVMD was selected for the modularisations as it is more robust than MQ and faster than EVM [3].

For the following formal definition of EVM, a clustering arrangement C of n items is defined as a set of sets $\{c_1, \ldots, c_m\}$, where each set (cluster) $c_i \subseteq \{1,...,n\}$ such that $c_i \neq \phi$ and $c_i \cap c_j = \phi$ for all $i \neq j$. Note that $1 \leq m \leq n$ and $n > 0$. Note also that $\bigcup_{i=1}^{m} c_i = \{1,..., n\}$. Let MDG M be an n by n matrix, where a '1' at row i and column j (M_{ij}) indicates a relationship between variable i and j, and '0' indicates that there is no relationship. Let c_{ij} refer to the jth element of the ith cluster of C. The score for cluster c_i is defined in Equation 2.

$$EVM(C,M) = \sum_{i=1}^{m} h(c_i, M) \qquad (1)$$

$$h(c_i, M) = \begin{cases} \sum_{a=1}^{|c_i|-1} \sum_{b=a+1}^{|c_i|} L(c_{ia}, c_{ib}) & , \text{if } |c_i| > 1 \\ 0 & , \text{Otherwise} \end{cases} \qquad (2) \qquad L(v_1, v_2, M) = \begin{cases} 0 & , v_1 = v_2 \\ +1 & , M_{v_1 v_2} + M_{v_2 v_1} > 0 \\ -1 & , \text{Otherwise} \end{cases} \qquad (3)$$

The objective of these heuristic searches is to maximise the fitness function. EVM has a global optimum which corresponds to all modules in a single cluster, where modules are all related to each other. The theoretical maximum possible value for EVM is the total number of links (relationships) in the graph, whereas the minimum value is simply the negative of the total number of links. EVM rewards maximising the cohesiveness of the clusters (presence of intra-module relationships), but it does not directly penalise inter-clustering coupling. In other words, it searches for all possible relationships within a cluster and rewards those that exist within the MDG and penalises those that do not exist within the MDG [12].

To speed up the process of the modularisation, EVMD was defined. It utilises an update formula on the assumption that one small change is being made between clusters. It is a faster way of evaluating EVM, where the previous fitness is known and the current fitness is calculated, without having to do the move. It produces the same results as EVM, but reduces the computational operations from $O(n\sqrt{n})$ to $O(\sqrt{n})$. For the formal definition of the EVMD we refer the reader to [3]. From this point forward EVM will be used when referring to the EVMD metric.

2.3   HS Metric

Homogeneity and Separation (HS) is an external coupling metric defined to measure the quality of the modularisation. HS is based on the Coupling Between Objects (CBO) metric, first introduced by Chidamber and Kemerer [6]. CBO (for a class) is defined as the count of the number of other classes to which it is coupled. It is based on the concept that if one object acts on another, then there is coupling between the two objects. Since the properties between objects of the same class are the same, the two classes are coupled when methods of one class use the methods defined by the other [6]. HS is a simple and intuitive coupling metric which calculates the ratio of the proportion of internal and external edges. HS is calculated by subtracting the number of links within clusters from the number of links that are between clusters, and then dividing the answer by the total number of links (to normalise it). The more links between the clusters the worse the modularisation, as only internal links are modularised (and not external ones). A value of +1 is returned if all the links are within the modules, a value of −1 is returned if all links are external coupling, and 0 is produced if there is an equal number. For the formal mathematical definition of the HS metric we refer the reader to [2].

3   Experimental Design

3.1   Data Creation

The large dataset used for this paper is from processed source code of an award win-ning product line architecture library, provided by Quantel. Quantel is one of the world's leading developers of high performance content creation and delivery systems across television and film post production. The dataset consists of information about different versions of a software system over time. The data source for this project is from processed source code of an award winning product line architecture library that has delivered over 15 distinct products. The entire code base currently runs to over 12 million lines of C++. The subset we are analysing for this paper is over 0.5 million lines of C++ code collected over the period 17/10/2000 to 03/02/2005, with 503 ver-sions in total. There are roughly 2-3 days' gaps between each check-in, giving a total timespan of 4 years and 4 months for the full dataset [4].

Table 1. Class Relation Types

Class relationship	Description
Attributes	Data members in a class
Bases	Immediate base classes
Inners	Any type declared inside the scope of a class
Parameters	Parameters to member functions of a class
Returns	Return from member functions of a class

A total of 6120 classes exist in the system; however, not all classes exist at the same time slice; there are between 434 and 2272 classes that exist at any particular point in time, referred to as "active" classes. The dataset consists of five time-series of un-weighted (binary) graphs. For this paper, graphs of the five types of relationship were merged together to form the 'whole system' for particular time slices. The relationship of how each graph represented between classes is shown in Table 1. Furthermore, the MDGs were significantly reduced. All modules that were not produced by Quantel and were not active at the time slice were removed. This has reduced the size of the graphs and the running time of the modularisation considerably. There are now between 194 and 1164 active classes at any one point, refer to Fig. 1. From the plot it can be ob-served that the number of active classes increases throughout the project.

3.2   Absolute Value Difference (AVD)

From experimentations conducted in [2] and [3] we predicted and showed that be-tween each software version there were no significant changes to the source code that made two successive versions very different (for seeding not to be possible). We pro-duced a set of results showing the similarity between the graphs Equation 4 shows how the AVD was calculated for each graph, where X and Y are two n by n binary matrices (MDGs). An AVD value of 0 indicates that two matrices are identical, whereas a large positive value indicates that they are different. A value between 0 and a large number gives a degree of similarity. Fig. 2 shows the AVDs of the full dataset.

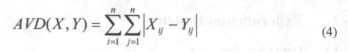

$$AVD(X,Y) = \sum_{i=1}^{n} \sum_{j=1}^{n} \left| X_{ij} - Y_{ij} \right| \tag{4}$$

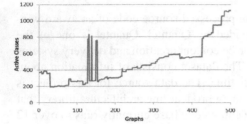
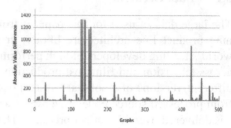

Fig. 1. Active classes at each software check-in

Fig. 2. Plot showing the AVDs of the full dataset

4   Modelling the Move Operator

For the following section, let MDG_1 and MDG_2 be an n by n matrix, G_1 be the optimal clustering arrangement, M_1 be the MDG associated with the clustering arrangement, E_1 be the optimal EVM for MDG_1 and E_2 be the optimal EVM for MDG_2. A difference of 1 between two MDGs indicates that one edge is being added or deleted. Assume that E_1 is the optimal EVM applied to M_1 and G_1 associated modularisation, and also that the data is of solid and dense clusters. In addition, from the literature we estimate and assume that the size and the number of clusters is $\sqrt{n}$ [14]. Finally, we hypothesise that only one move is needed to make the fitness function change.

When an edge is added or deleted, the difference in MDG is either going to be between two different clusters or between the same cluster. Thus, there are four possibilities that would result in a fitness change and thus impact the EVM value, refer to Table 2.

If an edge is added to the same cluster then the fitness function, EVM, will be incremented by one. But, if an edge was deleted from the same cluster, then EVM will be decremented by one, the edge will no longer be there and thus will be penalised. If an edge is deleted between two different clusters, EVM will not change. This is because EVM only looks at intra-clusters - there is no penalisation between clusters. On the other hand, if they are in different clusters and an edge is added, either the EVM does not change or the best EVM is attained by moving the variable into the cluster. If we assume that the size of the first cluster is $\sqrt{n}$ and the size of the second cluster is $\sqrt{n}$, this indicates that EVM will be incremented by one.

Table 2 shows the change to EVM, where E_1 is the old fitness and E_2 is the new fitness. From the table it can be seen that the worst case scenario involves choosing the correct variable and placing it in the correct cluster, to account for the one difference in the MDG, which will be the probable one difference in the EVM. Thus, now we compute the probability of a move occurring, which is linked to the iterations attempts in a Hill Climb.

Table 2. Implications of a move

	Same cluster	Different clusters
Add edge	$E_2 = E_1 + 1$	$E_2 = E_1$ OR $E_2 = E_1 + 1$
Delete edge	$E_2 = E_1 - 1$	$E_2 = E_1$

For each one difference between the MDGs, the correct variable needs to be selected. Normally, if a wrong move is made, there would either be no effect on the fitness or the fitness would be decremented by one. However, since we are using a Hill Climbing algorithm, if a wrong move is made a worst fitness would not be accepted.

Let n be the number of variables (classes) in an MDG. Let d be the AVD between two MDGs, and T be the number of iterations we are running the process for. There is a 1 in n chance of selecting the right variable, and to move it to the correct cluster there are $\sqrt{n}$ clusters. There are n variables to choose from and they can be moved to $\sqrt{n}-1$ clusters, as one cluster can be ruled out and that is the cluster it originated from.

Assume that $\Pr(\text{correct move}) = P = 1/(n\sqrt{n})$, Let $Q = 1-P$
The chance a single move occurs after T iterations is as follows:

$$\Pr(T=1) = P, \Pr(T=2) = PQ, \Pr(T=3) = PQ^2 \ \dots \ \Pr(T=i) = PQ^{i-1}$$

If we have d moves to make, then the probability that all of the d moves are made after T iterations of the Hill Climbing algorithms is:
$\Pr(\text{All } d \text{ moves after } T \text{ iterations}) = (1-Q^T)^d$
Let us assume that there is some acceptable level of confidence α that all the moves have been made, then we wish to compute a T for which this might happen:

$$\alpha = (1-Q^T)^d \qquad T\ln(Q) = \ln(1-\alpha^{1/d})$$
$$\alpha^{1/d} = 1-Q^T$$
$$Q^T = 1-\alpha^{1/d} \qquad T = \frac{\ln(1-\alpha^{1/d})}{\ln(Q)} \qquad (5)$$

5 Experimental Procedure

Two experiments that modularise the dataset were designed for this paper. The main difference between the experiments is the number of iterations they run for and their starting clustering arrangements; otherwise it is the same program. The two experiments were repeated 25 times each as Hill Climb is a stochastic method and there is a risk of the search reaching only the local maxima and thus produce varying results.

For experiment 1 (C), we modularised the dataset for 10 million iterations each. The starting clustering arrangement consisted of every variable in its own cluster. It assumes that all classes are independent; there are no relationships.

For experiment 2 (S), we modularised the dataset using results of the previous clustering arrangement from C. Instead of creating a random starting arrangement for the modularisation, the clustering arrangement of the preceding graph (produced from C) was used to give it a head start. For experiment 2, we selected a number of strategies to try to estimate the stopping conditions and find the minimal runtime needed for the process, they are:

Strategy 1 - The number of iterations for this strategy was fixed at 100,000 iterations, representing 1 per cent of the full run, apart from the first graph which was run for 10 million iterations.

Strategy 2 - The number of iterations for this strategy varied depending on the similarity between graphs. The AVD was calculated for all graphs and was used as a scalar for calculating the number of iterations. The more similar two successive graphs (low AVD), the less runs needed; and the more different they are (high AVD), the higher the number of iterations needed. The following equation was used for calculating the number of iterations of each graph: ITER = AVD X 8000.

Strategy 3 - is an estimate based on the probability of making the right move, computed using Equation (5). Several acceptable level of confidence values that represent the likelihood of obtaining the correct answer were selected; they were, T_1 - 99%, T_2 - 95%, T_3 - 90% and T_4 - 70%.

The convergence points of the 3 strategies (6 policies above) were computed and the maximum of these at each time slice was calculated. Convergence point can be defined as the earliest point in the iterations of the heuristic search of when the fitness function no longer increases until the end of the run. An extra 5% of the estimated number of iterations was added to the iterations of all graphs, for each of the 6 policies. Results produced were used to run Experiment 2; graphs were modularised using these computed values apart from the first graph, which was run for the full 10 million iterations.

6   Results and Discussion

The amount of time it takes the modularisation program to run is proportional to the number of fitness function calls. In our experiments, the number of fitness function calls is referred to as the number of iterations, and the time it takes the program to run is proportional to the number of iterations

Strategy 1 and 2 were introduced in [2] and needed improvement as the software check-ins do not necessarily need to run for a set number of iterations. The process might continue to run even when the algorithm has converged. Thus, Strategy 3 was introduced in order to estimate the number of iterations needed for each graph. The average fitness function calls for Strategy 1 and 2 are 100,000 and 464,956 iterations, respectively. The fitness function calls for Strategy 3 range from 12,825 iterations for T_4 to 23,162 iterations for T_1. From the results it can be seen that there is a large efficiency improvement using the new strategy compared to previous strategies.

Fig. 3 shows a count of the closest strategy estimate to the converged point. From the plot it can be seen that T_4 is the most accurate estimate as it is the closest/nearest to the converged point for most graphs. Even though the new strategy was based on a broad estimate of the number of average clusters [14], it still produced better estimates than the old strategies. Results show that 71 graphs from the dataset were modularised using the old strategies, whereas 260 graphs were modularised using Strategy 3. 171 of the graphs were omitted from the plot as they are zeros for all of the strategies. Currently, we are only looking at the most accurate strategy and thus we did not account for whether it is an underestimate or overestimate of the convergence point. We look to investigate this further as part of future work.

Given that the Munch algorithm runs for T, iterations, the fitness function is $O(\sqrt{n})$, and that the fitness function is where all of the computational complexity of the Hill Climbing algorithm is, then the overall complexity of the run is $O(T\sqrt{n})$. Thus, the smaller the value of T the faster the algorithm runs. Table 3 shows the time savings under each scheme compared to the full run of 10 million iterations. From the table it can be seen that the least amount of saving in terms of runtime is 92.88%, this is for Strategy 2. Results show that T4 has the highest percentage of saving in terms of runtime. We have also compared and computed how fast each of the strategies compared to C (iterations reductions factor). The results show that T_4 is 509 times faster than the full iterations, more than 5 times faster than Strategy 1 and 36 times faster than Strategy 2.

Table 3. Time saving under all schemes

Strategy	Time Saving %	Iter.
Strategy 1– 1%	99.00	100
Strategy 2 – 8000D	92.88	14
T_1– 99%	99.65	282
T_2– 95%	99.72	360
T_3– 90%	99.76	412
T_4– 70%	99.80	509

Table 4. Count of highest

Strategy	Count of highest
Strategy 1– 1%	167
Strategy 2– 8000D	159
T_1– 99%	5
T_2– 95%	0
T_3– 90%	0
T_4– 70%	0

Table 4 displays a frequency count of the largest iterations all the time for each of the strategies. It can be clearly seen that Strategy 1 and Strategy 2 are nearly the highest for all graphs. This illustrates that the previous strategies had a higher running time for 326 graphs compared to only 5 graphs for the new strategy.

Fig. 4 shows a plot of the convergence points of C and S for the full datasets. The convergence points indicate that the EVM is at a maximum. A gradually increasing trend can be observed for C, which indicates that a longer running time is needed for later graphs. The general trend of the results correlate with Fig. 1, which shows a gradual increase of the number of active classes throughout the project. Results of S are considerably lower than C throughout the full dataset, which indicates that the seeding technique works well. This is particularly true when comparing the results with Fig. 6 and 7, as they produce the same EVM and HS values for the majority of the graphs.

Fig. 5 shows a plot of the EVM of experiments C and S for the full dataset. It is not possible to differentiate C from the plot, as it overlaps with S. S produces the same results as C despite the fact that S was ran for a fraction of the original time of C. This proves that the seeding technique works and to a fair degree of accuracy.

Fig. 6 shows a plot of HS values of experiments C and S for the full dataset. It is not possible to differentiate C from the plot, as it overlaps with S. The same results are produced despite the fact that S was ran for considerably less time than C. It can be observed that HS results are gradually getting worse throughout the life of the project. We believe that when the system was designed there was more coupling than cohesion in the modules and as a result the internal structure of the system design was deteriorating over time. The negative HS values indicate that the inter-modules are more than the intra-module edges. In addition, it seems that large changes events occurred a number of times throughout the life of the system. There seem to be a reduction of coupling to a certain degree during these events.

Fig. 7 shows a plot of HS against EVM for the whole system. To find out whether there is a relationship between HS and EVM they were correlated. A value of -0.791 is produced, which indicates that the correlation is highly significant. For over 500 pairs of observations the 1 per cent significance level is at 0.115. It is interesting to observe that this strong correlation illustrates the credibility of EVM as a good metric. The plot shows that EVM is a good predictor for HS. HS cannot be used as a fitness function, as it would re-arrange all clusters into one (HS value of 1.0); since there would be no coupling. Despite the fact that EVM is not a measure of coupling or cohesion, it was still strongly correlated with HS. Thus, the metric is performing as desired, achieving low coupling and high cohesion.

Fig. 8 shows a plot of the AVD against the convergence points of the full datasets. The correlation of the AVD and the convergence points is 0.658, which indicates a very high correlation. From the plot, it can be seen that the lower the difference between subsequent graphs the quicker it will converge. This is good evidence in support of our hypothesis that the larger the difference the more iterations are needed.

7   Conclusion and Future Work

This paper presents a heuristic search technique where a large and gradually evolving industrial software system is evaluated. The evaluation shows that the modularisation technique introduced runs much faster than prior modularisation techniques. Code structure and sequence was used to improve the efficiency and effectiveness of software module clustering. This paper builds on previous work which demonstrated that the seeded process of the modularisation works well. Previous work introduced Strategy 1 and Strategy 2 which resulted in 99% and 93.88% time saving in terms of runtime. However, using a scalar to control the number of iterations is a much more robust way of conducting the experiment than running the process for a fixed length.

For this paper we attempted to improve the efficiency and convergence of the search process by introducing a strategy based on probability values of the significance of the seeded graphs. Using the new seeding strategy, we managed to produce results identical to the full modularisation of graphs while reducing the running time by more than 500 times. Thus, from the results produced if we were to choose a scheme for running the seeded modularisation then we would select T_4 as the scheme to use. The same theory applies when modularising the dataset using the preceding results of the modularisation.

Although, the estimate is fundamentally based on the assumption that the average number of clusters is $\sqrt{n}$, the results of the new strategy clearly demonstrate a significantly better limit than Strategy 2, evidently revealing that the approximation method works. However, for future work, we look to obtain a better estimate of the number of clusters. We also look to obtain the number of clusters from the dataset and compute the probabilities whilst running the process. In addition, in this paper, we have demonstrated that our hypothesis, introduced in Section IV, works empirically; however for future work we will include the formalised mathematical proof of the claim. Furthermore, we aim to compare the techniques and approaches proposed in this paper against more systems and perform a more systematic comparison.

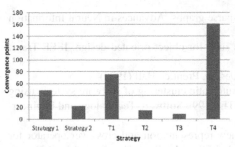

Fig. 3. Plot showing the rankings of the 6 policies

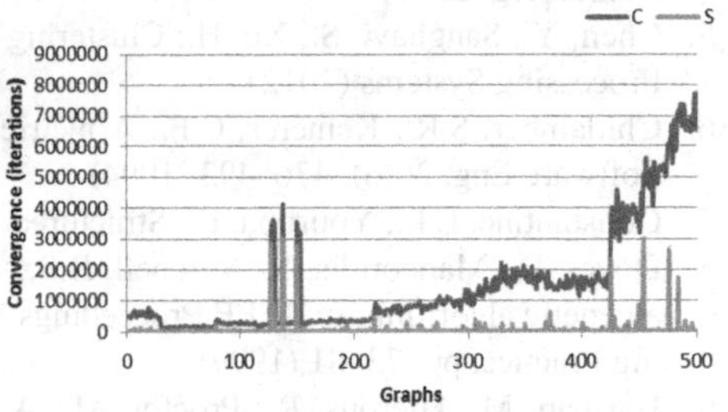

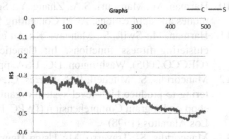

Fig. 4. Plot showing the convergence points of C and S for the full dataset

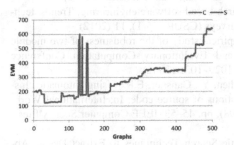

Fig. 5. Plot showing the EVM of C and S for the full dataset

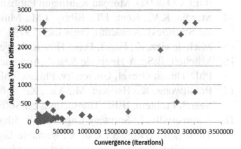

Fig. 6. Plot showing the HS of C and S for the full dataset

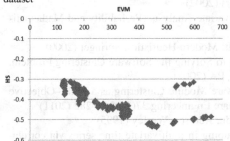

Fig. 7. Plot showing the HS against EVM for the full dataset

Fig. 8. Plot showing the AVDs against convergence points for the full dataset

References

1. Altman, D.G.: Practical Statistics for Medical research. Chapman and Hall (1997)
2. Arzoky, M., Swift, S., Tucker, A., Cain, J.: Munch: An Efficient Modularisation Strategy to Assess the Degree of Refactoring on Sequential Source Code Checkings. In: IEEE Fourth International Conference on Software Testing, Verification and Validation Workshops, pp. 422–429 (2011)

3. Arzoky, M., Swift, S., Tucker, A., Cain, J.: A Seeded Search for the Modularisation of Sequential Software Versions. Journal of Object Technology 11(2), 6:1–6:27

4. Cain, J., Counsell, S., Swift, S., Tucker, A.: An Application of Intelligent Data Analysis Techniques to a Large Software Engineering Dataset. In: Adams, N.M., Robardet, C., Siebes, A., Boulicaut, J.-F. (eds.) IDA 2009. LNCS, vol. 5772, pp. 261–272. Springer, Heidelberg (2009)

5. Chen, Y., Sanghavi, S., Xu, H.: Clustering sparse graphs. Advances in Neural Information Processing Systems (2012)

6. Chidamber, S.R., Kemerer, C.F.: A metrics suite for object oriented design. IEEE Trans, Software Eng. 20(6), 476–493 (1994)

7. Constantine, L.L., Yourdon, E.: Structured Design. Prentice Hall (1979)

8. Doval, D., Mancoridis, S., Mitchell, B.S.: Automatic clustering of software systems using a genetic algorithm. In: IEEE Proceedings STEP 1999 Software Technology and Engineering Practice, pp. 73–81 (1999)

9. Harman, M., Hierons, R., Proctor, M.: A new representation and crossover operator for search based optimization of software modularization. In: Proc. Genetic and Evolutionary Computation Conference, pp. 1351–1358. Morgan Kaufmann Publishers (2002)

10. Harman, M., Mansouri, S.A., Zhang, Y.: Search-based software engineering: Trends, techniques and applications. ACM Computing Surveys (CSUR) 45(1), 11 (2012)

11. Harman, M., Swift, S., Mahdavi, K.: An empirical study of the robustness of two module clustering fitness functions. In: Genetic and Evolutionary Computation Conference (GECCO 2005), Washington, DC, USA, pp. 1029–1036 (2005)

12. Mancoridis, S., Mitchell, B.S., Rorres, C., Chen, Y., Gansner, E.R.: Using automatic clustering to produce high-level system organizations of source code. In: International Workshop on Program Comprehension (IWPC 1998), pp. 45–53. IEEE Computer Society Press, Los Alamitos (1998)

13. Mancoridis, S., Traverso, M.: Using Heuristic Search Techniques to Extract Design Abstractions from Source Code. In: Proc. Genetic and Evolutionary Computation Conference (GECCO 2002). Morgan Kaufmann Publishers (2002)

14. Mardia, K.V., Kent, J.T., Bibby, J.M.: Multivariate Analysis (Probability and Mathematical Statistics). Academic Press Inc. (1979)

15. Michalewicz, Z., Fogel, D.B.: How to Solve It: Modern Heuristics. Springer (2000)

16. Mitchell, B.S.: A Heuristic Search Approach to Solving the Software Clustering Problem, PhD Thesis, Drexel, University, Philadelphia, PA (2002)

17. Praditwong, K., Harman, M., Yao, X.: Software Module Clustering as a Multi–Objective Search Problem. IEEE Transactions on Software Engineering 37(2), 264–282 (2011)

18. Sommerville, I.: Software Engineering, 5th edn. Addison-Wesley (1995)

19. Tucker, A., Swift, S., Liu, X.: Variable Grouping in multivariate time series via correlation. IEEE Transactions on Systems, Man, and Cybernetics, Part B: Cybernetics 31(2), 235–245 (2001)

Simple Pattern Spectrum Estimation
for Fast Pattern Filtering with CoCoNAD

Christian Borgelt and David Picado-Muiño

European Centre for Soft Computing
Gonzalo Gutiérrez Quirós s/n, 33600 Mieres, Spain
{christian.borgelt,david.picado}@softcomputing.es

Abstract. CoCoNAD (for *Continuous-time Closed Neuron Assembly Detection*) is an algorithm for finding frequent parallel episodes in event sequences, which was developed particularly for neural spike train analysis. It has been enhanced by so-called Pattern Spectrum Filtering (PSF), which generates and analyzes surrogate data sets to identify statistically significant patterns, and Pattern Set Reduction (PSR), which eliminates spurious induced patterns. A certain drawback of the former is that a sizable number of surrogates (usually several thousand) have to be generated and analyzed in order to achieve reliable results, which can render the analysis process slow (depending on the analysis parameters). However, since the structure of a pattern spectrum is actually fairly simple, we propose a simple estimation method, with which (an approximation of) a pattern spectrum can be derived from the original data, bypassing the time-consuming generation and analysis of surrogate data sets.

1 Introduction

About a year ago we presented CoCoNAD (for *Continuous-time Closed Neuron Assembly Detection*) [4], an algorithm for finding frequent parallel episodes in event sequences, which are defined over a continuous (time) domain. The name of this algorithm already indicates that the application domain motivating our investigation is the analysis of *parallel spike trains* in neurobiology: sequences of points in time, one per neuron, that represent the times at which an electrical impulse (*action potential* or *spike*) is emitted. Our objective is to identify *neuronal assemblies*, intuitively understood as groups of neurons that tend to exhibit synchronous spiking. Such cell assemblies were proposed as a model for encoding and processing information in biological neural networks [8]. As a (possibly) first step in the identification of neuronal assemblies, we look for (significant) *frequent neuronal patterns*, that is, groups of neurons that exhibit *frequent synchronous spiking* that cannot be explained as a chance occurrence [13,16]. In this paper we draw on this application domain for the parameters of the (artificially generated) data sets with which we tested the proposed pattern spectrum estimation, but remark that our method is much more widely applicable.

The CoCoNAD algorithm differs from other approaches to find frequent parallel episodes in event sequences, like those, for example, in [12,6,10] or [15]

H. Blockeel et al. (Eds.): IDA 2014, LNCS 8819, pp. 37–48, 2014.
© Springer International Publishing Switzerland 2014

(some of which are designed for discrete item sequences, although a transfer to a continuous (time) domain is fairly straightforward), by the support definition it employs. While the mentioned approaches define the support of a parallel episode as the (maximal) number of non-overlapping minimal windows covering instances of the episode, CoCoNAD relies on a maximum independent set (MIS) approach. This allows to count different instances of a parallel episode even though the windows covering them overlap, thus leading to a potentially higher support count. Nevertheless the resulting support measure remains anti-monotone, because no spike is contained in more than one counted instance [4].

Furthermore, in order to single out significant frequent patterns from the output, while avoiding the severe multiple testing problem that results from the usually very large number of frequent patterns, we proposed *pattern spectrum filtering* (PSF) in [13].[1] This method relies on generating and analyzing surrogate data sets as an implicit representation of the null hypothesis of items occurring independently. It eliminates all patterns found in the original data, for which a analogous pattern was found in a surrogate data set (since then the pattern can be explained as a chance event, cf. Section 3). This method was further detailed in [16], where it was also extended with *pattern set reduction* (PSR), which strives to eliminate spurious patterns that are merely induced by an actual pattern (that is, subset, superset and overlapping patterns) with a preference relation.[2]

These methods (PSF and PSR) proved to be very effective in singling out patterns from artificially generated data. However, the need to generate and analyze a sizable number of surrogate data sets (usually several thousand) can render the mining process slow, especially if the data exhibits high event frequencies and the analysis window width (maximum time allowed to cover an occurrence of a parallel episode) is chosen to be large. To overcome this drawback, we strive in this paper to exploit the fact that a pattern spectrum actually has a fairly simple structure and thus allows for an (at least approximate) estimation from the original data, bypassing surrogate data generation. The core idea is to count, based on the user-specified analysis window width, the possible "slots" for patterns of different sizes and to estimate from these counts the (expected) pattern support distribution with a Poisson approximation.

The remainder of this paper is structured as follows: Section 2 briefly reviews how (frequent) parallel episodes are mined with the CoCoNAD algorithm and Section 3 how the output is reduced with pattern spectrum filtering (PSF) and pattern set reduction (PSR) to significant, non-induced patterns. Section 4 describes the simple, yet effective method with which we estimate a pattern spectrum from the original data. In Section 5 we report experiments on artificially generated data sets and thus demonstrate the quality of pattern spectrum estimation. Finally, in Section 6 we draw conclusions from our discussion.

---

[1] Even though pattern spectrum filtering was presented for time-binned data in [13] (which reduces the problem to classical frequent item set mining: each time bin gives rise to one transaction), it can easily be transferred to the continuous domain.

[2] Although time-binned data was considered in [16], the idea of pattern set reduction can easily be transferred to continuous time, requiring only a small adaptation.

2   Mining Parallel Episodes with CoCoNAD

We (partially) adopt notation and terminology from [12]. Our data are (finite) *sequences of events* of the form $S = \{\langle i_1, t_1 \rangle, \ldots, \langle i_m, t_m \rangle\}$, $m \in \mathbb{N}$, where i_k in the *event* $\langle i_k, t_k \rangle$ is the *event type* or *item* (taken from an item base B) and $t_k \in \mathbb{R}$ is the time of occurrence of i_k, $k \in \{1, \ldots, m\}$. Note that the fact that S is a set implies that there cannot be two events with the same item occurring at the same time: events with the same item must differ in their occurrence time and events occurring at the same time must have different types/items. Note also that in our motivating application (i.e., spike train analysis), the items are the neurons and the events capture the times at which spikes are emitted.

Episodes (in S) are sets of items $I \subseteq B$ that are endowed with a partial order and usually required to occur in S within a certain time span. *Parallel episodes*, on which we focus in this paper, have no constraints on the relative order of their elements. An *instance (or occurrence) of a parallel episode* $I \subseteq B$, $I \neq \emptyset$, (or a *(set of) synchronous event(s)* for I) in an event sequence S with respect to a (user-specified) time span $w \in \mathbb{R}^+$ can be defined as a subsequence $\mathcal{R} \subseteq S$, which contains exactly one event per item $i \in I$ and which can be covered by a (time) window of width at most w. Hence the set of all instances of a parallel episode $I \subseteq B$, $I \neq \emptyset$, in S is

$$\mathcal{E}_S(I, w) = \{\mathcal{R} \subseteq S \mid \{i \mid \langle i, t \rangle \in \mathcal{R}\} = I \wedge |\mathcal{R}| = |I| \wedge \sigma(\mathcal{R}, w) = 1\},$$

where the operator σ captures the (approximate) synchrony of the events in $\mathcal{R}$:

$$\sigma(\mathcal{R}, w) = \begin{cases} 1 & \text{if } \max\{t \mid \langle i, t \rangle \in \mathcal{R}\} - \min\{t \mid \langle i, t \rangle \in \mathcal{R}\} \leq w, \\ 0 & \text{otherwise.} \end{cases}$$

That is, $\sigma(\mathcal{R}, w) = 1$ iff all events in $\mathcal{R}$ can be covered by a (time) window of width at most w. We then define the support of a parallel episode $I \subseteq B$ in S as

$$s_S(I, w) = \max\{|\mathcal{U}| \mid \mathcal{U} \subseteq \mathcal{E}_S(I, w) \wedge \forall \mathcal{R}_1, \mathcal{R}_2 \in \mathcal{U}; \mathcal{R}_1 \neq \mathcal{R}_2 : \mathcal{R}_1 \cap \mathcal{R}_2 = \emptyset\},$$

that is, as the size of a maximum independent set of the instances of I. Although in the general case the maximum independent set problem is NP-complete [9] and even hard to approximate [7], the problem instances we are facing here are constrained by the underlying one-dimensional time domain, which makes it possible to devise an efficient greedy algorithm that solves it exactly [14]. Pseudo-code of the support counting procedure can be found in [4].

Frequent parallel episodes are then mined, based on this support definition, with a standard recursive divide-and-conquer scheme that enumerates candidate item sets, which may also be seen as a depth-first search. The search is pruned, as in all such algorithms, with the so-called *apriori property*: *no superset of an infrequent parallel episode can be frequent*, since the support measure defined above can be shown to be *anti-monotone* (see, for example, [17,5]). Pseudo-code of the mining procedure including efficient event filtering can be found in [4].

3   Pattern Spectrum Filtering and Pattern Set Reduction

Trying to single out significant patterns proves to be less simple than it may appear at first sight, since one has to cope with the following two problems: in the first place, one has to find a proper statistic that captures how (un)likely it is to observe a certain pattern under the null hypothesis that items occur independently. Secondly, the huge number of potential patterns causes a severe multiple testing problem, which is not easy to overcome with standard methods. In [13] we provided a fairly extensive discussion and concluded that a different approach than evaluating individual patterns with statistics is needed.

As a solution, *pattern spectrum filtering* (PSF) was proposed in [13] based on the following insight: even if it is highly unlikely that a *specific group* of z items co-occurs c times, it may still be likely that *some group* of z items co-occurs c times, even if items occur independently. The reason is simply that there are so many possible groups of z items (unless the item base B as well as z are tiny) that even though each group has only a tiny probability of co-occurring c times, it may be almost certain that *one of them* co-occurs c times.[3] As a consequence, since there is no *a-priori* reason to prefer certain sets of z items over others (even though a refined analysis, on which we are working, may take individual item frequencies into account), we should not declare a pattern significant if the occurrence of a counterpart (same size z and same or higher support c) can be explained as a chance event under the null hypothesis of independent items.

As a consequence, we pool patterns with the same *pattern signature* $\langle z, c \rangle$, and collect for each signature the (average) number of patterns that we observe in surrogate data. This yields what we call a *pattern spectrum* (see Figures 2 and 3). Pattern spectrum filtering consists in keeping only such patterns found in the original data for which no counterpart with the same signature (or a signature with the same z, but larger c) was observed in surrogate data, as such a counterpart would show that the pattern can be explained as a chance event.

The essential part of this procedure is, of course, the generation of surrogate data, for which we rely on a simple permutation procedure: the occurrence times of the events are kept and the items (the event types) are randomly permuted. This destroys any co-occurrence of items that may be present in the data and thus produces data that implicitly represent the null hypothesis of independently occurring items. A discussion of other surrogate data generation approaches that are common in the area of neural spike train analysis can be found in [11].

Note that pattern spectrum filtering still suffers from a certain amount of *multiple testing*: every pair $\langle z, c \rangle$ that is found in the original data gives rise to one test. However, the pairs $\langle z, c \rangle$ are *much fewer* than the number of specific item sets. As a consequence, simple approaches like *Bonferroni correction* [2,1] become feasible, with which the number of needed surrogate data sets can be computed [13]: given a desired overall significance level α and the number k of

---

[3] This is actually the case for, say, $z = 5$ and $c = 4$ in our data, for which patterns are essentially certain to occur, see Figures 2 and 3, although the probability of observing a specific set of 5 items co-occurring 4 times is extremely small ($< 10^{-8}$).

pattern signatures to test, at least k/α surrogate data sets have to be analyzed. With the common choice $\alpha = 1\%$ and usually several dozen pattern signatures being observed, this rule recommends to generate several thousand data sets. In our experiments we always chose 10,000, regardless of the actual number of pattern signatures, in order to ensure a uniform procedure for all data sets.

As a further filtering step, *pattern set reduction* was proposed in [16], which is intended to take care of the fact that an actual pattern induces other, spurious patterns that are subsets or supersets or overlap the actual patterns. These spurious patterns are reduced with the help of a preference relation between patterns and the principle that only patterns are kept to which no other pattern is preferred. A simple heuristic, but very effective preference relation is the following: let $X, Y \subseteq B$ be two patterns with $Y \subseteq X$ and let $z_X = |X|$ and $z_Y = |Y|$ be their sizes and c_X and c_y their support values. The pattern X is preferred to Y if $z_X \cdot c_X \geq z_Y \cdot c_Y$. Otherwise Y is preferred to X. The core idea underlying this method is that under certain simplifying assumptions the occurrence probability of a pattern is inversely proportional to the number of individual events underlying it, that is, to the product $z \cdot c$. Intuitively, the above preference relation therefore prefers the less probable pattern. Alternatives to this preference relation and a more detailed discussion can be found in [16].

4   Pattern Spectrum Estimation

As already mentioned in the introduction, pattern spectrum filtering suffers from the problem that a sizeable number of surrogate data sets (usually several thousand) need to be generated and analyzed, which can render the analysis process slow, especially if due to high event frequencies and a large window width w an individual run already takes some time. Even though pattern spectrum generation lends itself very well to parallelization (since each surrogate data set can be generated and analyzed on a different processor core), it is desirable to find a faster way of obtaining (at least an approximation of) a pattern spectrum.

As a solution, we propose *pattern spectrum estimation* in this paper. This method draws on the idea that by counting the "slots" for patterns of different sizes, we can estimate the support distribution of the patterns via a standard Poisson approximation of the actual binomial distribution. By a "slot" for a pattern size z we mean any collection of z events in the event sequence $\mathcal{S}$ to analyze that can be covered by the chosen analysis window width w. Each such slot can hold an instance of a specific parallel episode $I \subseteq B, |I| = z$. With the probability of a pattern instance occurring in such a slot, that is, the probability that the z items constituting the parallel episode are chosen in a random selection (since we want to mimic independent items, as this is the implicitly represented null hypothesis), we obtain a probability distribution over the different numbers of occurrences of the parallel episode in the counted number of slots. This distribution is actually binomial, but it can be approximated well by a Poisson distribution, because the number of slots is usually very large while the occurrence probability of a specific parallel episode is very small.

By scaling the resulting probability distribution over the possible support values to the total number of patterns that can occur, we obtain expected counts for the different pattern signatures with size z. Executing the process for all sizes $z \in \{1, \ldots, |B|\}$ then yields the desired pattern spectrum.

Formally, the number of slots for each pattern size z is defined as

$$\forall z \in \{1, \ldots, |B|\}: \quad N_S(z, w) = \left| \{ \mathcal{R} \subseteq \mathcal{S} \mid |\mathcal{R}| = z \wedge \sigma(\mathcal{R}, w) = 1 \} \right|.$$

However, this formula does not lend itself well to implementation. Therefore, to count the slots for each pattern size, we first pass a sliding window over the event sequence $\mathcal{S}$, stopping at each event $\langle i, t \rangle \in \mathcal{S}$, and collecting the events in the (time) window $[t, t + w]$. That is, we consider the set of event sequences

$$\mathcal{W}_S(w) = \{ \mathcal{R}_e \mid e = \langle i, t \rangle \in \mathcal{S} \wedge \mathcal{R}_e = \{ \langle i', t' \rangle \in \mathcal{S} \mid t' \in [t, t + w] \} \}.$$

Using the mentioned sliding window method, this set is easy to enumerate.

From this set we then obtain the slot counts per pattern size z as

$$\forall z \in \{1, \ldots, |B|\}: \quad N_S(z, w) = \sum_{\mathcal{R} \in \mathcal{W}_S, |\mathcal{R}| \geq z} \binom{|\mathcal{R}| - 1}{z - 1}.$$

This formula can be understood as follows: only subsequences in $\mathcal{W}_S$ that contain at least z events can contain slots for a pattern of size z and therefore the sum considers only $\mathcal{R}$ with $|\mathcal{R}| \geq z$. In principle, all subsets of z events in a given $\mathcal{R}$ have to be considered. However, the event sequences in $\mathcal{W}_S$ overlap, and thus summing $\binom{|\mathcal{R}|}{z}$ could count the same slot multiple times. We avoid this by counting for each $\mathcal{R}$ only those subsets of size z that contain the first event in $\mathcal{R}$ (that is, the event at which the window defining $\mathcal{R}$ is anchored). Of the remaining $|\mathcal{R}| - 1$ events in $\mathcal{R}$ we then choose $z - 1$ to obtain a slot of size z.

For the support distribution estimation let us first assume that all items (and thus all parallel episodes) are equally likely (we abandon this assumption later, but it simplifies the explanation here) and occur independently (as required by the null hypothesis). Then the probability that a specific parallel episode $I \subseteq B$, $|I| = z$, occurs in a slot of size z is $P_S(I) = 1/\binom{|B|}{z}$. The probability distribution over the support values c can thus be approximated by a Poisson distribution as

$$P_S(\langle z, c \rangle) = \frac{\lambda^c}{c!} e^{-\lambda} \quad \text{with} \quad \lambda = N_S(z, w) / \binom{|B|}{z},$$

because $N_S(z, w)$ is (very) large and $1/\binom{|B|}{z}$ is (very) small and thus the standard conditions for a Poisson approximation are met. Multiplying this probability distribution by the number of parallel episodes of size z yields the expected number of patterns with signature $\langle z, c \rangle$, namely

$$E(\langle z, c \rangle) = \binom{|B|}{z} \frac{\lambda^c}{c!} e^{-\lambda},$$

and thus the desired pattern spectrum. To account for the finite number M of surrogate data sets that would have been generated otherwise, one may threshold it with $1/M$ and thus obtain an equivalent to a surrogate data pattern spectrum.

It should be noted, though, that this derivation is only an approximation in several respects. Apart from the Poisson approximation (which, however, is the least harmful, since the conditions for its application are met), it suffers from neglecting the following: in the first place, the support distributions for parallel episodes of the same size are negatively correlated, since more occurrences of one pattern must be compensated by fewer occurrences of other patterns. Hence simply multiplying the individual distributions by the number of possible parallel episodes is not quite correct. Secondly, the "slots" for a given size z overlap, that is, the same event can contribute to multiple slots for a given size. However, in the above derivation the slots are treated as if they are independent. Both of these issues can be expected to lead to an overestimate of the average number of patterns for a signature $\langle z, c \rangle$. The overestimate can be expected to be small, though, because the correlation is small due to the large number of parallel episodes and the amount of overlap is small relative to the total number of slots. Finally, the overlap actually increases the occurrence probability of a pattern, since a slot overlapping one that contains an instance of a pattern has a higher probability of containing the same pattern than an independent slot. This is less relevant, though, because CoCoNAD does not count both of two overlapping instances (see the support definition in Section 2).

However, the most serious drawback of the method as we described it up to now is the assumption that all items (and thus all parallel episodes) are equally likely. This assumption is rarely satisfied in practice, as the firing rates of recorded neurons tend to differ (considerably). Therefore we remove this assumption as follows: since for any practically relevant size of the item base B it is impossible to enumerate all parallel episodes of size z, we draw a sample of K subsets of the item base B having size z (we chose $K = 1,000$ to cover sufficiently many configurations), using equal probabilities for all items. For each drawn parallel episode $I \subseteq B$, $|I| = z$, we compute the Poisson distribution over the support values c as described above and sum these distributions over the elements of the sample. The result, a sum of Poisson distributions with different parameters λ (which take take of the different occurrence probabilities of the items), is then scaled to the total number of possible parallel episodes of size z. That is, the distribution is multiplied with $\binom{|B|}{z}/K$ (and thresholded with $1/M$ where M is the number of surrogate data sets that would have been generated otherwise) to obtain the pattern spectrum.

In this computation one has to take care that the probability of a parallel episode $I \subseteq B$, $|I| = z$, cannot simply be computed as $P_{\mathcal{S}}(I) = \prod_{i \in I} p_i$, where

$$\forall i \in B: \quad p_i = |\{t \mid \langle i, t \rangle \in \mathcal{S}\}| \, / \, |\mathcal{S}|$$

is the probability that a randomly chosen event has item i. The reason is that a chosen item cannot be chosen again and therefore the probability should rather be computed, using an order $i_1, \ldots, i_z$ of the items in I, like

$$P_{\mathcal{S}}(I) = \prod_{k=1}^{z} \frac{p_{i_k}}{1 - \sum_{j=1}^{k-1} p_{i_j}}.$$

However, there is no reason to prefer any specific order of the items over any other. To handle this problem, we draw a small sample of orders (permutations) for each chosen parallel episode and average over these orders as well as their reversed forms (unless $z \leq 4$, for which we simply enumerate all orders, since their number is manageable). We consider both a generated item order as well as its reverse, because the computed probabilities are certain to lie on opposite sides of the mean probability. In this way the average over the considered item orders can be expected to yield a better estimate of the mean probability.

In our experiments we found that if the item probabilities actually differed, this approach produced a better pattern spectrum estimate than assuming equal item probabilities. However, it tended to overestimate the occurrence frequencies of the pattern signatures. On the other hand, using equal item probabilities for the estimation (even though the probabilities actually differed) tended to produce underestimates. As a straightforward heuristic to correct these effects, we introduced a factor that contracts the probability dispersion of the items, thus reducing the overestimate. That is, before the support distribution estimation we transform the item probabilities computed above according to

$$p'_i = \bar{p} + \varrho(p_i - \bar{p}) \qquad \text{where} \qquad \bar{p} = 1/|B| \qquad \text{and} \qquad \varrho \in [0, 1].$$

By evaluating the quality of an estimated pattern spectrum relative to one derived from surrogate data, focusing on the expected signature counts close to the decision border for rejecting a found pattern (technically: expected counts $E(\langle z, c \rangle) \in [0.0001, 0.1]$) and using a logarithmic error measure (that is, computing differences of logarithms of pattern counts rather than differences of the counts directly), we found that $\varrho \in [0.4, 0.5]$ is a good choice for basically all parameter combinations that we tested. We only observed a slight dependence on the window width w: for larger values of w, smaller values of ϱ appear to produce better results. For the experiments reported in the next section we used the fixed value $\varrho = 0.5$, but results for other values did not differ much.

5   Experiments

We implemented our pattern spectrum estimation in both Python and C (see below for the source code) and applied it to a variety of data sets that were generated to resemble the data sets we meet in neural spike train analysis (our motivating application area). In total we generated 108 data sets, each of which represented 3 seconds of recording time. We varied the number of neurons (or items, $n \in \{40, 60, 80, 100\}$, which are typical numbers that can be recorded with state of the art equipment), the averaging firing rate ($r \in \{10, 20, 30\}$Hz), the firing rate variation over the neurons (either the same for all neurons or linearly increasing from the lowest to the highest, which was chosen to be 2 or 3 times the lowest rate) and the firing rate variation over time (using either a flat rate profile or a burst profile that mimics presenting and removing a stimulus three times, where the highest rate was chosen to be 2 or 3 times the lowest rate). As an illustration, dot displays of some of these data sets are shown in Figure 1.

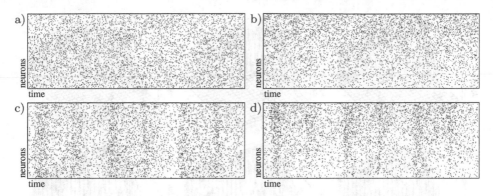

Fig. 1. Some examples of test data sets: a) stationary Poisson processes, same firing rate for all neurons; b) stationary Poisson processes, different firing rates (3:1 highest to lowest); c) burst profile (3:1 highest to lowest rate), same for all neurons; d) burst profile (3:1 highest to lowest rate), different average firing rates (3:1 highest to lowest)

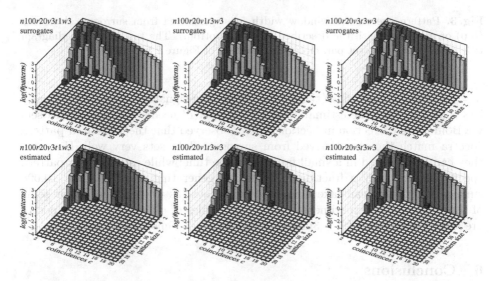

Fig. 2. Pattern spectra for window width 3ms, generated from surrogate data sets (top) or estimated with the described method (bottom). The top word in a diagram title encodes the data set parameters: n—number of neurons, r—firing rate, v—firing rate variation over neurons as $x : 1$ (highest to lowest), t—firing rate variation over time as $x : 1$ (highest to lowest), w—analysis window width. Grey bars extend beyond the top of the diagram, white squares represent zero occurrences. Note the logarithmic scale.

Each data set was then analyzed with four different window widths ($w \in \{2, 3, 4, 5\}$ms), yielding a total of 432 configurations. In each configuration a pattern spectrum was obtained by generating and analyzing 10,000 surrogate data sets and by estimating it with the described method. With this number of surrogate data sets we can be sure to meet an overall significance level of

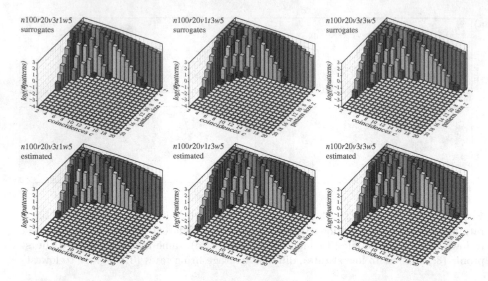

Fig. 3. Pattern spectra for window width 5ms, generated from surrogate data sets (top) or estimated with the described method (bottom). The top word in a diagram title encodes the data set parameters (cf. caption of Figure 2 for details).

$\alpha = 1\%$ or even lower, since the number of pattern signatures was always clearly less than 100. (See the estimation of the number of needed surrogate data sets via Bonferroni correction in Section 3.) We observed that the estimated pattern spectra match the ones derived from surrogate data sets very well. However, they can be obtained in a small fraction of the time: while estimating a pattern spectrum takes only a fraction of a second, generating and analyzing 10,000 surrogate data sets can take hours and even days (as we experienced for some of the data sets we experimented with). Examples of obtained pattern spectra are shown in Figure 2 ($w = 3$ms) and Figure 3 ($w = 5$ms).[4]

6   Conclusions

Although in several respects a (coarse) approximation, the pattern spectrum estimation we presented in this paper proved to produce very usable pattern spectra for the (artificially generated) data sets on which we tested it. The speed-up that can be achieved by estimation is substantial (often orders of magnitude). This speed-up can be exploited, for example, to automatically determine a proper window width w by trying different values and evaluating the result. Doing the same with surrogate data sets can turn out to be tedious and time-consuming, since each window width requires a new set of surrogates to be generated and analyzed. We are currently in the process of applying our pattern mining method

---

[4] Diagrams of the full set of pattern spectra can be found here:
http://www.borgelt.net/docs/spectra.pdf

(CoCoNAD + PSF + PSR, with pattern spectrum estimation as well as deriving a pattern spectrum by generating and analyzing surrogate data sets) to real-world data sets. Preliminary results look very promising.

Software and Source Code

Python and C implementations of the described estimation procedure as well as a Java based graphical user interface can be found at these URLs:

www.borgelt.net/pycoco.html   www.borgelt.net/cocogui.html

Acknowledgments. The work presented in this paper was partially supported by the Spanish Ministry for Economy and Competitiveness (MINECO Grant TIN2012-31372).

References

1. Abdi, H., Bonferroni, Šidák: Corrections for Multiple Comparisons. In: Salkind, N.J. (ed.) Encyclopedia of Measurement and Statistics, pp. 103–107. Sage Publications, Thousand Oaks (2007)
2. Bonferroni, C.E.: Il calcolo delle assicurazioni su gruppi di teste. Studi in Onore del Professore Salvatore Ortu Carboni, pp. 13–60. Bardi, Rome (1935)
3. Borgelt, C.: Frequent Item Set Mining. Wiley Interdisciplinary Reviews (WIREs): Data Mining and Knowledge Discovery 2, 437–456 (2012)
4. Borgelt, C., Picado-Muiño, D.: Finding Frequent Synchronous Events in Parallel Point Processes. In: Proc. 12th Int. Symposium on Intelligent Data Analysis (IDA 2013), pp. 116–126. Springer, Heidelberg (2013)
5. Fiedler, M., Borgelt, C.: Subgraph Support in a Single Graph. In: Proc. IEEE Int. Workshop on Mining Graphs and Complex Data, pp. 399–404. IEEE Press, Piscataway (2007)
6. Gwadera, R., Atallah, M., Szpankowski, W.: Markov Models for Identification of Significant Episodes. In: Proc. 2005 SIAM Int. Conf. on Data Mining, pp. 404–414. Society for Industrial and Applied Mathematics, Philadelphia (2005)
7. Høastad, J.: Clique is Hard to Approximate within n^{1e}. Acta Mathematica 182, 105–142 (1999)
8. Hebb, D.: The Organization of Behavior. J. Wiley & Sons, New York (1949)
9. Karp, R.M.: Reducibility among Combinatorial Problems. In: Miller, R.E., Thatcher, J.W. (eds.) Complexity of Computer Computations, pp. 85–103. Plenum Press, New York (1972)
10. Laxman, S., Sastry, P.S., Unnikrishnan, K.: Discovering Frequent Episodes and Learning Hidden Markov Models: A Formal Connection. IEEE Trans. on Knowledge and Data Engineering 17(11), 1505–1517 (2005)
11. Louis, S., Borgelt, C., Grün, S.: Generation and Selection of Surrogate Methods for Correlation Analysis. In: Grün, S., Rotter, S. (eds.) Analysis of Parallel Spike Trains, pp. 359–382. Springer, Berlin (2010)
12. Mannila, H., Toivonen, H., Verkamo, A.: Discovery of Frequent Episodes in Event Sequences. Data Mining and Knowledge Discovery 1(3), 259–289 (1997)

13. Picado-Muiño, D., Borgelt, C., Berger, D., Gerstein, G.L., Grün, S.: Finding Neural Assemblies with Frequent Item Set Mining. Frontiers in Neuroinformatics, 7: article 9 (2013) doi:10.3389/fninf.2013.00009
14. Picado-Muiño, D., Borgelt, C.: Frequent Itemset Mining for Sequential Data: Synchrony in Neuronal Spike Trains. In: Intelligent Data Analysis. IOS Press, Amsterdam (to appear, 2014)
15. Tatti, N.: Significance of Episodes Based on Minimal Windows. In: Proc. 9th IEEE Int. Conf. on Data Mining (ICDM 2009), pp. 513–522. IEEE Press, Piscataway (2009)
16. Torre, E., Picado-Muiño, D., Denker, M., Borgelt, C., Grün, S.: Statistical Evaluation of Synchronous Spike Patterns Extracted by Frequent tem Set Mining. Frontiers in Computational Neuroscience 7, article 132 (2013), doi:10.3389/fninf.2013.00132
17. Vanetik, N., Gudes, E., Shimony, S.E.: Computing Frequent Graph Patterns from Semistructured Data. In: Proc. IEEE Int. Conf. on Data Mining, pp. 458–465. IEEE Press, Piscataway (2002)

From Sensor Readings to Predictions: On the Process of Developing Practical Soft Sensors

Marcin Budka[1], Mark Eastwood[2], Bogdan Gabrys[1],
Petr Kadlec[3], Manuel Martin Salvador[1], Stephanie Schwan[3],
Athanasios Tsakonas[1], and Indrė Žliobaitė[4]

[1] Bournemouth University, UK
{mbudka,bgabrys,msalvador,atsakonas}@bournemouth.ac.uk
[2] Coventry University, UK
ab3276@coventry.ac.uk
[3] Evonik Industries, Germany
{petr.kadlec,stephanie.schwan}@evonik.com
[4] Aalto University and HIIT, Finland
indre.zliobaite@aalto.fi

Abstract. Automatic data acquisition systems provide large amounts of streaming data generated by physical sensors. This data forms an input to computational models (soft sensors) routinely used for monitoring and control of industrial processes, traffic patterns, environment and natural hazards, and many more. The majority of these models assume that the data comes in a cleaned and pre-processed form, ready to be fed directly into a predictive model. In practice, to ensure appropriate data quality, most of the modelling efforts concentrate on preparing data from raw sensor readings to be used as model inputs. This study analyzes the process of data preparation for predictive models with streaming sensor data. We present the challenges of data preparation as a four-step process, identify the key challenges in each step, and provide recommendations for handling these issues. The discussion is focused on the approaches that are less commonly used, while, based on our experience, may contribute particularly well to solving practical soft sensor tasks. Our arguments are illustrated with a case study in the chemical production industry.

1 Introduction

Automatic data acquisition systems, which are common nowadays, generate large amounts of streaming data. This data, provided by various physical sensors is used for monitoring and control of industrial processes, traffic patterns, environment and natural hazards to name a few. Soft sensors are computational models that aggregate readings of physical sensors to be used for monitoring, assessing and predicting the performance of the system. They play an increasingly important role in management and control of production processes [3, 7]. The popularity of soft sensors is boosted by increasing availability of real sensors,

H. Blockeel et al. (Eds.): IDA 2014, LNCS 8819, pp. 49–60, 2014.

data storage and processing capacities, as well as computational resources. Soft sensors operate online using streams of sensor readings, therefore they need to be robust to noise and adaptive to changes over time. They also should use a limited amount of memory and be able to produce predictions in at most linear time with respect to data arrival.

Building soft sensors for streaming data has received a lot of attention in the last decade (see e.g. [7,8]), often focusing on algorithmic aspects of the computational models, while the process of data preparation receives less attention in research literature [15]. Evidently, building a soft sensor is not limited to selecting the right model. In practice data preparation takes a lot of effort and often is more challenging than designing the predictive model itself. This paper discusses the process of building soft sensors with a focus on data preparation along with the case study from chemical industry. Our goal is to discuss the major issues of data preparation and experimentally evaluate the contribution of various data preparation steps towards the final soft sensor performance.

The main contribution of our study is a framework - a systematic characterization of data preparation process for developing industrial predictive models (soft sensors). Data preparation issue has received little attention in the research literature, while in industrial applications data preparation takes majority of the modelling time. In line with the framework we present our recommendations for data preparation that are based on our experience in building soft sensors within the chemical industry, and are illustrated with real data examples.

The paper is organised as follows. In Section 2 we discuss the requirements and expectations for soft sensors in chemical industry. Section 3 presents a framework for developing data driven soft sensors. In Section 4 we experimentally illustrate the role of three selected data preparation techniques in building accurate predictive models via a case study in the chemical production domain. Section 5 concludes the study, and discusses directions for future research.

2   Requirements and Expectations for Predictive Models in the Process Industry

In the process industry soft sensors are used in four main applications: (1) *online prediction* of a difficult-to-measure variable from easy-to-measure variables; (2) *inferential control* in the process control loop; (3) *multivariate process monitoring* for determining the process state from observed measurements; and (4) as a *hardware sensor backup* (e.g. during maintenance).

This study focuses on the data-driven soft sensors for online predictions of difficult-to-measure variables. Many critical process values (e.g. the fermentation progress in a biochemical process, or the progress of polymerisation in a batch reactor) are difficult, if not impossible to measure in an automated way at a required sampling rate. Sometimes the *first-principle* models, that are based on the physical and chemical process knowledge, are available. Although such models are preferred by practitioners, they are primarily meant for planning and design of the processing plants, and therefore usually focus on the ideal states of the process. Thus, such models can seldom be used in practice in a wide range of operating conditions. Moreover, often the process knowledge for modelling is not available at all. In such cases data-driven models fill the gap and often play an important role for the operation of the processes as they can extract the process knowledge automatically from the provided data.

A successful soft sensor is a model, which has been implemented into the process online environment and accepted by the process operators. In order to gain acceptance the soft sensor has to provide reasonable performance, be stable and predictable. This means that performance has to be immune to changes often happening in the production plants. It is not uncommon that physical sensors fail, drift or become unavailable. Hence, the soft sensor needs to have some kind of automated performance monitoring and adaptation capability. The predictive performance is not the only success criterion however.

Transparency is another important property for model success. It is essential for the process operators to understand, how the soft sensor came to its conclusions. This becomes even more critical if the predictions deviate from the true value. In such cases, it is of utmost importance to be able to backtrack the wrong prediction to its cause. For this reason pure black-box methods like certain types of Artificial Neural Networks may have problems with gaining acceptance.

Another challenge is that after the model is deployed, it is often used by personnel with limited background in machine learning. Therefore, the operation of the model has to be completely automated, and as simple as possible in order to avoid frustration and resistance from the operating personnel.

A systematic approach for soft sensor development has been proposed in [9]. The authors present it as a four step process consisting of handling missing data, detecting and handling outliers, deriving a regression model and validating it on independent data. An alternative methodology, presented in [16], focuses on three steps: data collection and conditioning, selection of influential features and correlation building. In [12], in addition to a general three-step methodology a more specialised one, based on multivariate smoothing procedure, is also discussed. Its distinguishing feature is the focus on the collection of process knowledge, which is not evident in other approaches. Other general methodologies for soft sensor development have been proposed in [3, 6, 7] and [4], with the latter based on the Six-Sigma process management strategy.

The methodology presented in this paper builds upon some ideas proposed in the literature, augmented by our own experience and knowledge gained during interaction with process experts, plant operators and soft sensor practitioners.

3 A Framework for Developing Data Driven Soft Sensors

The framework describes soft sensor development process in four steps:
1. setting up the performance goals and evaluation criteria;
2. data analysis (exploratory);
3. data preparation and preprocessing:

(a) data acquisition (deciding which data to collect and from which sources);
(b) data preprocessing (de-noising, handling outliers and missing values);
(c) data reduction (extracting relevant representation);
4. training and validating the predictive model.

We focus on the first three steps that have been understudied in data analysis literature. For model training and validation an interested reader is referred to one of the classical data mining or machine learning textbooks (e.g. [18]).

3.1 Setting Up Performance Goals and Evaluation Criteria

When starting a soft sensor project we first need to define what the soft sensor is needed for and what will be the quantitative and qualitative evaluation criteria.

Qualitative Evaluation. Many models are so called black-boxes, where it is difficult or impossible to track back the relation between the inputs and predictions. Knowing the effects of input features to the target is particularly important for controlling the process. Moreover, transparent models are typically more trusted and better accepted by the operators. Classification trees, regression trees, and nearest neighbour approaches are among the most transparent.

Computational requirements of the model need to be taken into account particularly in high throughput or autonomous systems operating on batteries.

Quantitative evaluation. The choice of appropriate error measure is critical. Not only it is important to choose a criterion, that is possible to optimize [1]. Even more important is that the criterion measures the performance aspects, that are practically relevant. The Root Mean Squared Error (RMSE) is very popular in research due to convenient analytical properties: $RMSE = \sqrt{1/n \sum_{i=1}^{n} (\hat{y}_i - y_i)^2}$, where y is the true target, $\hat{y}$ is the prediction and n is the size of the testing data. It punishes large deviations from the target, which is often very relevant for industrial applications, however, the meaning of this error may not be straightforward to interpret for the process experts and operators. The Mean Absolute Error (MAE) is often considered a more natural measure of average error [17]: $MAE = 1/n \sum_{i=1}^{n} |\hat{y}_i - y_i|$, but is more difficult to optimize.

Often, particularly in the control applications, predicting the direction of a signal change may be more important than low absolute deviation from the true value. In such cases it is useful to optimize the classification accuracy (CE). The accuracy is measured as a fraction of times the true signal from time $t - 1$ to t goes to the same direction (up or down) as the prediction.

Variability of the predictions is critical in process control applications. A flat prediction is preferred to spiky, since following the latter would require very frequent process adjustments. Jitter (J) measures an average distance that one would travel if the prediction signal was followed: $J = \frac{1}{n-1} \sum_{i=2}^{n} |\hat{y}_t - \hat{y}_{t-1}|$, where n is the number of observations, observations need to be ordered in time.

Robustness and prediction confidence is understood as resistance of the predictor to impurities in the data, such as noisy, outlying or missing observations. If the system was to provide a completely wrong prediction, it should rather not produce any prediction at all. While some predictive methods, e.g. Gaussian Processes,

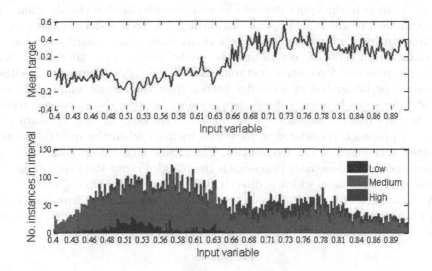

Fig. 1. An example of a profile plot — "low", "medium" and "high" are the target intervals

inherently provide a confidence value, most of the them do not. A common approach is to generate an ensemble of models and use the disagreement of the individual models as a confidence estimate. Such an approach however, would carry higher computational costs and lower transparency of the predictions. Alternatively, one can define a domain, where the model is applicable, and relate confidence values to the locations of test observations in this domain [11].

It is critical to understand the process and the potential role of the soft sensor as much as possible before deciding on which evaluation criteria to use.

3.2 Data Analysis

When the objectives of a soft sensor have been decided upon, the next step is data understanding. The goal is to discover characteristic properties of the data that would help to build a more effective predictive model.

Exploratory data analysis can help to discover anomalies in data, define necessary data preparation approaches and determine potentially useful model classes. Scatterplots of variable against variable or against target are commonly used in exploratory analysis. In addition, we recommend to plot variables over time and to construct variable profiles by defining a small number of intervals on the target y and plotting a visualisation of where datapoints in each interval lie along a given input variable x.

To construct a profile we divide an input variable into a number of bins and find how many datapoints in each bin fall in each interval of the target. These are plotted in a stacked bar chart as in the lower part of Figure 1. This makes a crude representation of the density functions $P(y|x)$, and $P(x)$. Plotting the mean of

y in each bin as in the upper part of Figure 1 provides further visualisation – in this example a profile in which the relationship is approximately monotonic.

Time series analysis involves the identification of stationarity in time series data, which assists in estimating the predictability of the time series, and can imply preferred forecasting methods. Any data-generating process is either stationary or integrated in an order higher than zero. Some empirically observed time series however exhibit dependencies between distant observations, and they are referred as *fractionally integrated processes* or *long-memory processes*. A process is considered to have long memory when the spectral density becomes unbounded at low frequencies. The most common approach to detect the presence of long-memory processes is the rescaled range statistic, commonly known as *R/S statistic*, which is directly derived by the *Hurst coefficient* [10].

We recommend to include domain experts in the data analysis from the beginning of the process in order to quickly detect potential problems of the available data.

3.3 Data Preparation and Preprocessing

Preprocessing transforms raw data into a format that can be more effectively used in training and prediction. Examples of preprocessing techniques include: outlier detection and removal, missing values replacement, data normalisation, data rotation, or feature selection. We leave out description of techniques, which can be found in machine learning textbooks, for instance [13, 18].

In industrial processes typically real time data processing is required. For autonomous operation preprocessing needs to be performed online, systematically, and design decisions need to be verifiable, therefore, the procedure and order of preprocessing actions need to be well defined. We propose a seven-step data preprocessing process, as highlighted in Figure 2.

The first step defines the data *design decisions*, such as, how data is queried from databases, how data sources are synchronized, at what sampling rate data arrives, whether any filtering (de-noising) is used. This step produces raw data in a matrix form, where each row corresponds to an observation of sensor readings at one time. Typically, the design choices remain fixed during the project.

In industrial processes data often comes from multiple sources, which may be separated physically (e.g. a long flow pipe) or virtually (e.g. data is stored in different ways), and need to be synchronized. Synchronization of virtual sources requires consolidating the data into a single database or stream, and is relatively straightforward. Synchronizing data from different physical locations is usually more challenging, and can be estimated based on the physical properties of the process (e.g. speed of flow), or approached as as a computational feature selection problem, where for each sensor different time lags are tried as candidate features.

The second step *filters* out irrelevant data. For example, we can discard data during plant shut-down times and non-steady operation states. Rules for detecting such periods can be defined by experts during the design step, or statistical change detection techniques could be used to identify them automatically.

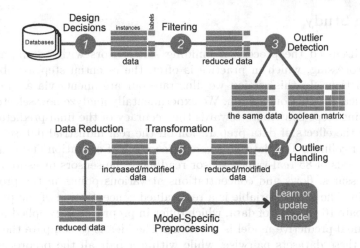

Fig. 2. The proposed order of data preprocessing steps

The third step *detects outliers* in four stages. Firstly, recordings are checked against physical constraints (e.g. water temperature cannot be negative). Secondly, univariate statistical tests can detect outliers in individual sensors. Thirdly, multivariate statistical tests on all variables together (see e.g. [2]) can detect outliers at an observation level. Finally, we check consistency of the instances with the target values. At this step outliers are flagged as missing values.

In the fourth step we *handle the identified outliers*, and/or missing values. In industrial applications predictions are needed continuously, therefore, removing observations with missing values is typically not an option. Standard missing value imputation techniques can be used, ranging from computationally light last observed value, or mean imputation, to various model based imputation methods (see e.g. [5]). The result is a modified data matrix, data size is still the same as produced in step 1.

The fifth step performs data *transformations*, which can modify the existing features (e.g. discretisation), derive new features (e.g. an indicator if the production is running), scale the data or rotate the data for de-correlation purposes. The result of this step is a data matrix that can have more or the same number of features than before and the same number of observations.

The sixth step *reduces data* by feature selection (or extraction) and observation selection (subsampling). As a result the data matrix will decrease in size.

The seventh step performs *model specific* preprocessing, e.g. further removing of outliers or undesired observations. This completes data preprocessing and we can proceed to model training.

While the design decisions (step 1) must be made, other steps (2-7) are optional, and it is up to data scientist to decide, which particular techniques to use. However, we recommend to keep the order as suggested. It enables reproducibility, allows easier documentation, and, most importantly, easier automation when it comes to implementation.

4   Case Study

We have discussed the process of building soft sensors with an emphasis on data preprocessing, which in practice is often the essential step for obtaining good predictions. In this section we illustrate our arguments via a case study from chemical production domain. We experimentally analyze how selected data preprocessing steps contribute towards the accuracy of the final predictor.

We test the effects of data preparation on the real industrial dataset from a debutanizer column. The dataset covers three years of operation and consists of 189 193 records of 85 real-valued sensor readings. The sensors measure temperatures, pressures, flows and concentrations at various points in the process at every 5 min. The target variable is a real-valued concentration of the product.

We prepare 16 versions of data, which differ in preprocessing applied to them, while the final predictive model is the same. The idea is to compare the performance on these datasets pairwise, while within a pair all the preprocessing is the same, but one step is different. That allows to asses the contribution of that particular preprocessing step in isolation, other factors held constant.

Table 1. Versions of data with different preprocessing

	Training size	Testing size	Subsampling	Synchronization	Feature selection	Fractal features	Difference data
RAW	188 752	21 859					
RAW-SYN	188 752	21 859		✓			
RAW-FET	188 752	21 859			✓		
RAW-SYN-FET	188 752	21 859		✓	✓		
SUB	15 611	1 822	✓				
SUB-SYN	15 611	1 822	✓	✓			
SUB-FET	15 611	1 822	✓		✓		
SUB-SYN-FET	15 611	1 822	✓	✓	✓		
SUB-DIF-FRA	15 610	1 822	✓			✓	✓
SUB-DIF-SYN-FRA	15 610	1 822	✓	✓		✓	✓
SUB-DIF	15 610	1 822	✓				✓
SUB-DIF-SYN	15 610	1 822	✓	✓			✓

The datasets are summarized in Table 1. **RAW** and **SUB** refer to different sampling rates. RAW uses data sampled at every 5 min, while SUB at every 1 h. **SYN** means that the input features of the data are synchronised by moving the features along the time axes to better reflect the physical process, as described in Section 3.3. **FET** indicates feature selection. We select 20 features that have the highest absolute correlation with the target variable from the original 85 features. We explore two options: early and late selection. Early (E) selection means that we select features from the first 1000 training examples. Late (L) selection means that we select features from the latest 1000 data points in the training set. If data is changing over time, we expect L to be more accurate. **FRA** means that the space of the dataset has been complemented with the features derived by computing the fractal dimension, as presented in Section 3.2. Fractal features describe the Hurst exponent for each input and the output, calculated over the last 128 measurements. Finally, **DIF** refers to transformation of the input features and the target variable. Differenced data replaces the original values with the first derivative with respect to time. It describes how much the values are changing in comparison to the previous time step. For example, suppose y_t is the original variable at time t, then the differenced data is $r_t = y_t - y_{t-1}$.

The experimental protocol is as follows. Size of the training and testing sets for each dataset are reported in Table 1. For testing we use a hold out set, which did not participate in the parameter tuning of the preprocessing methods. The performance is evaluated using the mean absolute error (MAE). The predictions on DIF datasets are transformed back to the original space before measuring the error. We use the Partial Least Squares regression (PLS) [14] as the predictive model with the number of hidden variables set to 10.

Tables 2 and 3 present the results in MAE, the lower - the better. Preprocessing actions are assessed pairwise - the baselines are on the left, and datasets with additional preprocessing are on the right. Table 2 covers non-differentiated data, and Table 3 presents the results on differentiated data. We treat these cases separately, since differentiating allows to capture different properties of the signal (autoregressive properties), and hence leads to different errors.

Table 2. Testing errors (MAE) on non-differentiated data (• - superior performance)

preprocessing#1	MAE#1	MAE#2	preprocessing#2	improvement
RAW	225	222 •	RAW-SYN	3 (1%)
SUB	227	221 •	SUB-SYN	6 (3%)
RAW-FET-E	228	198 •	RAW-FET-L	30 (13%)
RAW-SYN-FET-E	245	201 •	RAW-SYN-FET-L	44 (18%)
SUB-FET-E	236	193 •	SUB-FET-L	43 (18%)
SUB-SYN-FET-E	215	185 •	SUB-SYN-FET-L	30 (14%)

Table 3. Testing errors (MAE) on differentiated data (• - superior performance)

preprocessing#1	MAE#1	MAE#2	preprocessing#2	improvement
SUB-DIF	41.8	35.3 •	SUB-DIF-SYN	6.5 (16%)
SUB-DIF	41.8	32.4 •	SUB-DIF-FRA	9.4 (22%)

We see that each selected preprocessing action improves the predictive performance (the right approaches are better than the left). The largest improvement is achieved by late feature selection (RAW-SYN-FET-L and SUB-FET-L), as compared to early feature selection. This is an interesting observation. This suggests, that feature relevance is changing over time, and we can achieve as much as 18% reduction in the prediction error only by making the feature selection adaptive over time using the simplest fixed sliding window strategy.

Table 3 presents the mean absolute errors (MAE) of the predictions.

Comparing the results on non-differentiated data in Table 2 and differentiated data in Table 3 suggests that taking into account self-similarity is very beneficial. That is not surprising, considering that in chemical production processes operating conditions do not jump suddenly, hence, the concentration of the output also remains similar to what has been observed in the recent past, therefore, methods from time series modeling contribute well.

Experiments show that preprocessing actions consistently improve the predictive performance, with adaptive feature selection making the largest impact.

In Section 3.1 we also discussed qualitative criteria, namely transparency and computational load. In terms of computational load, PLS regression that was used can be updated recursively using analytical solutions, it does not require optimization loops, and is easy to handle on a commodity hardware.

PLS regression is one of the most transparent models available. While fitting the model is somewhat more involved, the result is a linear model. The coefficients at the inputs can be interpreted as the importance weights, hence, PLS regression provides good transparency and interpretability, easy to understand even for non-experts. All the tested preprocessing actions, except maybe the fractal dimension, do not reduce transparency and interpretability in any substantial way. For example, synchronization of the features (SYN) shifts observations in time, but the regression coefficients remain as interpretable, as before.

5   Conclusion

We analyzed data preparation process for building soft sensors in three main steps: establishing the evaluation framework, exploratory data analysis and data preparation. We recommended a sequence of data preparation techniques for building soft sensors. We illustrated our propositions with a case study with real data from industrial production process. The experiments showed that the selected preprocessing actions consistently improve the predictive performance, and adaptive feature selection makes the largest contribution towards improving the prediction accuracy.

This study opens several interesting directions for further research. Firstly, since relational and autoregressive data representations capture different patterns in data, combining the two approaches suggests a promising research direction. A straightforward way to combine would be to extend the feature space with autoregressive features. Alternatively, we could combine those different types of approaches into an ensemble for the final decision making.

Secondly, it would be interesting to explore how to filter out the effects of data compression when evaluating predictive models. One direction is to identify the real sensor readings and treat the compressed readings as missing values. It would be also interesting to analyse theoretically what compression algorithms would be the most suitable for streaming data, which is later intended to be used for predictive modelling.

Finally, despite thorough synchronisation of preprocessing steps we encountered different data representations (after each preprocessing step), not straightforward to integrate. Thus, our study confirmed the intuition that automating and combining multiple data preparation methods into a single autonomous system that would use a feedback loop to update itself is urgently needed.

Acknowledgment. Thanks Evonik Industries for providing the data. The research leading to these results has received funding from the European Commission within the Marie Curie Industry and Academia Partnerships & Pathways (IAPP) programme under grant agreement no. 251617.

References

1. Budka, M.: Clustering as an example of optimizing arbitrarily chosen objective functions. In: Advanced Methods for Comp. Collective Intell., pp. 177–186 (2013)
2. Chandola, V., Banerjee, A., Kumar, V.: Anomaly detection: A survey. ACM Computing Surveys 41(3), 1–58 (2009), doi:10.1145/1541880.1541882
3. Fortuna, L.: Soft sensors for monitoring and control of industrial processes. Springer (2007)
4. Han, C., Lee, Y.: Intelligent integrated plant operation system for six sigma. Annual Reviews in Control 26, 27–43 (2002)
5. Little, R.J.A., Rubin, D.B.: Statistical Analysis with Missing Data, 2nd edn. Wiley (2002)
6. Kadlec, P., Gabrys, B.: Architecture for development of adaptive on-line prediction models. Memetic Computing 1(4), 241–269 (2009)
7. Kadlec, P., Gabrys, B., Strandt, S.: Data-driven soft sensors in the process industry. Computers and Chemical Engineering 33(4), 795–814 (2009)
8. Kadlec, P., Grbic, R., Gabrys, B.: Review of adaptation mechanisms for data-driven soft sensors. Computers & Chemical Engineering 35(1), 1–24 (2011)
9. Lin, B., Recke, B., Knudsen, J., Jorgensen, S.: A systematic approach for soft sensor development. Computers & chemical engineering 31(5-6), 419–425 (2007)
10. Mandelbrot, B.: The fractal geometry of nature. W.H. Freeman (1983)
11. Netzeva, T., Worth, A., Aldenberg, T., Benigni, R., Cronin, M., Gramatica, P., Jaworska, J., Kahn, S., Klopman, G., Marchant, C., et al.: Current status of methods for defining the applicability domain of (quantitative) structure-activity relationships. Alternatives to Laboratory Animals 33(2), 1–19 (2005)
12. Park, S., Han, C.: A nonlinear soft sensor based on multivariate smoothing procedure for quality estimation in distillation columns. Computers & Chemical Engineering 24(2-7), 871–877 (2000)
13. Pearson, R.K.: Mining imperfect data. Society for Industrial and Applied Mechanics, USA (2005)
14. Qin, J.: Recursive PLS algorithms for adaptive data modeling. Computers & Chemical Engineering 22(4-5), 503–514 (1998)
15. Žliobaitė, I., Gabrys, B.: Adaptive preprocessing for streaming data. IEEE Trans. on Knowledge and Data Engineering 26, 309–321 (2014)
16. Warne, K., Prasad, G., Rezvani, S., Maguire, L.: Statistical and computational intelligence techniques for inferential model development: a comparative evaluation and a novel proposition for fusion. Eng. Appl. of Artif. Intell. 17, 871–885 (2004)

17. Willmott, C., Matsuura, K.: Advantages of the mean absolute error (MAE) over the root mean square error (RMSE) in assessing average model performance. Climate Research 30, 79–82 (2005)
18. Witten, I., Frank, E., Hall, M.: Data Mining: Practical Machine Learning Tools and Techniques, 3rd edn. Morgan Kaufmann (2011)

Comparing Pre-defined Software Engineering Metrics with Free-Text for the Prediction of Code 'Ripples'

Steve Counsell[1], Allan Tucker[1], Stephen Swift[1], Guy Fitzgerald[2], and Jason Peters[1]

[1] Dept. of Computer Science, Brunel University, Uxbridge, UK
[2] School of Business and Economics, Loughborough University, UK
{steve.counsell@brunel.ac.uk}

Abstract. An ongoing issue in industrial software engineering is the amount of effort it requires to make 'maintenance' changes to code. An equally relevant research line is determining whether the effect of any maintenance change causes a 'ripple' effect, characterized by extra, unforeseen and wide-ranging changes in other parts of the system in response to a single, initial change. In this paper, we exploit a combination of change data and comment data from developers in the form of free text from three 'live' industrial web-based systems as a basis for exploring this concept using IDA techniques. We explore the predictive power of change metrics *vis-à-vis* textual descriptions of the same requested changes. Interesting observations about the data and its properties emerged. In terms of predicting a ripple effect, we found using either quantitative change data or qualitative text data provided approximately the same predictive power. The result was very surprising; while we might expect the relative vagueness of textual descriptions to provide less explanatory power than the categorical metric data, it actually provided the approximate same level. Overall, the results have resonance for both IT practitioners in understanding dynamic system features and for empirical studies where only text data is available.

Keywords: Software maintenance, web-based systems, prediction, metrics.

1   Introduction

In the development of industrial software, it is well-recognized that software maintenance accounts for at least seventy-five percent of overall development costs [15]. Maintenance of code (both through fault fixing and for other perfective reasons) is still the subject of intense research activity and any study which provides insights into our understanding of this facet of software is potentially of use [4]. One under-researched aspect of the software maintenance process is the notion of a 'ripple' effect [3]. In other words, if we make one change to a computer system, does this cause 'knock-on' wider (and hence more costly and time-consuming) changes to be required elsewhere in the same system? In this paper, we explore three industrial web-based systems to provide insights into this phenomenon using IDA techniques as a basis [2, 14]. We use data about requests for changes made by system end-users after the systems had been made 'live' (i.e., put into production) as a basis of our analysis;

H. Blockeel et al. (Eds.): IDA 2014, LNCS 8819, pp. 61–71, 2014.

this data includes the type of maintenance change requested, affected parts of the system, developer effort data as well as whether that change had caused a ripple effect. Using Bayesian Networks (BNs) [14], we first explore a number of relationships in the data that inform an understanding of a ripple effect. We also compare the predictive power of numerical data for predicting a ripple effect against short textual descriptions of the actual requested changes to the same system. Hypothetically, we might expect numerical data to be a better indicator of a system feature than textual descriptions of the same changes, since natural language is usually more subjective and vague. Results showed however, that this was not the case - textual descriptions of changes provided the same approximate predictive power.

The remainder of the paper is organized as follows. In the following section, we describe related work. In Section 3, we provide data definitions and the organization context of the study including the data collected. We then describe results using BNs and text analysis using a word tool (Section 4) before concluding and suggesting further work (Section 5).

2   Related Work

Research into the area of software maintenance and its associated problems has been ongoing for over forty years by the software engineering (SE) community. The maintenance 'iceberg' is a phrase often used to describe the hidden cost and effort of making changes to software [5]. In this paper, we explore a relatively recent maintenance phenomenon of a 'ripple' effect. The concept of a ripple effect was first investigated by Black [3] using C code as a basis and has been widely cited since. A tool was developed which used an algorithm to compute the ripple effect for the C programming language and the research explored its theoretical and empirical properties. Despite its importance, examination of the ripple effect has been under-researched in the past and so the study presented is an opportunity to understand this characteristic of changes in software in a little more detail. In terms of the SE-IDA crossover, application of BNs to the prediction of fault-proneness in software has been proposed in the past by Fenton et al., [8]; the authors suggested that BNs could be applied effectively in numerous applications and, as such, have been applied to areas as diverse as football result prediction and 'agile' software development processes [6, 10]. The research presented thus informs the growing area of search-based software engineering [9]. A cornerstone of search-based software engineering is the robustness with which the IDA-based search algorithms can be applied [11]; it is the marriage of SE and IDA in this paper which allows novel and interesting information about existing software artifacts to be generated.

3   Study Context

The context of the case study used in this paper was an IT development division of a small to medium-sized enterprise in London, UK. The company was an established web services development consultancy with over fifteen years experience in the design and production of commercial products. All systems described in the next section were of roughly similar size in their functionality and the same core team of four to six developers in the company produced each system. A waterfall development methodology was adopted for each system. Each change to a system requested by an end-user of the developers generated a Change Request Form (CRF) on which details of the request were formally stated.

3.1   The Three Systems Studied

System A was commissioned by a client providing business services and support to organizations providing human resource applications, for example, payroll systems. The project required an integrated corporate on-line system which disseminated and marketed human resource products. The specification included a back-end database, a content management system and an open forum communication facility. **System B** involved the on-line implementation of an end client's portfolio management service which would allow their customers to delegate management of their assets. The developed product would allow a fully active, real-time system for customers; it provided a calculation facility for computing complex financial queries and the production of reports based on customers current portfolio positions. **System C** was to develop an online implementation of financial services; in particular, the investment and management of customer's portfolios. This included services such as loans, treasury-based transactions and pensions. The project required the inter-connectivity between corporate-wide sub-systems linked to a web application.

3.2   Data Collected

The following items were collected manually (and then transcribed to file) from each of the three systems by one of the authors who spent 6 months at the company as part of a research project and with recourse and access to the CRFs of each of the three systems:

The architectural layer of a system: Separating a system into architectural layers is considered good practice in the design of computer systems since it logically delineates the boundaries of a system making it easier to develop and maintain 'partitioned' systems. Defined as either: presentation, system or data. The *presentation* layer is the uppermost level of the software application and displays information related to services on the client side. It communicates with other layers by outputting results to the browser and all other tiers in the network. The *system* layer (or business-logic layer) controls an application's functionality by implementing detailed processing and logic. The *data* layer consists of database servers and is where the data about the system is stored and retrieved.

Evidence of a ripple effect: In this paper, a ripple effect is more formally defined as a change made at one architectural layer requiring a change (or number of changes) at another architectural layer.

The maintenance type: In this paper, we adopt the well-known maintenance categories of Swanson [17]. The three categories of are: *Corrective*: Performed in response to processing, implementation features that are faulty and includes emergency fixes and routine debugging. *Perfective*: Performed to eliminate processing inefficiencies, enhance performance or improve maintainability. *Adaptive*: Performed in response to changes in the data and processing environment; examples are changes to organizational processes or the tax rules of a country.

Effort: Measured by the number of developer hours required to satisfy a specific CRF, rounded to the nearest half hour. The effort value is inclusive of any added effort incurred by a change as a result of a ripple effect.

A textual description of the change as part of the CRF: A short sentence in English describing the requested change and produced by the developer as part of the CRF in consultation with the user. This was the text used as a basis of predicting a ripple effect described later. An example is *"Error message when secondary pages accessed via navigation box"* (a corrective change in System A). Equally, *"Remove bullet point numbering system for text content"* (an adaptive change in System B).

4   Data Analysis

4.1   Summary Analysis

A total of 425 CRFs were collected from the live phase of each project and were collected and verified by one of the authors. Table 1 shows the frequency of ripple effects in each of the three systems (i.e., 'Ripple' or 'No ripple'). System B is clearly the system with the greatest number of ripple effects and System C the lowest. Interestingly, System B was the only system delivered on time (the other two systems were delivered late). This statement might seem paradoxical. In other words, why would a system delivered on time be more prone to a ripple effect? The answer is relatively straightforward: the system was delivered to the client on time, but with known bugs and inadequacies. In other words, much of the high-intensive effort (usually associated with a ripple) was required *after* the system had been shipped to the client. Often the end client would request and then receive the system even though they knew that it contained bugs.

Table 1. Distribution of CRFs

System	CRFs	Ripple	No ripple
A	134	62	72
B	128	90	38
C	163	55	108
Total	425	207	218

4.2   Bayesian Network Analysis

We developed a Bayesian Network (BN) with the following five nodes: effort (in hours), ripple ('y' or 'n'), architectural layer (presentation (p), system (s) or data (d)), the maintenance type (adaptive (a), corrective (c) or perfective (p)) and system (A, B or C). As a preliminary analysis, one relationship that we might expect from the developed BN is that: *A low effort value is usually associated with the absence of a ripple effect (and vice versa) since the incidence of a ripple effect, by its nature, will usually cause a higher effort value than if there had been no ripple effect.* BNs were learnt using the change data where the ripple node was predicted, based upon evidence from the other nodes; the K2 learning algorithm was used to learn the structure [7] and a ten-fold cross-validation approach used to score the ability of the model to predict ripples. An identical approach was then applied to the free text which was pre-processed into a document term matrix; any word which appeared in the comments was treated as a variable once it was stemmed [18]. This resulted in twenty-two variables (terms) and hence a wrapper feature selection approach was used [12] using BN classifiers to score combinations of terms. This process ensured that we would identify *combinations* of terms rather than single words. The selected features were then used to build a BN to predict ripples. Again, ten-fold cross validation was used to compare the predictive capabilities of the pre-processed text to the change data. Standard inference was used with the junction tree algorithm [13] to explore "what if?" scenarios with respect to ripple effects. We also explored Agrawal's association rule analysis [1] when applied to this data as a comparison.

Fig. 1 shows the BN settings when effort is set to low (here, '-inf' denotes minus infinity and '+inf', positive infinity). A probability of 'no' ripple effect with that low effort is reported as 0.98, confirming the intuitive relationship between effort and ripple. However, we note an unexpected feature not so evident from this initial exploration - the probability of the 'presentation' layer prevailing as the source of a change with low effort (and no ripple) was 0.94. One explanation for this result may be that changes at the presentation layer generally tend to be cosmetic, requiring relatively simple changes only. An example might be to change or remove some text from a web page. Such a change is unlikely to cause a ripple effect and that may explain the high probability value.

Association rule analysis [1] revealed a similar relationship (rule 1) and it was System C where this relationship was most prevalent (rule 2), see Fig.2. Another interesting feature derived from both of these association rules was that the adaptive category of change was most evidently the source of a non-ripple effect (from both rules). Inspection of the raw data for adaptive changes (all three systems) with no ripple revealed typical change text to be of the form: *'remove text'*, *'change text'* and *'line space required between text'*. All of these types of change are low effort and again are unlikely to cause a ripple effect. (Note: Fig. 2 details the top two rules extracted. Rule 1 is interpreted as: "if effort is "less than 1.75", there is no ripple, change type is' 'a', then layer will be p with a confidence of 0.98 (very high), lift of 1.6 and leverage of 0.08".)

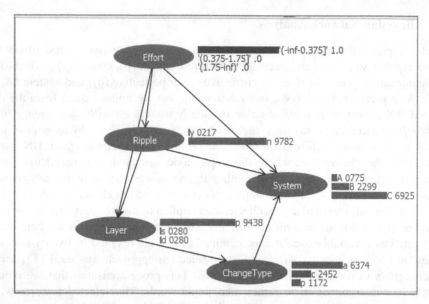

Fig. 1. Relationship between effort (low: -inf-0.375) and ripple effect (y)

Rule 1: Effort='(-inf-1.75]' Ripple=n ChangeType=a 92 ==> Layer=p 90
<conf:(0.98)> lift:(1.6) lev:(0.08) [33] conv:(11.91)
Rule 2: System=C Effort='(-inf-1.75]' Ripple=n ChangeType=a 53 ==>
Layer=p 51   <conf:(0.96)> lift:(1.57) lev:(0.04) [18] conv:(6.86)

Fig. 2. Association rule outputs (top two rules generated)

One other aspect of the analysis which emerged from exploration of the BN was the relationship between change type, ripple and layer. While the presentation layer appears to be associated with low effort (see Fig.1), the combination of a ripple effect set to 'yes' and perfective maintenance change ('p') appears to be prevalent at the system layer (probability 0.82 - see Fig. 3). By way of explanation, perfective maintenance can often be quite complex in nature and involve manipulation of significant amounts of complex program logic. This process often results in a ripple effect as other parts of the system are affected. Over the three systems, the average perfective effort for a change was 3.31; this compares with 1.62 for the adaptive and 2.08 for the corrective categories.

Finally, it was noticeable from the dataset how relatively few data layer CRFs there were in the dataset used for the analysis. This layer accounted for only 26 of the 425 CRFs. (6.1%). The BN showed that the adaptive layer (probability of 0.81) was most prevalent when a CRF related directly to a data layer change (set to d) with ripple effect (y). To explain, inspection of the CRF data revealed that many of the data layer changes were to include new data (in the form of new graphs incorporating new elements) and these required adaptation of the underlying database; this is likely to have caused a range of further changes (ripples) at both the presentation and system layer;

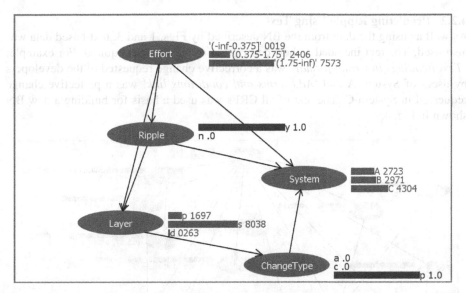

Fig. 3. Relationship between ripple and change type

in fact, 17 of the 26 CRFs for the data layer caused a ripple effect (65.38%). This compares with just 38% for the presentation layer and 64.75% for the system layers. Some examples of data layer change requests were: *"Investment challenge section text content addition"* and *"About the game section text content"*.

4.2.1 Predicting Ripple Using Data

To predict a ripple effect using numerical data, we first used the data of the BN (i.e., system, layer, effort, ripple and maintenance type) to predict whether a ripple effect was evident. Ten-fold cross validation was used to create different training and test sets. The model used was the training set and, thereafter, how well it predicts ripple on the test set. In other words, evidence is entered for all the other variables and the most probable answer for ripple used. Table 2 shows the confusion matrix for this analysis and also the true positive (sensitivity) and true negative (specificity) rates.

Table 2. Confusion matrix (numerical change data)

	a (ripple = 'y')	b (ripple = 'n')
a (ripple = 'y')	166	41
b (ripple = 'n')	77	141
	Sensitivity = 0.80	Specificity = 0.77

The number of correctly classified instances was 307 and incorrectly classified 118, giving a prediction rate of 72.24% (weighted average of 0.742). A Kappa statistic of 0.45 and ROC of 0.74 for ripple=y and 0.74 for ripple=n was reported.

4.2.2 Predicting Ripple Using Text

As well as using the data from the BN described by Figs. 1 and 3, text-based data was also used. The text included a short description of the change required. For example: *"Text headings in wrong format"* was a corrective change requested of the developers by users of System A and *'Add terms and conditions link'* was a perfective change requested in System C. The text of all CRFs was used a basis for building a new BN shown in Fig. 4.

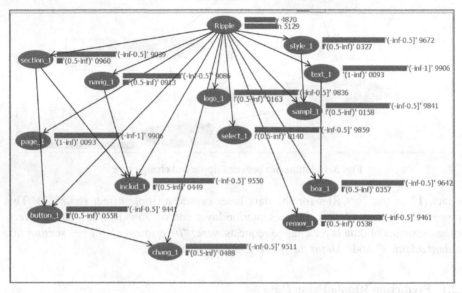

Fig. 4. BN using text as a basis for predicting a ripple effect

The nodes of the BN have 'ripple' at their root; the remaining nodes are the most popular words derived from the set of CRFs (i.e., {text, page, section, navig, button, remov, change, admin include, size, box, update, file, form, addition, database, paragraph, head, list, sampl, select, ripple}). Table 3 shows the confusion matrix for the predictions using the twenty-two selected words; again, it also includes the sensitivity and specificity values.

Table 3. Confusion matrix (text based prediction)

	a (ripple = 'y')	b (ripple = 'n')
a (ripple = 'y')	104	103
b (ripple = 'n')	19	199
	Sensitivity = 0.50	Specificity = 0.66

The number of correctly classified instances was 303; incorrectly classified instances were 122, giving a prediction rate of 71.29% (weighted average of 0.754). A Kappa statistic of 0.42 and ROC values of 0.75 for ripple=y and 0.75 for ripple=n reported.

The data clearly shows that using textual data as a mechanism for prediction is not significantly worse than using concrete numerical data (72.24% for numerical data compared with 71.29% for textual data). These results are highly significant on a methodological basis and for future software engineering studies. In many scenarios, only qualitative text data, as opposed to concrete categorical data, is available. If the evidence from this study is a guide, and there is no loss in terms of the conclusions we can draw from text-based data, then new opportunities arise for data analysis and potential time and cost savings in not having to collect numerical data where textual data only is available (and numerical data is difficult to collect).

4.3   Text Analysis

One final part of the analysis was to try and understand the key characteristics of the CRF text descriptions. In other words, could we determine, from the text of each CRF, the reasons why a ripple effect occurred? This would give us added insights into firstly, the cause of most effort across the three systems and secondly, if a source of common ripple effect could be identified, a set of criteria for a project manager to explore ways of avoiding future ripple effects. "Wordle" is a tool for generating "word clouds" from text [19]. The clouds give greater prominence to words which appear more with a higher probability, based on the posterior distributions in the text BN having observed ripple or not. We generated word clouds for text-based changes where a ripple effect occurred and where it did not. Fig.4 shows the two clouds for the non-ripple text (left cloud) and the ripple text (right cloud).

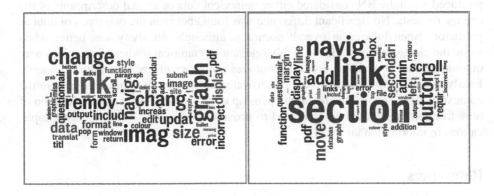

Fig. 5. Word clouds for non-ripple text (left) and ripple text (right)

While both cloud diagrams in Fig. 5 have several common words, for example 'link' and 'navig', two key features from the figures stand out. For the non-ripple cloud diagram (left), 'change/chang', 'graph', 'imag' (for image) and 'link' largely dominate and this reflects the large number of CRFs with simple descriptions such as change text, remove text or to change the wording of an navigational link. On the other hand, the ripple text cloud diagram (right) is dominated by the words 'link' and, in particular, the word 'section'. Typical of the type of change that included this word was: "*Charges section requires database and admin*". This type of change was non-trivial in the sense that it required other parts of the application to be changed and hence a ripple effect was caused. Often, this was an addition to the database (at the data layer) and this required change at the system layer to accommodate the new functionality. Interestingly, a distinction can therefore be made between the two types of 'link' in each diagram. Simply changing the wording of a link caused no ripple effect – a simple textual change. Adding new links or errors in the links *did* cause a ripple effect. Cloud diagrams could thus be used by a project manager to target areas that are clearly ripple-prone, at the same time overcoming issues with synonyms in the text and complementing a BN analysis.

5   Conclusions and Future Work

Most systems suffer from constant requests for changes to systems after they are in operation. The nature of these changes varies widely and some may cause   unforeseen effort in unanticipated parts of the system – a so called 'ripple effect'. In this paper, we explored the ripple effect in three industrial, web-based systems. Our objective was to gain insights into the features of 425 change requests and the relationship they had with a ripple effect. We used a combination of text mining and Bayesian Networks (BNs) to determine relationships between changes that caused a ripple and interesting results emerged which demonstrated the dependencies between system features (effort, architectural layer, for example). Of equal interest was the result produced when the BN considered either numerical data or textual descriptions of the change requests. No significant difference was found between the two types of data as predictors when looking at overall accuracy, although sensitivity was better when using the change metrics. This is highly relevant to empirical studies where often only qualitative data is available and it encourages a number of lines for future research. Firstly, we will explore other areas of software engineering (e.g. bug analysis) in the context of IDA. Second, the research opens up opportunities for further exploration of how free-form text could be optimized to provide maximum information with implications in information theory [16].

References

1. Agrawal, R., Imieliński, T., Swami, A.: Mining association rules between sets of items in large databases. In: Intl. Conference on Management of Data, Washington, USA, p. 207 (1993)
2. Berthold, M., Klawonn, F., Hoppner, F., Borgelt, C.: Guide to Intelligent Data Analysis: How to Intelligently Make Sense of Real Data. Springer (2010)
3. Black, S.: Computing ripple effect for software maintenance. Journal of Software Maintenance and Evolution: Research and Practice 13(4), 263–279 (2001)
4. Brooks, F.: The Mythical Man-Month: Essays on Soft. Eng. Addison-Wesley (1975)
5. Canning, R.: The Maintenance 'Iceberg'. EDP Analyzer 10(10), 1–14 (1972)

6. Constantinou, A., Fenton, N., Neil, M.: pi-football: A Bayesian network model for forecasting Association Football match outcomes. Know. Based Syst. 36, 322–339 (2012)
7. Cooper, G., Herskovits, E.: A Bayesian Method for the Induction of Probabilistic Networks from Data. Machine Learning 9 (1992)
8. Fenton, N., Neil, M., Hearty, P., Marsh, W., Marquez, D., Krause, P., Mishra, R.: Predicting Software Defects in Varying Development Lifecycles using Bayesian Nets. Information & Software Technology 49, 32–43 (2007)
9. Harman, M.: The current state and future of search based software engineering. In: Future of Software Engineering. IEEE Computer Society Press, Los Alamitos (2007)
10. Hearty, P., Fenton, N., Marquez, D., Neil, M.: Predicting Project Velocity in XP Using a Learning Dynamic Bayesian Network Model. IEEE Trans. Soft. Eng. 35(1), 124–137 (2009)
11. Jain, A.: Data Clustering: 50 Years Beyond K-Means. Pattern Recognition Letters 31(8), 651–666 (2010)
12. Kohavi, R., John, G.: Wrappers for feature subset selection. Artificial Intelligence 97(1-2), 273–324 (1997)
13. Lauritzen, S., Spiegelhalter, D.: Local computations with probabilities on graphical structures and their application to expert systems. Journal of the Royal Statistical Society, Series B Methodological) 50(2), 157–224 (1988)
14. Pearl, J.: Bayesian Networks: A Model of Self-Activated Memory for Evidential Reasoning. In: Proceedings of the 7th Conference of the Cognitive Science Society, University of California, Irvine, CA, pp. 329–334 (1985)
15. Pressman, R.: Software Engineering, A Practitioner's Approach. McGraw Hill (1982)
16. Shannon, C.: A Mathematical Theory of Communication. Bell System Technical Journal 27, 379–423 (1948)
17. Swanson, E.: The dimensions of maintenance. In: Proceedings of the 2nd International Conference on Software Engineering, San Francisco, US, pp. 492–497 (1996)
18. http://www.r-project.org (last accessed: May 18, 2014)
19. http://www.wordle.net (last accessed: May 18, 2014)

ApiNATOMY: Towards Multiscale Views
of Human Anatomy

Bernard de Bono[1], Pierre Grenon[1],
Michiel Helvensteijn[2], Joost Kok[2], and Natallia Kokash[2]

[1] University College London (UCL), United Kingdom
[2] Leiden Institute of Advanced Computer Science (LIACS), The Netherlands

Abstract. Physiology experts deal with complex biophysical relationships, across multiple spatial and temporal scales. Automating the discovery of such relationships, in terms of physiological meaning, is a key goal to the physiology community. ApiNATOMY is an effort to provide an interface between the physiology expert's knowledge and all ranges of data relevant to physiology. It does this through an intuitive graphical interface for managing semantic metadata and ontologies relevant to physiology. In this paper, we present a web-based ApiNATOMY environment, allowing physiology experts to navigate through circuitboard visualizations of body components, and their cardiovascular and neural connections, across different scales. Overlaid on these schematics are graphical renderings of organs, neurons and gene products, as well as mathematical models of processes semantically annotated with this knowledge.

1   Introduction

Knowledge of physiology is extensive and complex. To provide software support for using and manipulating physiology data, formalization of the knowledge is required. An *ontology* consists of a set of terms, and their relations, representing a specific domain of knowledge. They are created and maintained by knowledge domain experts, and are used as computer-readable taxonomies by software tools to support knowledge management activities in that domain. When knowledge is formalised in this way, it is possible to record explicit descriptions of data elements in the relevant domain using ontologies; this is the process of semantic annotation or the generation of *semantic metadata*.

For example, physiology experts deal with complex biophysical operations across multiple spatial and temporal scales, which they represent in terms of the transfer of energy from one form to another and/or from one anatomical location to another. Different kinds of descriptions of these biophysical operations are produced by different disciplines in biomedicine. For instance, (i) a medical doctor may describe the mechanism by which a stone in the ureter causes damage in the kidney; (ii) a pharmacologist may depict the process by which a drug absorbed from gut transits to the hip joints where it reduces inflammation; (iii) a molecular geneticist may trace the anatomical distribution of the expression of a particular gene to understand the cause of a skeletal malformation; and,

H. Blockeel et al. (Eds.): IDA 2014, LNCS 8819, pp. 72–83, 2014.

(iv) a bio-engineer may build a mathematical model to quantify the effect of hormone production by the small intestine on the production of bile by the liver. These descriptions take diverse forms, ranging from images and free text (e.g., journal papers) to models bearing well-defined data (e.g., from clinical trials) or sets of mathematical equations (which might be used as input for a simulation tool).

The physiology community is investing considerable effort in building ontologies for the annotation and semantic management of such resources. For example, a number of reference ontologies have been created to represent gene products [3], chemical entities [4], cells [5] and gross anatomy [6]. Together, these ontologies consist of hundreds of thousands of terms, such that the volume of semantic metadata arising from resource annotation is considerable.

Unfortunately, conventional technology for the visualization and management of ontologies and metadata is not usefully accessible to physiology experts, as they involve unfamiliar, abstract technicalities. A number of *generic* ontology visualization tools have been developed to assist knowledge acquisition, browsing and maintenance of ontologies [2]. Such tools, however, put considerable and unrealistic demands on the users' familiarity and expertise in semantic web technologies and the design principles of ontologies. Having to become technically proficient with such technologies is a burden few physiology experts can bear without losing touch with their long term goals. Rather, domain experts should be able to manage data based on a familiar perspective, in which its meaning is made explicit in terms of the expert's own knowledge; a long standing challenge in knowledge engineering.

In this paper, we present ApiNATOMY, a web-based environment that allows physiology experts to navigate through circuitboard visualizations of body components, and their cardiovascular and neural connections, across different scales. It supports a plugin infrastructure to overlay graphical renderings of organs, neurons and gene products on these schematics, in support of biomedical knowledge management use cases discussed in the next section.

The remainder of the paper is structured as follows: Section 2 gives an overview of the ontology-, metadata- and data-resources that we focused on for the ApiNATOMY prototype, and outline key use-case scenarios that motivate our work. Sections 3 and 4 then discuss the visualization techniques we applied to arrange and display those resources. Section 5 provides some insight into the implementation of the prototype. Finally, Sections 6 and 7 offer an overview of related efforts in the field, and conclude the paper with a discussion of the anticipated implications of our tool, as well as planned future work.

2   Use Cases and Data Resources

In this section, we briefly discuss a core use case for the ApiNATOMY application: the generation of interactive schematics in support of genomics and drug discovery studies. We introduce some of the key ontology- and data-resources required in this case. In so doing, we set the stage for an exposition of our early-stage results in the ApiNATOMY application effort.

The domains of genomics and drug discovery are heavily dependent on physiology knowledge, as both domains take into account the manufacture of proteins in different parts of the body and the transport of molecules that interact with those proteins, such as drugs, nutrients, and other proteins. We aim to provide an interactive, schematic overview of data resources important to these domains. This includes gene expression data (e.g., [7]), and data on the transport routes taken by molecular interactors (e.g., [8]). Such data may be usefully depicted in the form of a physiology *circuitboard*.

In ApiNATOMY, a physiology circuitboard schematic consists of an *anatomical treemap* and an overlay of *process graphs*. Our earlier prototypes [9,10] presented treemaps of the Foundational Model of Anatomy (FMA) ontology [6]. Nesting of one treemap tile inside another indicated that the term associated with the child tile is either a mereotopological *part* or a *subclass* of the term associated with the parent tile. Our newest prototype also adopts this convention.

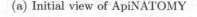

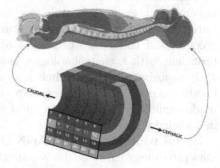

(a) Initial view of ApiNATOMY

(b) Longitudinal section through the male human body, justifying the layout [9]

Fig. 1. The main 24-tile layout of the ApiNATOMY circuitboard

The ApiNATOMY graphical user interface (Figure 1(a)) supports user interaction with circuitboard schematics via point-and-click navigation of the treemap content. The upper level of the anatomical treemap is arranged to resemble the longitudinal section through the middle of the human body (Figure 1(b)). Each of the organs in the plan is composed of multiple tissues and sub-organs. The GUI supports filtering across multiple levels and zooming into selected areas.

This type of interaction extends also to the overlayed process graphs. These graphs project routes of blood flow processes linking different regions of the human body —using data generated in [11]—, as well as transport processes along neurons of the central nervous system (i.e., the brain and spinal cord) — using data obtained via the Neuroscience Information Framework [12].

The ApiNATOMY GUI is built from inception as a three-dimensional environment. This facilitates interaction not only with 3D renderings of the circuit boards themselves, but also with a wide range of geometry/mesh formats for volumetric models of biological structure across scales. For instance, it is already possible to overlay Wavefront .obj data from BodyParts3D [13] as well as .swc data provided by neuromorpho.org [14]. Easy access to such visual resources is critical to the understanding of long-range molecular processes in genomics and drug discovery research.

In the next two sections, we discuss our techniques for constraining treemap layouts to generate stable anatomical treemaps (Section 3.1), designing and overlaying physiological communication routes for the cardiovascular and neural systems (Sections 3.2 and 3.3), and depicting three-dimensional models of organs and protein architecture diagrams for the anatomical overview of gene expression data (Section 4).

3 Visualizing Ontologies and Connectivity Data

In this section we discuss our considerations in the visualization of ontological hierarchies using treemaps, and connectivity data using graph overlays.

3.1 Treemaps

Treemaps [15] visualize hierarchical data by using nested shapes in a space-filling layout. Each shape represents a geometric region, which can be subdivided recursively into smaller regions. The standard shape is a rectangle. Nodes in a treemap, also called *tiles*, represent individual data items. Node size, color and text label can be used to represent attributes of the data item. In interactive environments such as ApiNATOMY, it is possible to navigate between different layers and zoom into selected tiles [16].

We do not use node size to represent information. We focus on tile color and *position*. Tiling algorithms used for typical applications of treemaps (e.g., visualizing the structure of a computer file system) do not usually associate tile positions with any characteristic of the data, and as such, it does not matter if tiles shift around arbitrarily. But this is not the case in our scenario. As shown in Figure 1, relative tile positions are quite relevant, and should be kept stable while the user filters data and zooms in and out. Otherwise, their perception of the data could be easily disrupted. Moreover, the user should be able to enforce constraints on (relative) tile positions to make the treemap views structurally resemble body regions. Hence, we developed a stable and customizable *tiling algorithm* that arranges tiles according to a given template [10].

The schematic body plans created using template-based treemaps can be seen in Figure 1. Figure 1(a) shows the top level 24 tile body anatomy plan. The choice of this layout is explained by Figure 1(b), which shows how it can conceptually wrap around the longitudinal axis of the human body. The treemap layout is controlled by the (default) templates and remains stable during navigation.

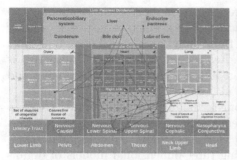 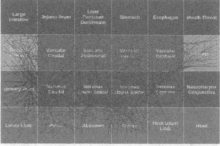

(a) Selected blood vessel connections   (b) Arterial connections from left ventricle

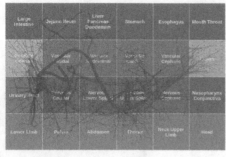 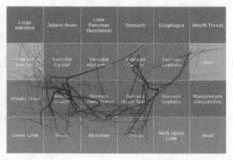

(c) Bundled connections from left ventricle (d) Bundled connections to right atrium

Fig. 2. Overlaying cardiovascular connections. A straightforward approach works well when the number of connections is limited (a). When many connections need to be displayed, we quickly lose overview (b). This is mitigated by employing edge-bundling techniques (c,d).

3.2   Process Graphs

With the treemap-based body plans as background, we overlay the schematic representation of *body systems* such as circulatory, respiratory, or nervous systems. Body systems are essentially graphs with nodes corresponding to body parts (treemap tiles) or entities inside of body parts (e.g., proteins, cells), and edges corresponding to organ system compounds such as blood vessels or nervous connections that pass through such body parts or sub-parts. They may also contain auxiliary nodes that are not represented on the treemap but still carry important biomedical information.

Body systems are intrinsically complex and require efficient data visualization techniques to help avoid clutter induced by the large amount of graph edges and their crossings. Our users need to trace individual connections of body systems, as well as view large parts at once. Edge bundling techniques [17,18,19] have been proposed to improve perception of large, dense graphs. Such techniques generally rely on edge rerouting strategies that are either solely targeted at improving visual perception (by using the positions of nodes) or exploit the relationships among connectivity data as guidelines for a more natural allocation of graph edges and nodes. Our application requires a mixture of these techniques.

If there are too many edges to get a clear overview of the data —as in Figure 2(b), which shows the full connectivity graph for the left ventricle (7101) on the top-level body plan— we can apply hierarchical edge bundling techniques that use path structure to bundle common sub-paths. The result for the left ventricle is shown in Figure 2(c), which gives a much nicer overview. The result for the right atrium is shown in Figure 2(d).

After a one-time pre-processing to import data from available external sources, we store connectivity data in a convenient format. A user can interact with and edit this data using the tool.

3.3 Analyzing the Connectivity Data: An Example

Consider the blood vessels in the human body. Our initial dataset on this is a graph based on the FMA ontology, and consists of approximately 11,300 edges and over 10,000 distinct nodes. In this graph, an edge represents a flow process over an unbranched blood-vessel segment. Nodes represent blood vessel junctions and end-points. Samples of records from the dataset are shown in Table 1.

Table 1. Vascular connectivity data from the FMA ontology. The first column is a unique segment identifier. The second shows the type of a segment (1: arterial, 2: microcirculation, 3: venous, and 4: cardiac chamber). The third contains FMA IDs. The fourth and fifth contain identifiers of the two connected nodes.

Segment	T.	FMA	Node 1	Node 2	Description
121a	2	62528	62528_2	62528_4	Arterioles in Microcirculation segment of Wall of left inferior lobar bronchus
121c	2	62528	62528_4	62528_5	Capillaries in Microcirculation segment of Wall of left inferior lobar bronchus
121v	2	62528	62528_3	62528_5	Venules in Microcirculation segment of Wall of left inferior lobar bronchus
⋮	⋮	⋮	⋮	⋮	⋮
8499	1	69333	8498_0	62528_2	Arterial Segment 8499 of Trunk of left second bronchial artery from origin of supplying terminal segment to the arteriolar side of the Wall of left inferior lobar bronchus MC
9547	3	66699	9546_0	62528_3	Venous Segment 9547 of Trunk of left bronchial vein from origin of supplying terminal segment to the venular side of the Wall of left inferior lobar bronchus MC

A *microcirculation (MC)* is represented by three edges connected in series: one representing tissue arterioles, a second for the bed of capillaries, and a third for the venules. In Table 1, the anatomical entity in which the MC is embedded is 62528 ("Wall of left inferior lobar bronchus"). The topology of its MC segment connectivity is as follows:

$$62528\_2 \xrightarrow{121a} 62528\_4 \xrightarrow{121c} 62528\_5 \xleftarrow{121v} 62528\_3.$$

MC segment 121a is supplied with blood by the arterial segment 8499, and MC segment 121v is drained of blood by the venous segment 9547.

The accurate and comprehensible visualization of the cardiovascular system requires complex pre-processing; a biomedical expert in our team identified about 12 rules for the extraction of relevant data from the full dataset. For illustration purposes, Figure 2 shows only paths connecting MCs of the walls of the heart to MCs belonging to the sub-organs of the tiles in our upper level 24 tile body plan. To obtain this view, we looked for the shortest paths — due to the way the data is represented in the initial data set, cycles are possible. For example, the path from the left ventricle to the wall of left inferior lobar bronchus MC is $7101 \rightarrow 2406 \rightarrow \cdots \rightarrow 8499 \rightarrow 62528$, and the path from there to the right atrium is $7096 \leftarrow 771 \leftarrow \cdots \leftarrow 9546 \leftarrow 9547 \leftarrow 62528$.

The first and the last IDs in this path correspond to the tiles in the treemap, while the intermediate IDs are represented by auxiliary nodes. One of the issues we encountered is the need to determine optimal positions for these nodes. Since several paths can have common sub-paths, as shown in Figure 2(a), the intermediate nodes should be positioned so as to minimize the overall path length. This motivates our application of the *sticky force-directed graph visualization* method [20,21] in which a sub-set of nodes have fixed coordinates, and the other nodes are positioned by simulating imaginary forces applied by their edges.

4   Visualizing Models and Metadata

The entities in the ApiNATOMY ontologies have various data associated with them, to which they are explicitly linked via semantic metadata annotations. This includes static and dynamic 3D models of body organs and their subsystems. For instance, we extract and display neuronal reconstructions and associated metadata from http://neuromorpho.org [14]. Figure 3(a) shows a sample neuron model associated with the neocortex (reached through "Nervous Cephalic" → "Region of cerebral cortex" → "Neocortex"). ApiNATOMY allows users to show multiple 3D objects together in their proper context. For example, Figure 3(b) shows a screenshot including the "Neocortex" neuron, as well as 3D models of the "Liver" and "Stomach", retrieved from BodyParts3D [13].

ApiNATOMY also supports the visualization of protein- and drug-interaction networks (Figure 4(a)) that are represented as graphs on top of treemap tiles. We are in the process of acquiring and integrating relevant data from the Ensembl genomic database [22]. In Ensembl, gene models are annotated automatically using biological sequence data (e.g. proteins, mRNA). We query this database to extract genes, transcripts, and translations with related protein features, such as PFAM domains. ApiNATOMY generates diagrams of protein-interactions and positions them on tiles where the corresponding genes are expressed. For easy access, protein diagrams are able to represent domain features in the form of color-coded 3D shapes extending from the circuitboard (Figure 4(b)).

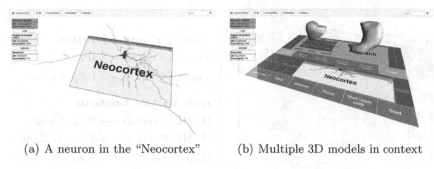

(a) A neuron in the "Neocortex"   (b) Multiple 3D models in context

Fig. 3. Visualizing static 3D models

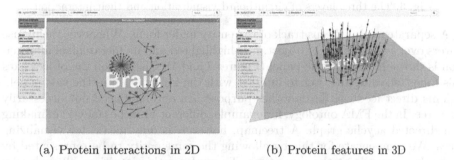

(a) Protein interactions in 2D   (b) Protein features in 3D

Fig. 4. Visualizing protein expression, protein interaction and protein features

5 Implementation

In this section we discuss a number of implementation aspects of ApiNATOMY. For maximum compatibility across operating systems as well as handheld devices, the whole application is written in Javascript. The main framework in use is AngularJS, which provides a Model-View-Controller architecture, as well as two-way databinding. Connectivity- and protein-protein interaction diagrams (Figures 2 and 4(a)) are generated using D3.js, and all 3D functionality (Figures 3 and 4(b)) is implemented using Three.js, which provides a convenient abstraction layer over WebGL.

The circuit-board is rendered with essentially three layers, which are shown in Figure 5. The treemap is generated with plain HTML. On top of this, a partly transparent diagram layer is rendered by D3.js. The positions of the tiles and the positions of the diagram nodes are synchronized with AngularJS two-way databinding. When 3D mode is activated, Three.js takes control of both layers. Besides rendering 3D objects with WebGL, it can manipulate HTML elements using CSS 3D transforms. When using both rendering engines in conjunction, Three.js can keep WebGL and HTML perfectly synchronized. Together with AngularJS two-way databinding, we get very fine control of positioning. This is demonstrated particularly well in Figure 4(b). To render .swc neuron files (Figure 3), ApiNATOMY uses SharkViewer, an open source Three.js library developed by the Howard Hughes Medical Institute [23].

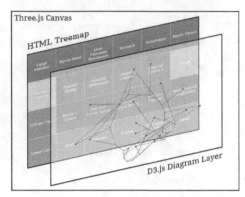

Three.js Canvas

HTML Treemap

D3.js Diagram Layer

Able to Control Position:
Three.js Canvas $\longrightarrow$ HTML Treemap
Three.js Canvas $\longrightarrow$ D3.js Layer

Able to Add Content:
HTML Treemap $\longrightarrow$ D3.js Layer
HTML Treemap $\longrightarrow$ Three.js Canvas
D3.js Layer $\qquad\longrightarrow$ Three.js Canvas

Fig. 5. The three layers of circuitboard visualization and their interaction

A separate module keeps track of the entity under focus. Whenever the mouse hovers over a specific tile or object, it is highlighted and its hierarchical information is shown in the left side-panel (Figures 3 and 4). Clicking on the object fixes this focus, allowing the user to interact with the information in the side-panel.

This direct feedback has another purpose. An ontology need not necessarily be a tree. In the FMA ontology, for example, different branches may join, making it a directed acyclic graph. A treemap, however, is only meant for visualizing trees. We compensate for this by allowing the same entity to be represented by more than one tile at the same time. To reinforce this intuition, all such tiles are highlighted in unison when the mouse hovers over any one of them. Only one visible tile per entity may be opened up to show its children. Such a tile is considered 'active', and only active tiles participate in the visualization of cross-tile connectivity data.

6 Rationale for Our Approach and Related Work in Anatomy and Physiology Knowledge Visualization

The need for the multi-scale visualization and analysis of human body systems is well recognized by biomedical communities. For example, the 3D Multiscale Physiological Human initiative deals with combinations of physiological knowledge and computational approaches to help scientists in biomedicine to improve diagnostics and treatments of various disorders [24,25]. In addition, numerous anatomy-related taxonomies and databases have been created and are widely used by researchers in the biomedical field [26]. While various generic visualization techniques can be used to display biomedical ontologies [2], to the best of our knowledge, ApiNATOMY is the first systematic approach to integrate such knowledge in one extensible and configurable framework.

Among the most effective taxonomy visualization techniques are space-filling diagrams, and in particular, treemaps. de Bono et al. [9] describes limitations of existing treemapping tools for biomedical data visualization. To overcome these limitations, we introduced a generic method to build custom templates which is applied in our tool to control layout of ApiNATOMY body tissues. Among the advantages of the proposed treemapping method are customizable layouts, visualization stability and multi-focus contextual zoom. The detailed comparison of our method with existing treemaping algorithms can be found in [10]. Burch and Diehl [27] discuss the ways to display multiple hierarchies and conclude that overlaying connectors on top of treemaps is the most visually attractive and easy to follow approach. Among the alternative options they considered are separate, linked and colored tree diagrams, sorted and unsorted matrices and sorted parallel coordinate views. Regarding the way to layout the connectors, two naive methods were considered: straight connections and orthogonal connections.

Our application requires multiple taxonomies consisting of thousands of items to be displayed on relatively small screens of handhold devices. We employ the same visualization technique with more advanced treemapping and connector layout algorithms. Due to the potentially large amount of vascular connectivity data, we employ the hierarchical edge bundling technique [17] that results in an intuitive and realistic depiction of blood flow across a treemap-based plan of the human body. In contrast to the scenarios in the aforementioned work, not every node in our vascular connection dataset has a corresponding node in the treemap. Thus, force-directed graph drawing method [28] is added to the scene to find optimal positions of intermediate junctions on the paths that connect the root of the taxonomy (i.e. in the heart) with its leaves (body tissues shown as treemap tiles). The variation of the force-bundling method suitable for our application is known as sticky force-directed placement [20] which allows to fix the positions of certain nodes and allocate other nodes to achieve mechanical equilibrium between forces pulling the free nodes towards fixed positions.

Other potentially useful methods did not provide the desired result. The first approach we tried consists of applying the force-directed edge bundling method [29] to bundle entire paths among the heart chambers and body tissues, but this does not reflect the hierarchical structure of vascular connectivity graph. The second approach, force-based edge bundling over a graph produced by sticky force-directed node allocation algorithm results into unnatural distortion of short edges towards each other. Other edge-bundling methods(e.g., [18,19,30]) operate on graphs with known node positions and thus would produce visualizations on our data that suffer from similar problems.

7   Conclusions and Future Work

The core goal for ApiNATOMY is to put clinicians, pharmacologists, basic scientists and other biomedical experts in direct control of physiology knowledge management (e.g. in support of integrative goals outlined in [31]). As the domain of physiology deals with processes across multiple anatomical scales, the schematic ApiNATOMY approach provides a more flexible and customizable depiction of process participants, and the routes they undertake, compared to conventional methods of anatomy navigation that constrain visualization to regional views of very detailed and realistically proportioned 3D models (such as Google Body [32]). In this paper, we presented our initial results in the development of a generic tool that creates an interactive topological map of physiology communication routes. These routes are depicted in terms of (i) treemaps derived from standard reference anatomy ontologies, as well as (ii) networks of cardiovascular and neural connections that link tiles within these treemaps. These topological maps, also known as circuitboard schematics, set the stage for the visual management of complex genomic and drug-related data in terms of the location of gene products and the route taken by molecules that interact with them. While the implementation of our tool is still in its early stages, we have already started taking steps in preparation for future developments, supporting:

- the visually-enhanced construction of mathematical models in systems biology (e.g., as discussed in [33]),
- the collaborative graphical authoring of routes of physiology communication (e.g., brain circuits) and, crucially,
- the automated discovery of transport routes given (i) a fixed- location receptor and (ii) its corresponding ligand, found elsewhere in the body.

Above all, our aim is to ensure that ApiNATOMY is easy to use for biomedical professionals, and available across a wide range of platforms, to foster collaborative exchange of knowledge both within, and between, physiology communities.

References

1. de Bono, B., Hoehndorf, R., Wimalaratne, S., Gkoutos, G., Grenon, P.: The RICORDO approach to semantic interoperability for biomedical data and models: strategy, standards and solutions. BMC Research Notes 4, 313 (2011)
2. Katifori, A., Halatsis, C., Lepouras, G., Vassilakis, C., Giannopoulou, E.: Ontology visualization methods - a survey. ACM Comput. Surv. 39(4) (2007)
3. Blake, J.A., et al.: Gene ontology annotations and resources. Nucleic Acids Res. 535, D530–D535 (2013)
4. Hastings, J., de Matos, P., Dekker, A., Ennis, M., Harsha, B., Kale, N., Muthukrishnan, V., Owen, G., Turner, S., Williams, M., Steinbeck, C.: The ChEBI reference database and ontology for biologically relevant chemistry: enhancements for 2013. Nucleic Acids Res. 41(D1), D456–D463 (2013)
5. Bard, J., Rhee, S.Y., Ashburner, M.: An ontology for cell types. Genome Biol. 6(2), R21 (2005)
6. Rosse, C., Mejino Jr., J.L.V.: A reference ontology for biomedical informatics: the foundational model of anatomy. J. Biomed. Inform. 36(6), 478–500 (2003)
7. EBI: Arrayexpress home, EBI (2012), http://www.ebi.ac.uk/arrayexpress/
8. Harnisch, L., Matthews, I., Chard, J., Karlsson, M.O.: Drug and disease model resources: a consortium to create standards and tools to enhance model-based drug development. CPT Pharmacomet. Syst. Pharmacol. 2, e34 (2013)
9. de Bono, B., Grenon, P., Sammut, S.: ApiNATOMY: A novel toolkit for visualizing multiscale anatomy schematics with phenotype-related information. Hum. Mutat. 33(5), 837–848 (2012)
10. Kokash, N., de Bono, B.J.K.: Template-based treemaps to preserve spatial constraints. In: Proc. IVAPP 2014 (2014)
11. de Bono, B.: Achieving semantic interoperability between physiology models and clinical data. In: Proc. of IEEE Int. Conf. on e-Science Workshops, pp. 135–142 (2011)

12. Gardner, D., et al.: The neuroscience information framework: A data and knowledge environment for neuroscience. Neuroinformatics 6(3), 149–160 (2008)
13. Mitsuhashi, N., Fujieda, K., Tamura, T., Kawamoto, S., Takagi, T., Okubo, K.: Bodyparts3d: 3d structure database for anatomical concepts. Nucleic Acids Res. 37, D782–D785 (2009)
14. Ascoli, G.A.: Mobilizing the base of neuroscience data: the case of neuronal morphologies. Nat. Rev. Neurosci. 7(4), 318–324 (2006)
15. Johnson, B., Shneiderman, B.: Tree-maps: a space-filling approach to the visualization of hierarchical information structures. In: Proc. of the 2nd Conference on Visualization 1991, pp. 284–291. IEEE (1991)
16. Blanch, R., Lecolinet, E.: Browsing zoomable treemaps: Structure-aware multiscale navigation techniques. TVCG 13, 1248–1253 (2007)
17. Holten, D.: Hierarchical edge bundles: Visualization of adjacency relations in hierarchical data. IEEE Transactions on Visualization and Computer Graphics 12(5), 741–748 (2006)
18. Gansner, E.R., Hu, Y., North, S.C., Scheidegger, C.E.: Multilevel agglomerative edge bundling for visualizing large graphs. In: Battista, G.D., Fekete, J.D., Qu, H. (eds.) Proce. of PacificVis, pp. 187–194. IEEE Computer Society (2011)
19. Hurter, C., Ersoy, O., Telea, A.: Graph bundling by kernel density estimation. Comp. Graph. Forum 31, 865–874 (2012)
20. Fruchterman, T., Reingold, E.: Graph drawing by force-directed placement. Software Practice and Experience 21(11), 1129–1164 (1991)
21. Bostock, M.: Sticky force layout. Online visualization tool, http://bl.ocks.org/mbostock/3750558 (accessed on May 20, 2014)
22. EBI: Ensemble. Online web page (2014) (accessed on May 20, 2014)
23. Weaver, C., Bruns, C., Helvensteijn, M.: SharkViewer. Howard Hughes Medical Institute, Janelia Farm Research Campus. doi:10.5281/zenodo.10053 (2014)
24. Magnenat-Thalmann, N., Ratib, O., Choi, H.F. (eds.): 3D Multiscale Physiological Human. Springer (2014)
25. Magnenat-Thalmann, N. (ed.): 3DPH 2009. LNCS, vol. 5903. Springer, Heidelberg (2009)
26. Burger, A., Davidson, D., Baldock, R. (eds.): Anatomy Ontologies for Bioinformatics. Computational Biology, vol. 6. Springer (2008)
27. Burch, M., Diehl, S.: Trees in a treemap: Visualizing multiple hierarchies. In: Proc. VDA 2006 (2006)
28. Battista, G.D., Eades, P., Tamassia, R., Tollis, I.G.: Graph Drawing: Algorithms for the Visualization of Graphs. Prentice Hall (1999)
29. Holten, D., van Wijk, J.J.: Force-directed edge bundling for graph visualization. Comput. Graph. Forum 28(3), 983–990 (2009)
30. Selassie, D., Heller, B., Heer, J.: Divided edge bundling for directional network data. IEEE Trans. Visualization & Comp. Graphics (Proc. InfoVis) (2011)
31. Hunter, P., et al.: A vision and strategy for the virtual physiological human in 2010 and beyond. Philos. Trans. A Math. Phys. Eng. Sci. 368(2010), 2595–2614 (1920)
32. Wikipedia: Zygote body (2014), http://en.wikipedia.org/wiki/Zygote_Body (accessed on May 20, 2014)
33. de Bono, B., Hunter, P.: Integrating knowledge representation and quantitative modelling in physiology. Biotechnol. J. 7(8), 958–972 (2012)

Granularity of Co-evolution Patterns
in Dynamic Attributed Graphs

Élise Desmier[1], Marc Plantevit[2], Céline Robardet[1],
and Jean-François Boulicaut[1]

[1] INSA Lyon, LIRIS CNRS UMR 5205, F-69621 Villeurbanne, France
[2] Université Claude Bernard Lyon 1, LIRIS CNRS UMR 5205, F-69621 Villeurbanne

Abstract. Many applications see huge demands for discovering relevant
patterns in dynamic attributed graphs, for instance in the context of social interaction analysis. It is often possible to associate a hierarchy on the
attributes related to graph vertices to explicit prior knowledge. For example, considering the study of scientific collaboration networks, conference
venues and journals can be grouped with respect to types or topics. We
propose to extend a recent constraint-based mining method by exploiting
such hierarchies on attributes. We define an algorithm that enumerates
all multi-level co-evolution sub-graphs, i.e., induced sub-graphs that satisfy a topologic constraint and whose vertices follow the same evolution
on a set of attributes during some timestamps. Experiments show that
hierarchies make it possible to return more concise collections of patterns
without information loss in a feasible time.

1 Introduction

Due to the success of social media and the ground-breaking discovery in experimental sciences, network data have become increasingly available in the last
decade. Consequently, graph mining is recognized as being one of the most studied and challenging tasks for the data mining community. Two different and
complementary ways have been considered so far: (1) analyzing graphs based on
macroscopic properties (e.g., degree distribution, diameter) [9] or partitioning
techniques [12], and (2) extracting more sophisticated properties within a pattern discovery setting. In particular, local pattern mining in graphs has received
much attention, leading to the introduction of new problems (e.g., mining collections of graphs [17,22] or single graphs [5,7]). The graph vertices are generally
depicted by additional information that form with the graph structure an attributed graph [15,16,18,20]. Such attributed graphs support advanced discovery
processes providing insightful patterns.

However, there exists other types of augmented graphs such as evolving [1,4,19]
or multidimensional graphs [2]. A growing body of literature has investigated
augmented graphs by only considering one of the above types at a time. In [11],
we proposed to tackle both dynamic and attributed graphs by introducing the
problem of trend sub-graphs in dynamic attributed graph discovery. This new

H. Blockeel et al. (Eds.): IDA 2014, LNCS 8819, pp. 84–95, 2014.

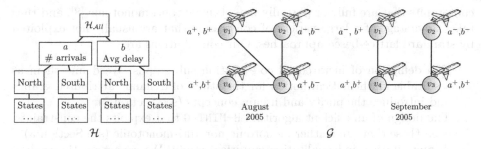

Fig. 1. US domestic flights dynamic graph

kind of patterns relies on the graph structure and on the temporal evolution of the vertex attribute values. In this paper, we go deeper in the analysis of dynamic attributed graphs by also examining the existence of a hierarchy over the vertex attributes. Indeed, we believe that the subsumption power of hierarchies is of most interest to summarize patterns and avoid unperceptive/useless/meaningless patterns. We propose to mine maximal dynamic attributed sub-graphs that satisfy some constraints on the graph topology and on the attribute values. To be more robust towards intrinsic inter-individual variability, we do not compare raw numerical values, but their trends, that is, their derivative at time stamp t. Let us consider the example in Fig. 1 that depicts a dynamic attributed graph describing the US domestic flights. The vertices stand for the airports and edges link airports that are connected by at least a flight during the time period of observation. Two attributes described the vertices of the graph: a is the number of flight arrivals and b is the average delay of arrival. At each time period of observation, we only consider the evolution or trend of the attribute values, and the value increases are denoted $+$, whereas their decreases is denoted $-$. The two attributes a and b can be specialized according to the geographical location of the airports where planes come from (see hierarchy $\mathcal{H}$ on Fig. 1). The incoming number of flights and their average delay can be decomposed into the ones coming from the North and South areas, as well as the distinct states of America. Hence, if a pattern describes a phenomenon that characterizes the whole airplane system, the most appropriate level of description is the first one (namely attributes a and b), whereas if the pattern is specific to a peculiar state, the involved attributes will be the ones at the leaves of the hierarchy.

The connectivity of the extracted dynamic sub-graphs is constrained by a maximum diameter value that limits the length of the longest shortest path between two vertices. Additional interestingness measures are used to assess the relevancy of the trend dynamic sub-graphs and guide their search with user-parametrized constraints. In this unified framework, these measures aim at evaluating (1) how the vertices outside the trend dynamic sub-graph are similar to the ones inside it; (2) the dynamic of the pattern through time; (3) the quality of the description of the pattern given the hierarchy over the vertex attributes. The algorithm designed to compute these patterns traverses the lattice of dynamic attributed sub-graphs in a depth-first manner. It prunes and propagates constraints that are fully or partially monotonic or anti-monotonic [8], and thus takes advantage of a large variety of constraints that are usually not exploited by standard lattice-based approaches. Our contributions are:

- The definition of hierarchical co-evolution sub-graphs: We define them as a suitable mathematical notion for the study of dynamic attributed graphs and introduce the purity and h-gain concepts (see Section 2).
- The design of an efficient algorithm H-MINTAG that exploits the constraints, even those that are neither monotonic nor anti-monotonic (see Section 3).
- A quantitative and qualitative empirical study. We report on the evaluation of the efficiency and the effectiveness of the algorithm on a real-world dynamic attributed graph (see Section 4).

2   Hierarchical Co-evolution Sub-graphs

A dynamic attributed graph $\mathcal{G} = (\mathcal{V}, \mathcal{T}, \mathcal{A})$ is a sequence over a time period $\mathcal{T}$ of attributed graphs $\{G_1, \ldots, G_{|\mathcal{T}|}\}$ where each attributed graph G_t is a triplet $(\mathcal{V}, E_t, A_t)$, with $\mathcal{V}$ a set of vertices that is fixed throughout the time, $E_t \subseteq \mathcal{V} \times \mathcal{V}$ a set of edges at timestamp t, and $\mathcal{A}$ a set of attributes common to all vertices at all times. $A_t(v) \in \mathbb{R}^{|\mathcal{A}|}$ are the values of vertex v at time t on $\mathcal{A}$.

A *vertex induced* dynamic sub-graph of $\mathcal{G}$ is an induced subgraph across a subsequence of $\mathcal{G}$, denoted by (V, T) with $V \subseteq \mathcal{V}$ and $T \subseteq \mathcal{T}$. In order to take into account both the fact that attributes can be expressed according to different levels of granularity and the end-user's prior knowledge, we assume that a hierarchy $\mathcal{H}$ is provided over the set of vertex attributes $\mathcal{A}$. A hierarchy $\mathcal{H}$ on $dom(\mathcal{H})$ is a tree whose edges are a relation is_a, a specialization (resp. generalization) relationship that corresponds to a path in the tree from the root node $\mathcal{H}_{All}$ to the leaves, that are the attributes of $\mathcal{A}$ (resp. from the leaves to the root). Different functions are used to run through the hierarchy:

- $parent(x)$ returns the direct parent of the node $x \in dom(\mathcal{H})$
- $children(x)$ returns the direct children of the node $x \in dom(\mathcal{H})$
- $leaf(x)$ returns all the leaves down from $x \in dom(\mathcal{H})$

We aim at identifying relevant sub-graphs that rely on the graph structure, the temporal evolution of attributes and the associated hierarchy. To this end, we define a new kind of pattern, the so-called *hierarchical co-evolution sub-graphs*. Intuitively, a hierarchical co-evolution sub-graph $P = \{V, T, \Omega\}$ is such that $V \subseteq \mathcal{V}$, $T \subseteq \mathcal{T}$ and $\Omega \subseteq \{dom(\mathcal{H}) \times \{+, -\}\}$, a set of signed attributes. Such a dynamic sub-graph of $\mathcal{G}$ is induced by (V, T) and its vertices follow the same trends defined by Ω. Such dynamic sub-graphs, whose attribute values increase or decrease at the same timestamps, may be unconnected. Therefore, to support analysis based on the graph structure, we introduce a structural constraint that is based on the diameter of the induced dynamic subgraph and provides relevant patterns.

A hierarchical co-evolution sub-graph is then defined as follows:

Definition 1 (Hierarchical co-evolution Sub-Graph). $P = (V, T, \Omega)$ *is a sequence of graphs* $G_t[V]$ *induced[1] by the vertices of* V *in the graphs* G_t, $t \in T$. *The sets* V, T *and* Ω *are such that* $V \subseteq \mathcal{V}$, $T \subseteq \mathcal{T}$ *and* $\Omega \subseteq \{dom(\mathcal{H}) \times \{+, -\}\}$. *The pattern* P *has to satisfy the two following constraints:*

1. *Each signed attribute* $(a, s) \in \Omega$ *defines a trend that has to be satisfied by any vertex* $v \in V$ *at any timestamp* $t \in T$. *Thus, if* (v, a, t) *is the value of attribute* a *at time* t *for vertex* v, $trend(v, a, t) = s$ *with:*

$$trend(v, a, t) = + \ iff \ \sum_{a_i \in leaf(a)} (v, a_i, t) < \sum_{a_i \in leaf(a)} (v, a_i, t+1)$$

$$trend(v, a, t) = - \ iff \ \sum_{a_i \in leaf(a)} (v, a_i, t) > \sum_{a_i \in leaf(a)} (v, a_i, t+1)$$

   Thus, if $\forall v \in V$, $\forall t \in T$ *and* $\forall (a, s) \in \Omega$, *we have* $trend(v, a, t) = s$, *then* $coevolution(P)$ *constraint is satisfied.*
2. *Given* Δ, *a user-defined threshold, and* $sp_G(v, w)$ *the length of the shortest path between vertices* v *and* w *in graph* G, *the constraint* $diameter(P)$ *is satisfied iff* $\forall t \in T$, $\max_{v,w \in V} sp_{G_t[V]}(v, w) \leq \Delta$.

The maximum diameter constraint makes it possible to focus on some specific graph structure within the discovery of hierarchical co-evolution sub-graphs. Indeed, it allows to check how far the vertices are from each other. $\Delta = 1$ implies that the sub-graph is a clique, $\Delta = 2$ implies that the vertices of the sub-graph have at least one common neighbor. More generally, the higher the maximum diameter threshold Δ, the sparser the sub-graphs can be. Until $\Delta = |V| - 1$, the sub-graphs have to be connected.

The attribute value of a parent node within the hierarchy is evaluated by adding the corresponding values of its children. Therefore, even if the trend conveyed by an attribute of the parent is true, it is important to check how this information is valid, i.e., if the trends associated to its children are similar. Indeed, if a children attribute has a large increase while the other children have a small decrease, the sum associated to the parent attribute may result in an increase that is not followed by most of its leaves. The purity measure evaluates the correlation between trends of the leaves of an attribute. Given the Kronecker function $\delta_{condition}$ and given a user-defined threshold $\psi \in [0, 1]$, the *purity* of a pattern returns the number of valid trends $trend(v, a, t) = s$ of the pattern compared to the total number of trends:

$$purity(V, T, \Omega) = \frac{\sum_{v \in V} \sum_{t \in T} \sum_{(a,s) \in \Omega} \sum_{\ell \in leaf(a)} \delta_{trend(v, \ell, t) = s}}{|V| \times |T| \times |leaf(\Omega)|}$$

---

[1] $G_t[V] = (V, E_t \cap \{V \times V\})$

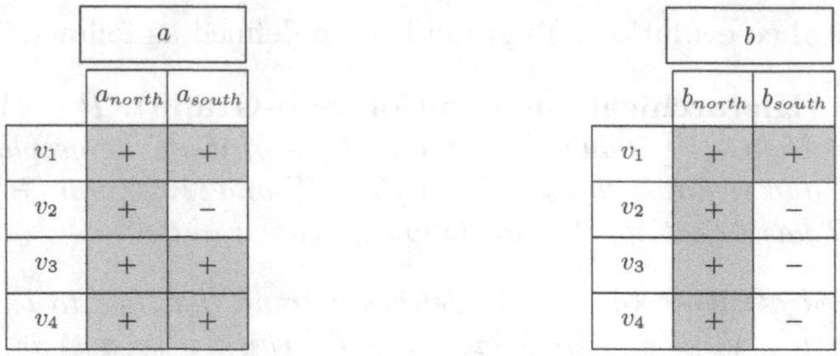

a		
	a_{north}	a_{south}
v_1	+	+
v_2	+	−
v_3	+	+
v_4	+	+

b		
	b_{north}	b_{south}
v_1	+	+
v_2	+	−
v_3	+	−
v_4	+	−

$P_1 = \{\{v_1, v_2, v_3, v_4\}, \{Aug.\ 2005\}, \{(a, +)\}\}$ $P_2 = \{\{v_1, v_2, v_3, v_4\}, \{Aug.\ 2005\}, \{(b, +)\}\}$

Fig. 2. Illustration of the purity values of two patterns extracted from dynamic graph presented in Fig. 1

From this measure, we can derive the predicate $purityMin(P)$ which is true iff $purity(P) \geq \psi$. For example of Fig. 2, the purity of the pattern $P_1 = \{\{v_1, v_2, v_3, v_4\}, \{Aug.\ 2005\}, \{(a, +)\}\}$ is equal to $\frac{7}{8} = 0.875$ whereas the one of $P_2 = \{\{v_1, v_2, v_3, v_4\}, \{Aug.\ 2005\}, \{(b, +)\}\}$ is equal to $\frac{5}{8} = 0.625$.

One inconvenient of hierarchy is that it may introduce redundancy among the hierarchical co-evolution sub-graphs. An important issue is thus to avoid this redundancy by identifying the good level of granularity of a pattern. The question is thus to determine whether the pattern is worth to be specialized. Fig. 2 illustrates this problem with two patterns in two dimensions, i.e., lines depict vertices, columns are related to attributes. A cell is coloured if the trend of the attribute is +. Considering the pattern $P_3 = \{\{v_1, v_2, v_3, v_4\}, \{Aug.\ 2005\}, \{(a_{north}, +)\}\}$, its purity is equal to 1, and the one of pattern P_1 is of 0.875. There is no much interest in specializing pattern P_3 into P_1: end-users may prefer to consider the pattern P_3 as it is more synthetic while having a similar purity. On the other hand, the pattern $P_4 = \{\{v_1, v_2, v_3, v_4\}, \{Aug.\ 2005\}, \{(b_{north}, +)\}\}$ as a purity of 1 while its parent P_2 has a purity of 0.625. Then it seems much more interesting to keep the "parent" attribute instead of producing redundant pieces of information.

To this end, we introduce the *gain* of purity that evaluates whether the purity of the pattern would increase if it gets specialized or not. To this aim, we compare the purity of the a pattern P with respect to the purity of its "parent" patterns, that is, all the patterns made by generalizing one of the attributes of P. Given a user-threshold $\gamma \geq 1$, the gain of purity is defined as the purity of the "children" pattern compared to the purity of its "parent" patterns:

$$gainMin(P) \text{ iff } \frac{purity(P)}{max_{P_i \in parent(P)}(purity(P_i))} \geq \gamma$$

where $(V, T, \Omega_i) \in parent(V, T, \Omega)$ if $\exists (a_i, s_i) \in \Omega_i$ and $\exists (a, s) \in \Omega$ s.t. $a \in children(a_i)$ and $(\Omega_i \setminus a_i) = (\Omega \setminus children(a_i))$. From Fig. 2, the pattern P_4 has a gain equal to 1.6, whereas P_3 has a gain equal to 1.14.

Before ending this Section, let us formalize the general problem we want to solve as follows:

Problem 1 (*Maximal hierarchical co-evolution sub-graph discovery*). Let $\mathcal{G}$ be a dynamic attributed graph, $\mathcal{H}$ be a hierarchy over the set of vertex attributes $\mathcal{A}$, Δ be a maximum diameter threshold, and γ be a minimum gain threshold. Additional quality measures Q can be used, as defined in [11] (e.g., volume, vertex specificity, temporal dynamic). Given a conjunction of constraints $\mathcal{C}_Q$ over Q, the maximal hierarchical co-evolution sub-graph mining problem is to find the set of all the patterns that satisfy the constraints *coevolution*, *diameter*, *gainMin* and $\mathcal{C}_Q$.

3   H-MINTAG Algorithm

Algorithm 1 presents the main steps of H-MINTAG. The search space of the algorithm can be represented as a lattice which contains all possible tri-sets from $\mathcal{V} \times \mathcal{T} \times (dom(\mathcal{H}) \times \{+, -\})$, with bounds $\{\emptyset, \emptyset, \emptyset\}$ and $\{\mathcal{V}, \mathcal{T}, dom(\mathcal{H}) \times \{+, -\}\}$. The enumeration of all the patterns by materializing and traversing all possible tri-sets from the lattice is not feasible in practice. Therefore, in the algorithm, all possibly valid tri-sets are explored in a depth-first search manner which allows to extract the whole collection of hierarchical co-evolution sub-graphs and the constraints are used to reduce the search space by using their properties to not develop tri-sets that can not be valid patterns. The enumeration can be represented as a tree where each node is a step of the enumeration. A node contains two tri-sets P and C. P is the pattern in construction and C contains the elements not yet enumerated and that can potentially be added to the pattern. At the beginning, P is empty and C contains all the elements of $\mathcal{G}$, i.e., $P = \emptyset$ and $C = \{\mathcal{V}, \mathcal{T}, dom(\mathcal{H}) \times \{+, -\}\}$. The extracted patterns are the ones that respect the *diameter*, the *coevolution*, the *gainMin*, the *maximality* constraints, and the other possible constraints as defined in [11].

At each step of the enumeration, either an element of C is enumerated (vertex, timestamp or attribute) (lines 18-27) or an attribute of P is specialized (lines 5-10) and an attribute from C is enumerated (lines 11-15) while keeping the non specialized attribute. At the beginning of the algorithm, one vertex, one timestamp and one attribute are enumerated to allow a better use of the constraints to prune the search space. At each step, the elements of C (vertices, timestamps and attributes) are deleted if they can not be added to P without invalidating it, i.e., if they cannot respect the different constraints (line 1). If P does not respect the constraints, the enumeration is stopped.

The constraints *coevolution*, *diameter*, *purityMin* are not anti-monotonic considering the algorithm. They cannot be used directly to prune the search space. However, some piecewise monotonic properties of these constraints can be used to reduce the search space.

The *coevolution* constraint is not anti-monotonic considering the specialization of an attribute: If a vertex v does not respect the trend a^s at time t, no conclusions can be derived for the trends of any of its leaf attributes a_i. Indeed, the trend associated to a is computed while summing the values of the $a_i \in leaf(a)$, so some a_i can have an opposite trend. However, considering the enumeration

---

Algorithm 1. H-MINTAG

Require: $P = \varnothing, C = (\mathcal{V}, \mathcal{T}, children(\mathcal{H}_{All})), attr, \mathcal{C}_\mathcal{Q}$
Ensure: Maximal hierarchical co-evolution sub-graphs
1. Propagation(C)
2. **if** $\neg empty(C)$ **and** $\mathcal{C}_\mathcal{Q}(P, C)$ **then**
3.   **if** $attr \neq \varnothing$ **then**
4. $child \leftarrow children(attr)$
5.   **for** i in $1..|child|$ **do**
6.   **if** $gainMin(P.V \cup C.V, P.T \cup C.T, P.A \setminus attr \cup child[i])$ **then**
7.   H-MINTAG$((P \setminus attr) \cup child[i], C \cup child[i+1..|child|], child[i])$
8. $hasSon \leftarrow true$
9.   **end if**
10.   **end for**
11.   **if** $hasSon$ **then**
12.   **for** i in $1..|C.A|$ **do**
13.   H-MINTAG$(P \cup C.A[i], C \setminus C.A[1..i], i)$
14.   **end for**
15.   **else**
16. $attr \leftarrow \varnothing$
17.   **end if**
18.   **else**
19. $E \leftarrow ElementTypeToEnumerate(P, C)$
20.   **for** i in $1..|C.E|$ **do**
21.   **if** $E = A$ **then**
22. $attr \leftarrow C.E[i]$
23.   **end if**
24.   H-MINTAG$(P \cup C.E[i], C \setminus C.E[1..i], attr)$
25.   **end for**
26.   H-MINTAG$(P, C \setminus C.E, \varnothing)$
27.   **end if**
28. **else if** $\mathcal{C}_\mathcal{Q}(P)$ output (P)
29. **end if**

---

of the proposed algorithm, the *coevolution* can be pruned if the next step is not a specialization step. Indeed, if the attributes of the pattern are leaves of $\mathcal{H}$ or if the attributes have already passed the specialization step, the constraint is anti-monotonic. Then enumeration can be stopped if *coevolution*(P) is false and elements e of C can be deleted if *coevolution*$(P \cup e)$ is false.

The *diameter* constraint is neither monotonic nor anti-monotonic. The addition of a vertex to a set of vertices can increase or decrease the diameter of the induced subgraph. Then, it is not possible to check strictly the diameter on P and C, however one can check if the induced graph can respect the *diameter* constraint while adding all or part of the vertices of C. Thus, during the algorithm, the following relaxed constraint $lightDiameter(P, C)$ is used:

$$\forall t \in T, \max_{v,w \in P.V} sp_{G_t[P.V \cup C.V]}(v, w) \leq \Delta$$

Otherwise, no valid pattern can be enumerated. Moreover only elements of C that can be added while respecting the diameter constraint are kept, i.e., $C.V = \{v \in C.V | \forall t \in T, \max_{w \in P.V} sp_{G_t[P.V \cup C.V]}(v, w) \leq \Delta\}$ and $C.T = \{t \in C.T | \max_{v,w \in P.V} sp_{G_t[P.V \cup C.V]}(v, w) \leq \Delta\}$.

The *purityMin* constraint is not anti-monotonic. Indeed, while specializing an attribute, the number of trends $trend(v, a, t) = s$, $v \in (P.V \cup Q.V)$, $t \in (P.T \cup Q.T)$, $(a, s) \in (P.\Omega \cup Q.\Omega)$ can increase or decrease if the leaf attributes do not follow the same trend as their parent. One must compute the number of trends that validate either s or $\bar{s}$ for at least one of the leaf attribute of a. Then the number of valid trends $\sum_{v \in P.V \cup C.V} \sum_{t \in P.T \cup C.T} \sum_{\ell \in leaf(P.\Omega \cup C.\Omega)} \delta_{trend(v,a,t)=s} + \delta_{trend(v,a,t) \neq s}$ is anti-monotonic and the number of possible trends $|P.V| \times |P.T| \times |leaf(P.\Omega)|$ is monotonic. The *lightPurity* relaxed constraint is anti-monotonic:

$$lightPurity(P, C) = \frac{\sum_{v \in P \cup C.V} \sum_{t \in P \cup C.T} \sum_{(a,s) \in leaf(P \cup C.\Omega)} \delta_{trend(v,a,t)=s} + \delta_{trend(v,a,t) \neq s}}{|P.V| \times |P.T| \times |leaf(P.\Omega)|}$$

If $lightPurity(P, C) < \psi$ is false, then the enumeration can safely be stopped.

4   Experiments

We carried out some experiments on a dynamic attributed graph built from the DBLP Computer Science Bibliography[2]. Vertices of the graph represent 2,145 authors who published at least 10 papers in a selection of 43 conferences and journals of the Data Mining and Database communities between January 1990 and December 2012. This time period is divided into 10 overlapping periods. A hierarchy over the 43 attributes is built considering the type of publications (e.g., journal, conference), the related area (e.g., database, machine learning,

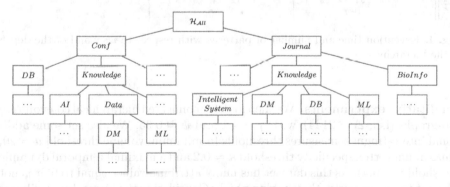

Fig. 3. Hierarchy of DBLP dataset

---

[2] http://dblp.uni-trier.de/db/

data mining, bioinformatics). This hierarchy contains 59 nodes and has a depth equal to 5, it is partly represented in Fig. 3. The default setting is $\Delta = 1, \Gamma = 1.1, \Psi = 0.2$ and two maximum threshold on the vertex specificity (κ) and the temporal dynamic (τ) set to 0.5.

Quantitative experiments. The impact of the hierarchy can be analyzed with respect to 3 parameters: the purity and the h-gain and the depth of the hierarchy. Fig. 4 reports the execution time of H-MINTAG and the number of patterns according to these parameters. The purity constraint has a significant and similar positive impact on both the execution time and the number of patterns. Increasing the h-gain enables to discard many patterns while the running time is marginally impacted. To study the impact of the depth of the hierarchy, we modified the hierarchy by deleting levels of abstractions. Hierarchy with depth equal to 0 is the dataset with no-hierarchy (i.e., only the 43 attributes). In our approach, the deeper the hierarchy, the lower the number of patterns. The execution time also decreases when the hierarchy becomes deeper.

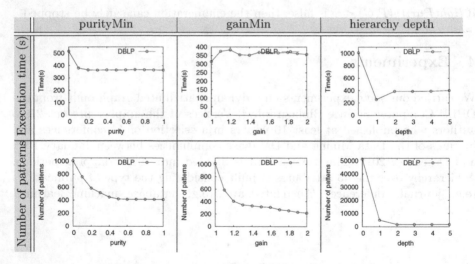

Fig. 4. Execution time and number of patterns with respect to ψ, γ and of the depth of the hierarchy

Qualitative experiments. We then look for connected hierarchical co-evolution sub-graphs (i.e., $\Delta = 2144$), with $\gamma = 1.1$ and $\psi = 0.35$. We also set some additional interestingness measures thresholds (a minimum volume threshold $\vartheta = 20$, a maximum vertex specificity threshold $\kappa = 0.2$ and a maximal temporal dynamic threshold $\tau = 0.4$). As this dataset has many attribute values equal to 0, it is not relevant to set the purity threshold too high. Considering the hierarchy, attributes too generalized as "conference" or "journal" are not really interesting, then γ was set to 1.1 to obtain patterns not too generalized. Two patterns were obtained in

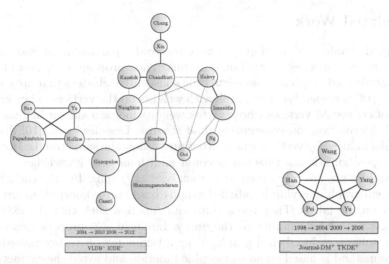

Fig. 5. First (on the left) and second (on the right) patterns extracted from DBLP
with the parameters: $\vartheta = 20$, $\Delta = -1$, $\gamma = 1.1$, $\psi = 0.35$ $\kappa = 0.2$ and $\tau = 0.4$

this extraction. The first pattern is presented in Fig. 5 (left). This pattern concerns 17 authors who decreased their number of publications in VLDB and ICDE
between 2004 and 2012. This pattern is relatively sparse, as the edges are dotted
when they exist only at one of the two timestamps, for instance "Raymond T.
Ng" is connected to "Beng Chin Ooi" at the first timestamp and to "Yannis E.
Ioannidis" at the second timestamp, but he is connected to none author at both
timestamps. It represents small groups of authors who work together occasionally. Moreover, if the decreasing of publication in VLDB seems logical considering
the new publication policy of the "VLDB endowment" it is noteworthy that it is
also true for the ICDE conference. This pattern has small outside densities with
$VertexSpecificity = 0.126$ and $TemporalDynamic = 0.118$. Since the decreasing in "VLDB" concerns many authors at this timestamp (not only those involved
in this pattern), we can conclude that the vertex specificity is mainly due to the
decreasing in "ICDE". The low temporal dynamic specificity means that they do
not decrease their number of publication in these conferences and that the pattern
can show that this small community changed its publication policy.

The second pattern is illustrated in Fig. 5 (right). It involves 5 authors that
increase their number of publication in the journal "IEEE-TKDE" and in the
data-mining journals between 1998 and 2006. This pattern reflects that even if
the journal "IEEE-TKDE" is considered as a database journal in the hierarchy,
it has a high attractiveness in data mining. This pattern has a purity of 0.417,
which means that they publish in a lot of data-mining journals; it seems to
make sense since these authors are well-known in the data mining community.
The vertex specificity is equal to 0.073 which depicts that this behavior is truly
specific to these authors. And the temporal dynamic is equal to 0.4 which shows
that their number of publications maybe oscillates. That points out that it is
difficult to publish regularly in this type of journals.

5   Related Work

Recently, dynamic attributed graphs have received a particular interest. Boden et al. [3] mine sequences of attributed graphs. They propose to extract clusters in each attributed graph and associate time consecutive clusters that are similar. Jin et al. [14] consider dynamic graph with weights on the vertices. They extract groups of connected vertices whose vertex weights follow a similar increasing or decreasing evolution, on consecutive time stamps. Desmier et al. [10] discover subgraphs induced by vertices whose attributes follow the same trends. However, these propositions do not take into account additional user knowledge.

Hierarchies are not often used in the analysis of graphs. In [21], the authors propose subgraph querying in labelled graphs based on isomorphisms using an ontology on the labels. They use a similarity function such that the extracted subgraphs have labels similar to the query. Inokuchi [13] propose generalized frequent subgraphs in labelled graphs using a taxonomy on vertex and edge labels.The method is based on an isomorphic function and avoid the extraction of over-generalized patterns. The authors of [6] defines the taxonomy-superimposed graph mining problem. They compute frequency based on generalized isomorphism with a one-to-one mapping function. These propositions treat labelled graphs instead of attributed graphs and do not deal with dynamic aspect of the graphs.

6   Conclusion

We propose to extract hierarchical co-evolution sub-graphs from a dynamic attributed graph and a hierarchy. These patterns are sets of vertices that are connected and that follow the same trends over a set of attributes over time, with attributes that are either those of the dataset or of the hierarchy. We also define some constraints to reduce the execution time and increase the relevancy of the patterns, in particular according to hierarchy. We design an algorithm H-MINTAG to compute the complete set of patterns. Experiments on a real-world dataset prove that this method extracts, in a feasible time, interesting patterns based on the user parametrized constraints.

Acknowledgements. The authors thank ANR for supporting this work through the FOSTER project (ANR-2010-COSI-012-02).

References

1. Berlingerio, M., Bonchi, F., Bringmann, B., Gionis, A.: Mining graph evolution rules. In: Buntine, W., Grobelnik, M., Mladenić, D., Shawe-Taylor, J. (eds.) ECML PKDD 2009, Part I. LNCS, vol. 5781, pp. 115–130. Springer, Heidelberg (2009)
2. Berlingerio, M., Coscia, M., Giannotti, F., Monreale, A., Pedreschi, D.: As time goes by: Discovering eras in evolving social networks. In: Zaki, M.J., Yu, J.X., Ravindran, B., Pudi, V. (eds.) PAKDD 2010, Part I. LNCS, vol. 6118, pp. 81–90. Springer, Heidelberg (2010)

3. Boden, B., Günnemann, S., Seidl, T.: Tracing clusters in evolving graphs with node attributes. In: CIKM, pp. 2331–2334 (2012)
4. Borgwardt, K.M., Kriegel, H.P., Wackersreuther, P.: Pattern mining in frequent dynamic subgraphs. In: Int. Conf. on Data Mining (ICDM), pp. 818–822 (2006)
5. Bringmann, B., Nijssen, S.: What is frequent in a single graph? In: Washio, T., Suzuki, E., Ting, K.M., Inokuchi, A. (eds.) PAKDD 2008. LNCS (LNAI), vol. 5012, pp. 858–863. Springer, Heidelberg (2008)
6. Cakmak, A., Özsoyoglu, G.: Taxonomy-superimposed graph mining. In: EDBT, pp. 217–228 (2008)
7. Calders, T., Ramon, J., van Dyck, D.: Anti-monotonic overlap-graph support measures. In: ICDM, pp. 73–82 (2008)
8. Cerf, L., Besson, J., Robardet, C., Boulicaut, J.-F.: Closed patterns meet n-ary relations. TKDD 3(1), 3:1–3:36 (2009)
9. Chakrabarti, D., Faloutsos, C.: Graph mining: Laws, generators, and algorithms. ACM Comput. Survey 38(1) (2006)
10. Desmier, E., Plantevit, M., Robardet, C., Boulicaut, J.-F.: Cohesive co-evolution patterns in dynamic attributed graphs. In: Discovery Science, pp. 110–124 (2012)
11. Desmier, E., Plantevit, M., Robardet, C., Boulicaut, J.-F.: Trend mining in dynamic attributed graphs. In: Blockeel, H., Kersting, K., Nijssen, S., Železný, F. (eds.) ECML PKDD 2013, Part I. LNCS, vol. 8188, pp. 654–669. Springer, Heidelberg (2013)
12. Ester, M., Ge, R., Gao, B.J., Hu, Z., Ben-moshe, B.: Joint cluster analysis of attribute data and relationship data. In: SIAM SDM, pp. 246–257 (2006)
13. Inokuchi, A.: Mining generalized substructures from a set of labeled graphs. In: ICDM, pp. 415–418 (2004)
14. Jin, R., McCallen, S., Almaas, E.: Trend Motif: A Graph Mining Approach for Analysis of Dynamic Complex Networks. In: ICDM, pp. 541–546. IEEE (2007)
15. Moser, F., Colak, R., Rafiey, A., Ester, M.: Mining cohesive patterns from graphs with feature vectors. In: SDM, pp. 593–604 (2009)
16. Mougel, P.N., Rigotti, C., Plantevit, M., Gandrillon, O.: Finding maximal homogeneous clique sets. Knowl. Inf. Syst. 39(3), 579–608 (2014)
17. Nijssen, S., Kok, J.N.: Frequent graph mining and its application to molecular databases. In: Systems, Man and Cybernetics (SMC), vol. 5, pp. 4571–4577 (2004)
18. Prado, A., Plantevit, M., Robardet, C., Boulicaut, J.F.: Mining graph topological patterns. IEEE TKDE, 1–14 (2013)
19. Robardet, C.: Constraint-based pattern mining in dynamic graphs. In: ICDM, pp. 950–955 (2009)
20. Silva, A., Meira Jr., W., Zaki, M.J.: Mining attribute-structure correlated patterns in large attributed graphs. PVLDB 5(5), 466–477 (2012)
21. Wu, Y., Yang, S., Yan, X.: Ontology-based subgraph querying. In: ICDE, pp. 697–708 (2013)
22. Yan, X., Han, J.: gSpan: Graph-Based Substructure Pattern Mining. In: ICDM, pp. 721–724 (2002)

Multi-user Diverse Recommendations through Greedy Vertex-Angle Maximization

Pedro Dias and Joao Magalhaes

Dep. Computer Science
Universidade Nova de Lisboa, Portugal
p.dias@campus.fct.unl.pt, jm.magalhaes@fct.unl.pt

Abstract. This paper presents an algorithm capable of providing meaningful and diversified product recommendations to small sets of users. The proposed approach works on a high-dimensional space of latent factors discovered by the bias-SVD matrix factorization techniques. While latent factor models have been widely used for single users, in this paper we formalize recommendations for multi-user as a multi-objective minimization problem. In the pursuit of recommendation diversity, we introduce a metric that explores the angles among product factor vectors and extracts from these a measurable real-life meaning in terms of diversity. In contrast to the majority of recommender systems for groups described in literature, our system employs a collaborative filtering approach based on latent factor space instead of content-based or ratings merging approaches.

1 Introduction

Recommender systems emerged with the purpose of providing personalized and meaningful content recommendations based on user preferences and usage history. In the context of recommendation for groups, where there is more than one user to please, recommendations must be provided in a different way so that the whole group of users is satisfied. The most successful approaches to recommender systems are commonly oriented to single users, providing these with highly personalized recommendations. In our implementation, we attempt to explore state-of-the-art latent factor collaborative filtering techniques in the pursuit of meaningful and diversified multi-user recommendations, by extending the potentialities of these techniques to a multi-user context. The purpose of latent factor approaches to recommender systems is to map both users and products onto the same latent factor space, representing these as vectors with k dimensions. That is, the user i factors vector, $u_i = (u_{i_1}, u_{i_2}, \cdots, u_{i_k})$ and the product j factors vector $p_j = (p_{j_1}, p_{j_2}, \cdots, p_{j_k})$. By representing users and products in such way, one can evaluate the extent to which users and products share common characteristics by comparing their k factors against each other. The principle underlying this approach is that both users and products can be represented under a common reduced dimensionality space of latent factors that

H. Blockeel et al. (Eds.): IDA 2014, LNCS 8819, pp. 96–107, 2014.

are inferred from the data and explain the rating patterns. Our algorithm operates exclusively in the latent-factor space, in which one can easily relate different users. Moreover, by clustering this space we obtain a set of interest-groups to which users belong, enabling us to experiment with different multi-group scenarios.

Our approach intends to find sets of products that satisfy different users at the same time, while pursuing maximum product diversity. To achieve such goal, we defined satisfaction metrics and diversity metrics, and designed optimization algorithms to maximize these indicators. Finding the set of products within the large scope of available products that yields the highest satisfaction and diversity presents a complex combinatorial problem, which raised some challenges. On the pursuit of multi-user diversified recommendations, we needed to deal with the non-convex nature of the objective functions we attempted to optimize, which led us to develop a deterministic greedy search algorithm that avoids local minima and returns a nearly optimal solution, while avoiding an exhaustive and computationally infeasible search through all possible solutions.

This paper is organized as follows: section 3 describes the matrix factorization implementation, section 4 presents our multi-user recommendation algorithms and section 5 presents an evaluation and discussion of our system's performance. Next, we discuss related work.

2   Related Work

Collaborative filtering approaches attempt to infer user preferences by analysing the patterns and historic of consumption of all users in the system, mining the relations between users and products based on their interactions. An early application of collaborative filtering was the open architecture GroupLens, implemented by Resnick et al.[1] with a similar purpose. This approach introduced the concept of user feedback provided explicitly by users in the form of ratings (explicit feedback) or extracted from user activity analysis (implicit feedback). Within collaborative filtering approaches, the latent factor approach alone has proven to yield state-of-the-art results [2]. Applications of such approach include neural networks [3], latent variable models [4] and Singular Value Decomposition (SVD)[5].

Although recommender systems have recently attracted a lot of attention from the scientific community, recommendation for multi-user groups has not been widely addressed, since most recommendation techniques are oriented to individual users and focus on maximizing the accuracy of their preference predictions. A. Jameson et al. [6] conducted an enlightening survey presenting the most relevant works on the field of recommendation for groups. The main challenges faced when providing recommendation for groups are (1) capturing user preferences, (2) combining user preferences into a representation of group preferences, (3) defining criteria to assess the adequacy of recommendations, and (4) delivering recommendations. Group recommender systems can be compared according to how they deal with these challenges. In 2002, the Flytrap system was proposed by A. Crossen et al. [7], presenting a simple system designed to build a soundtrack that would please all users within a group in a target environment. The Flytrap system, relied on the songs metadata and users' listening patterns. The CATS system, proposed McCarthy et al. [8], is designed to recommend travel packages to groups of users. It relies on the explicit feedback provided by the group users as a *more of this / less of that* fashion. This user feedback is recorded and linearly combined between all users within the group to be afterwards compared against the set of features that represent each travel package. Another example of group recommender systems is the system Bluemusic proposed by Mahato et al. [9]. In this approach users are detected via bluetooth and the awareness of their presence has direct influence on a playlist which is being played on a public place. While most recommender for group systems are more concerned with gathering data from users, in this paper we propose a specific algorithm and especially designed for recommending products for groups of users.

Baltrunas et al. [10] examined a late-fusion approach to recommendation for groups. Recommendations are computed individually and later combined into a single ranked list of recommendations. They assume that the order of the recommended items (independent of their position in the rank) is more important than optimizing the top elements of the rank for all users. Moreover, they do not explicitly tackle diversity as we do in this paper. Recognizing that product ratings and consumption patterns might differ, Seko et al. [11] proposed to extend the Power Balance Map (a distribution density of shared usage history) with new dimensions related to items metadata and user behavior. Different users are then related in this newly created space. Barbieri and Manco [12] argued that just optimizing RMSE might not lead to improvements in terms of recommendations or user satisfaction. This fact has been recognized at general by authors such as Ziegler et. al [13] who explicitly addressed the diversification recommendations. Following this idea we propose to maximize the diversity by predicting the missing product in a set of products for a multi-user scenario.

Thus, our approach is related to greedy search algorithms for multi-objective functions. This problem can be solved by combinatorial optimization, such as [14], or forward-backward greedy search algorithms [15]. The algorithm proposed in this paper first computes a joint set of recommendation for each individual function and incrementally searches for the best set of products that satisfy the sum of functions.

3   Collaborative Filtering with Bias-SVD

In the context of recommender systems, matrix factorization is mainly performed through methods that approximate Singular Value Decomposition (SVD). SVD is a technique to decompose a matrix into the product $Q\Sigma V$, where Q contains the left singular vectors, Σ contains the singular values and V contains the right singular vectors of the original matrix. The application of SVD to recommender systems is motivated by the desire of decomposing the ratings matrix into a

2-matrices representation $R = U \cdot P^T$. Where each vector (row) u_i of U represents a user i and each vector (row) p_j of P represents a product j.

The goal of using matrix factorization in recommendation problems is to enable the assessment of user preferences for products by calculating the dot product of their factor representations, as the predicted preference $\hat{r}_{ij} = u_i \cdot p_j^T$ of user i for product j.

A noticeable improvement proposed by Koren et al. [2] defines a baseline predictor and considers the deviations from the average rating for users and products, referred to as **user and product biases**. User and product biases can be taken into account to better capture the real essence of user preferences, minding the fact that different users tend to give higher or lower ratings and different products tend to get higher or lower ratings, as well. Thus, the plain-SVD model can be improved into a bias-SVD model by setting a global mean rating average baseline prediction and adding parameters to capture biases, resulting in the prediction rule defined by eq. 1.

$$\hat{r}_{ij} = \mu + u_i \cdot p_j^T + b_i + b_j \tag{1}$$

Here, μ is the mean rating average, b_i represents user i bias and b_j represents product j bias. Accordingly, the expression to be minimized corresponds to:

$$[U, P] = \arg\min_{u_i, p_j} \sum_{r_{ij} \in R} (\hat{r}_{ij} - \mu - b_i - b_j - u_i \cdot p_j^T)^2 + \tag{2}$$
$$\lambda \cdot (\|u_i\|^2 + \|p_j\|^2 + b_i^2 + b_j^2)$$

This expression accomplishes three goals: matrix factorization by minimization, biases compensation and the corresponding regularization for over-fitting control. The first part of eq. 2 pursues the minimization of the difference (henceforth referred to as error) between the known ratings present on the original R ratings matrix and their decomposed representation (U and P). The second part controls generality by avoiding over-fitting during the learning process, where λ is a constant defining the extent of regularization, usually chosen by cross-validation.

4   Multi-user Recommendations

Computing group-based preferences comprises combining the preferences of those users into a representation of group preferences. Since the system already has some knowledge regarding these target users' preferences, in the form of product ratings, such knowledge shall be used to produce a diversified list of recommendations for groups. For combining user preferences, an *early fusion* approach was taken. The *early fusion* approach consists in combining the factor vectors of target users in a linear combination,

$$g = \frac{\sum_i^m u_i}{m}. \tag{3}$$

In eq. 3, m is the number of target users, u_i is the factor vector representing user i and g is the resulting combined factor vector representing the preferences of the group. The intuition behind this choice for computing group preferences is the following: according to the intuition behind the latent factor representation of users and products, each latent factor represents a characteristic associated to users and products, meaning that each user's latent factor vector will represent the extent to which that user likes (positive factor values) or dislikes (negative factor values) a given characteristic in a product. By combining user's latent factor vectors as our early fusion approach suggests, we expect to neutralize contradicting user preferences and preserve common user preferences, representing these in a resulting group latent factor vector. Afterwards, the system attempted to produce a list of recommendations that would please the group in terms of user satisfaction and product diversity. In this section, these steps will be addressed in detail.

4.1   Group Bias-SVD: Maximizing Group Satisfaction

Group satisfaction is defined as the rating that a group would give to a product. Our system attempts to recommend lists of products that maximize the group satisfaction, by making use of the users' latent factor vectors and the group latent factor vector, obtained through the early fusion approach previously mentioned. Once obtained the latent factor vector representing group preferences, recommendations can be computed with the prediction rule introduced by the bias-SVD model in eq. 1, only this time using the group preferences factor vector as a super-user, as

$$\hat{r}_{gj} = \mu + g \cdot p_j^T + b_g + b_j \tag{4}$$

where the group bias b_g is computed by averaging all m users' individual biases. By using the prediction rule defined by eq. 4, one can predict which products are more likely to satisfy this artificial n-user-group: products with higher predicted score are expected to be more fit to recommend the group of users. However, although the system relies on the group factor vector to produce recommendations, individual user preferences are still used to assure a minimum degree of individual satisfaction, by setting a minimum individual predicted preference threshold $minUSat$. With such threshold, the system avoids recommending products that would significantly displease some of the target users, even if the overall predicted multi-user preference for those products is high. To achieve this goal, the system stores all products within the database that fulfil the $minUSat$ restriction and sorts them in descending order of predicted multi-user preference into a list L, so that the n top products can be recommended to the group. Thus, our objective function intends to find the set S of n products that maximizes the average predicted multi-user rating, as described by eq. 5.

$$S = \underset{\{p_1,\cdots,p_n\}}{\arg\max} \frac{\sum_{j=1}^n \hat{r}_{gj}}{n}, \quad \{p_1,\cdots,p_n\} \in L \tag{5}$$

4.2   Maximizing Product Diversity

In addition to aiming for maximizing the satisfaction of the group in terms of preference, the system also attempts to present a diversified list of products. In most cases, striving towards maximizing diversity compromises satisfaction. To deal with this satisfaction-diversity trade-off issue, the system takes as input the parameters $minGSat$ and $minUSat$, defining the minimum multi-user and individual user satisfaction respectively. Thus, the first thing the system does is to obtain a list E, containing only the top products that fulfil the minimum satisfaction restrictions imposed by the parameters $minGSat$ and $minUSat$. Once obtained the list E, the systems attempts to find the set S of n products contained in E that maximizes diversity.

Defining Diversity. To define diversity let us revisit the latent factor representation of users and products obtained through matrix factorization: each user and product is represented by a vector of latent factors, where these latent factors are abstract or real-life characteristics. This means that a product vector's direction within the latent factor high-dimensional space indicates which characteristics this product has and which it does not have. Hence, two products can be compared by comparing the direction to which their respective factor vectors point. In this sense, the straightforward choice for a comparison metric between products is the cosine of the angle formed by these products. This metric is well known as the cosine similarity/distance, defined as $cosSim_{a,b} = \frac{p_a \cdot p_b}{\|p_a\| \cdot \|p_b\|}$. Here, p_a and p_b denote products a and b factor vectors. Notice that cosine similarity values range from -1 to 1. Using the cosine similarity metric, our intuition is that a set of 2 products will be more diversified if the cosine similarity between its products is negative or close to zero. Fig. 1 illustrates the intuition behind cosine diversity. Accounting for vector directions by measuring diversity in terms of sums of the vertex angles avoids the *curse of high-dimensionality* problem and is more consistent with what we consider to be diversity. Since our goal is to maximize diversity among products, i.e., to minimize similarity, we define the cosine diversity metric as $cosDiv_{a,b} = -cosSim_{a,b}$. This cosine diversity metric measures the diversity between two products, but our goal is to find a diversity metric that measures the diversity of a set of n products. In that sense, we define

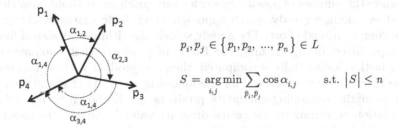

$$p_i, p_j \in \{p_1, p_2, ..., p_n\} \in L$$

$$S = \operatorname*{arg\,min}_{i,j} \sum_{p_i, p_j} \cos \alpha_{i,j} \quad \text{s.t.} \; |S| \leq n$$

Fig. 1. Maximizing the vertex-angle diversity

the diversity of the set S as the average cosine diversity between all products in the set and between all products and all users in the target group G, as in eq. 6.

$$cosDiv_S = \frac{\sum_{\forall a,b \in S} (cosDiv_{a,b}) + \sum_{\forall p \in S, \forall u \in G} (cosDiv_{p,u})}{|S|^2 + |S| \cdot |G|} \qquad (6)$$

Thus, the objective function becomes as described by eq. 7.

$$S = \underset{\{p_1, \cdots, p_n\}}{\arg\max} \ cosDiv_{\{p_1, \cdots, p_n\}}, \qquad \{p_1, \cdots, p_n\} \in E \qquad (7)$$

As we can observe from eq. 7 and fig. 1, the optimization problem we face is a combinatorial problem with a non-convex objective function, since the diversity each product adds to the set depends on which products have already been selected to be part of the product set to recommend. Thus, the universe of possible solutions is very wide, where finding the optimal solution constitutes a challenge in terms of computational efficiency.

Algorithm 1. Greedy vertex-angle maximization algorithm

E ← getTopProds(G,*minGSat*,*minUSat*)
S ← E.topSublist(n)
for t **do**
   seed ← newRandomProd(E)
   newS.add(seed)
   E.remove(seed)
   while |newS| < n **do**
      newProd ← getMostDivProd(E,G,newS)
      newS.add(newProd)
      E.remove(newProd)
   end while
   if newS.divScore > S.divScore **then**
      S ← newS
   end if
end for

Greedy Vertex-angle Maximization. To solve the combinatorial problem we face, computing all possible combinations of n products is not a feasible option, since the number of possibilities can easily reach many thousands. Hence, we opted by taking a greedy search approach which is dramatically cheaper in terms of computational effort. The greedy search algorithm we designed follows a few steps, listed in alg. 1: first, the list E of products that maximize each user satisfaction individually is computed, then one product is chosen randomly from the E list to be used as seed for the following steps. Afterwards, each new product from the remaining candidate products list is chosen to be included in the solution according to the cosine diversity value it carries: the candidate product with the highest cosine diversity score considering the target users and the already selected products, will be included in the selected products list.

This last step is repeated until the list of products to recommend reaches the desired number n of products selected for recommendation. These steps will produce one solution. Repeating these steps t times with t different seeds will produce t different solutions from which the algorithm picks the one that yields the highest cosine diversity score. In alg. 1, t is the pre-determined number of runs the algorithm should take and consequently the number of different solutions it will return, n is the number of products we intend to recommend the group and $minGSat/minUSat$ are the previously introduced parameters to define the minimum group satisfaction and the minimum user satisfaction, respectively. To avoid sacrificing satisfaction (i.e., group RMSE), products were selected to be seeds according to their order in the E list (recall that products in E are sorted by satisfaction: predicted group preference), i.e., if the greedy search takes t runs, the top t products on the E list will be sequentially selected as seeds for each run. Thus, in our case $t \leq |E|$.

5   Evaluation

5.1   Dataset

For the following experiments, the Movielens dataset was used. This dataset contains 10 million ratings on 0 to 5 scale with 0.5 point increments, given by 69878 users to 10681 movies. To obtain the latent factor space through matrix factorization, we considered only users with at least 20 ratings, so that we would have a reasonable confidence level about each user's preferences and a reasonable number of ratings in each subset. The dataset was further split into 3 subsets: training (65%), validation (15%), and test (20%), for cross-validation. At the matrix factorization stage, only the training and validation subsets were used, leaving the test subset for final evaluation of multi-user recommendation experiments.

5.2   Evaluation Protocol

The latent factor space was discovered using the bias-SVD matrix factorization model described in section 3, and the space dimensionality was set at 50 latent factors. Experiments were made with different scenarios: single-group, 2-group and 4-group. In each scenario, the system produced lists of products to a multi-user set of 4 randomly selected users that may belong to one, two or four different interest-groups. On the pre-processing stage, we also used a k-means clustering algorithm to find interest-groups among users. We set a minimum threshold of 150 users per cluster so that every user would be categorized as part of a representative and reasonably-sized interest-group. The recommendation lists produced in each scenario were then evaluated based on two criteria: user **satisfaction** and product **diversity**, which will be detailed in the next sections.

Target users must have at least 25 ratings on the test set. To assure there will be a reasonable number of ratings to evaluate group satisfaction we selected products containing ratings from at least 75% of the target users. Each recommendation list contains 5 products. Additionally, the $minUSat$ was set to 2.5.

5.3   Experiment: Group Satisfaction

In this experiment we assess the group bias-SVD model. For each of the three aforementioned scenarios, 50 different experiments were made to assure the results have statistical significance. In each individual experiment, a set of 5 products was recommended to a set of 4 users. When evaluating multi-user recommendations we face several challenges, in particular the absence of multi-user preference data. To overcome this limitation the test data followed a product eligibility criteria to select products that were commonly rated by all users.

Metrics for Measuring Multi-user Satisfaction. Once multi-user recommendations are produced, we measured some indicators associated with these recommendations to evaluate their quality. The group satisfaction value $avgSat$ is the average rating given by target users to recommended products on the test set. To help making a correct interpretation of the $avgSat$ value we registered the best and worst possible satisfaction values considering the set of products eligible for recommendation and the target users. This makes it possible to evaluate the obtained satisfaction levels according to how good or bad they could have been.

Results: Multi-user Satisfaction Experiments. Figure 2, illustrate the results obtained on the single-group, 2-group and 4-group scenarios, respectively. The experiments illustrated on these charts are sorted in descending order of the maximum possible satisfaction. As we can observe on all three figures, the multi-user satisfaction curve tends to be substantially closer to the best possible multi-user satisfaction curve than to the worst possible multi-user satisfaction curve. The average normalized multi-user satisfaction values ($gSat/maxSat$) obtained in all three scenarios ranged from 76% to 80%. These results mean that the recommender system is producing good multi-user recommendations. Finally, the normalized multi-user satisfaction values indicate no relevant differences among the system's performances for single-group, 2-group and 4-group scenarios, showing its versatility when recommending products to users from different interest-groups.

5.4   Experiment: Product Diversity

A set of experiments following the same protocol as those performed to evaluate multi-user satisfaction was made to evaluate product diversity. This time we made 100 experiments for each scenario instead of 50, produced recommendation lists with 15 products instead of 5, and decided not to enforce product eligibility standards. This way, we can recommend more products at once, make more unique experiments and explore a larger scope of products to obtain

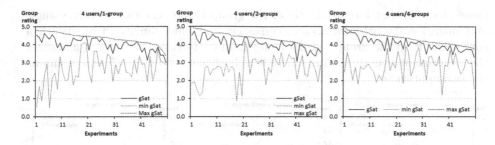

Fig. 2. Multi-user satisfaction evaluation scenarios: 4 users, single-group; 4 users, 2-groups; 4 users, 4-groups. Average normalized satisfaction: 76.0%, 80.5% and 80.4% respectively.

maximum diversity. To evaluate the performance of our diversity-oriented approach to multi-user recommendation we compared it against the performance of a satisfaction-oriented approach (group bias-SVD) where recommendation lists contain the 15 products that rank higher in terms of predicted multi-user satisfactions, regardless of how diversified this list is.

Metrics for Measuring Product Diversity. The metric used for testing the vertex-angle diversity algorithm is based on product genres. In this dataset, each product is associated to a list of genres. Since there is no universal definition for diversity, we consider that a list of products is diversified if there is a high number of different genres associated to those products and if the occurrences of the genres involved is balanced. A straightforward way of measuring the balance between genre occurrences is to calculate its variance: lower variance means more balanced genre occurrences. Thus, real product diversity for a list of products S is defined as eq. 8.

$$div_S = nGenres_S/varGenres_S \qquad (8)$$

Here, $nGenres_S$ represents the number of different genres related to the products in list S and $varGenres_S$ represents the variance in genre occurrences in that same product list S.

Results: Product Diversity Experiments. The top row of figure 3 illustrate the diversity scores obtained with both the diversity-oriented (greedy vertex maximization) and the satisfaction-oriented approach (group bias-SVD) in single-group, 2-group and 4-group scenarios. As we can observe, the normalized diversity-oriented approach produced more diversified recommendations in 77%-83% of the experiments without lowering multi-user satisfaction below the minimum 3.5 $minGSat$ threshold, which represents an overall successful performance of the proposed multi-user recommender system. The bottom row of figure 3 depicts 100 experiments in satisfaction-versus-diversity charts for the single-group, 2-group and 4-group scenarios. Each point corresponds to an experiment where 15 products were recommended to the group of users. The salient fact from these charts is that independently of the group heterogeneity, the vertex-angle maximization algorithm works on a consistent and wider range of diversity when comparing it to the group bias-SVD algorithm. Moreover, figure 4 summarizes the experiment overall results, where it is best observed the differences among the two approaches.

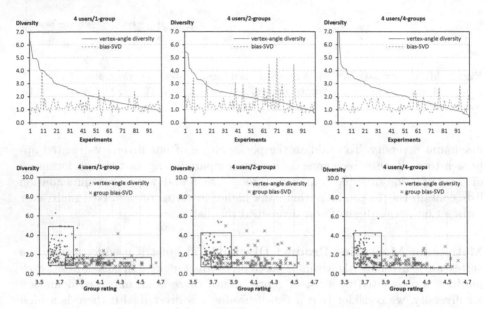

Fig. 3. Diversity evaluation scenarios: 4 users, single-group; 4 users, 2-groups; 4 users, 4-groups

	vertex-angle algorithm			group bias-SVD		
nr. of groups	1-group	2-group	4-group	1-group	2-group	4-group
Diversity	2.274	2.136	2.214	1.244	1.443	1.268
Satisfaction	3.696	3.710	3.692	4.118	4.085	4.040
Diversity	2.208			1.318		
Satisfaction	3.699			4.081		

Fig. 4. Summary results comparing the group bias-SVD to the vertex-angle maximization algorithms

6 Discussion

In this paper we addressed the problem of computing diverse but effective recommendations in multi-user scenarios. Previous approaches to recommendations for multi-users either followed a content-based method or a simple collaborative approach merging the ratings of users. In contrast, our method transfers the entire problem into a high-dimensional latent factor space. The proposed technique achieved a trade-off between accuracy (satisfaction) and diversity. We showed that the vertex-angle maximization algorithm can compute consistent and coherent recommendations across several heterogeneous groups of users. Moreover, the use of latent space to solve problems related to multi-user recommendation introduced in this paper was our most significant accomplishment, encouraging further research on the subject.

References

1. Resnick, P., Iacovou, N., Suchak, M., Bergstrom, P., Riedl, J.: Grouplens: An open architecture for collaborative filtering of netnews, pp. 175–186. ACM Press (1994)
2. Koren, Y., Bell, R.M., Volinsky, C.: Matrix factorization techniques for recommender systems. IEEE Computer 42(8), 30–37 (2009)
3. Salakhutdinov, R., Mnih, A., Hinton, G.: Restricted boltzmann machines for collaborative filtering. In: Proceedings of the 24th International Conference on Machine Learning, ICML 2007. pp. 791–798. ACM (2007)
4. Ovsjanikov, M., Chen, Y.: Topic modeling for personalized recommendation of volatile items. In: Balcázar, J.L., Bonchi, F., Gionis, A., Sebag, M. (eds.) ECML PKDD 2010, Part II. LNCS, vol. 6322, pp. 483–498. Springer, Heidelberg (2010)
5. Sarwar, B.M., Karypis, G., Konstan, J.A., Riedl, J.T.: Application of dimensionality reduction in recommender system – a case study. In: ACM Webkdd Workshop (2000)
6. Jameson, A., Smyth, B.: Recommendation to groups. In: Brusilovsky, P., Kobsa, A., Nejdl, W. (eds.) Adaptive Web 2007. LNCS, vol. 4321, pp. 596–627. Springer, Heidelberg (2007)
7. Crossen, A., Budzik, J., Hammond, K.J.: Flytrap: intelligent group music recommendation. In: Int'l Conference on Intelligent User Interfaces. IUI 2002, pp. 184–185. ACM (2002)
8. McCarthy, K., Salamó, M., Coyle, L., McGinty, L., Smyth, B., Nixon, P.: Cats: A synchronous approach to collaborative group recommendation. In: Int'l Florida Artificial Intelligence Research Society Conference, Melbourne Beach, Florida, USA, May 11-13, pp. 86–91. AAAI Press (2006)
9. Mahato, H., Kern, D., Holleis, P., Schmidt, A.: Implicit personalization of public environments using bluetooth. In: CHI 2008 Extended Abstracts on Human Factors in Computing Systems. CHI EA 2008, pp. 3093–3098. ACM (2008)
10. Baltrunas, L., Makcinskas, T., Ricci, F.: Group recommendations with rank aggregation and collaborative filtering. In: Proceedings of the Fourth ACM Conference on Recommender Systems, RecSys 2010, pp. 119–126. ACM (2010)
11. Seko, S., Yagi, T., Motegi, M., Muto, S.: Group recommendation using feature space representing behavioral tendency and power balance among members. In: ACM Conference on Recommender Systems, RecSys 2011, pp. 101–108. ACM (2011)
12. Barbieri, N., Manco, G.: An analysis of probabilistic methods for top-N recommendation in collaborative filtering. In: Gunopulos, D., Hofmann, T., Malerba, D., Vazirgiannis, M. (eds.) ECML PKDD 2011, Part I. LNCS, vol. 6911, pp. 172–187. Springer, Heidelberg (2011)
13. Ziegler, C.N., McNee, S.M., Konstan, J.A., Lausen, G.: Improving recommendation lists through topic diversification. In: Proceedings of the 14th International Conference on World Wide Web, pp. 22–32. ACM (2005)
14. Jaszkiewicz, A.: Genetic local search for multi-objective combinatorial optimization. European Journal of Operational Research 137(1), 50–71 (2002)
15. Zhang, T.: Adaptive forward-backward greedy algorithm for sparse learning with linear models, NIPS (2008)

ERMiner: Sequential Rule Mining
Using Equivalence Classes

Philippe Fournier-Viger[1], Ted Gueniche[1], Souleymane Zida[1],
and Vincent S. Tseng[2]

[1] Dept. of Computer Science, University of Moncton, Canada
[2] Dept. of Computer Science and Information Engineering,
National Cheng Kung University, Taiwan
{philippe.fournier-viger,esz2233}@umoncton.ca, ted.gueniche@gmail.com,
tsengsm@mail.ncku.edu.tw

Abstract. Sequential rule mining is an important data mining task with
wide applications. The current state-of-the-art algorithm (RuleGrowth)
for this task relies on a pattern-growth approach to discover sequential rules. A drawback of this approach is that it repeatedly performs a
costly database projection operation, which deteriorates performance for
datasets containing dense or long sequences. In this paper, we address
this issue by proposing an algorithm named ERMiner (Equivalence class
based sequential Rule Miner) for mining sequential rules. It relies on
the novel idea of searching using equivalence classes of rules having the
same antecedent or consequent. Furthermore, it includes a data structure
named SCM (Sparse Count Matrix) to prune the search space. An extensive experimental study with five real-life datasets shows that ERMiner
is up to five times faster than RuleGrowth but consumes more memory.

Keywords: sequential rule mining, vertical database format, equivalence classes, sparse count matrix.

1   Introduction

Discovering interesting sequential patterns in sequences is a fundamental problem in data mining. Many studies have been proposed for mining interesting
patterns in sequence databases [12]. Sequential pattern mining [1] is probably
the most popular research topic among them. It consists of finding subsequences
appearing frequently in a set of sequences. However, knowing that a sequence
appears frequently is not sufficient for making predictions [4]. An alternative
that addresses the problem of prediction is sequential rule mining [4]. A sequential rule indicates that if some item(s) occur in a sequence, some other item(s)
are likely to occur afterward with a given confidence or probability.

Two main types of sequential rules have been proposed. The first type is
rules where the antecedent and consequent are sequential patterns [11,15,13].
The second type is rules between two unordered sets of items [6,4]. In this paper
we consider the second type because it is more general and it was shown to

H. Blockeel et al. (Eds.): IDA 2014, LNCS 8819, pp. 108–119, 2014.

provide considerably higher prediction accuracy for sequence prediction in some domains [5]. Moreover, another reason is that the second type has been used in many real applications such as e-learning [6], manufacturing simulation [9], quality control [2], web page prefetching [5], anti-pattern detection in service based systems [14], embedded systems [10], alarm sequence analysis [3] and restaurant recommendation [8].

Several algorithms have been proposed for mining this type of sequential rules. CMDeo [6] is an Apriori-based algorithm that explores the search space of rules using a breadth-first search. A major drawback of CMDeo is that it can generate a huge amount of candidates. As as alternative, the CMRules algorithm was proposed. It relies on the property that any sequential rules must also be an association rule to prune the search space of sequential rules [6]. It was shown to be much faster than CMDeo for sparse datasets. Recently, the RuleGrowth [4] algorithm was proposed. It relies on a pattern-growth approach to avoid candidate generation. It was shown to be more than an order of magnitude faster than CMDeo and CMRules. However, for datasets containing dense or long sequences, the performance of RuleGrowth rapidly deterioates because it has to repeatedly perform costly database projection operations. Because mining sequential rules remains a very computationally expensive data mining task, an important research question is: "Could we design faster algorithms?"

In this paper, we address this issue by proposing the ERMiner (Equivalence class based sequential Rule Miner) algorithm. It relies on a vertical representation of the database to avoid performing database projection and the novel idea of explorating the search space of rules using equivalence classes of rules having the same antecedent or consequent. Furthermore, it includes a data structure named SCM (Sparse Count Matrix) to prune the search space.

The rest of the paper is organized as follows. Section 2 defines the problem of sequential rule mining and introduces important definitions and properties. Section 3 describes the ERMiner algorithm. Section 4 presents the experimental study. Finally, Section 5 presents the conclusion.

2   Problem Definition

Definition 1 (sequence database). Let $I = \{i_1, i_2, ..., i_l\}$ be a set of items (symbols). An *itemset* $I_x = \{i_1, i_2, ..., i_m\} \subseteq I$ is an unordered set of distinct items. The *lexicographical order* $\succ_{lex}$ is defined as any total order on I. Without loss of generality, it is assumed in the following that all itemsets are ordered according to $\succ_{lex}$. A *sequence* is an ordered list of itemsets $s = \langle I_1, I_2, ..., I_n \rangle$ such that $I_k \subseteq I$ ($1 \leq k \leq n$). A *sequence database* SDB is a list of sequences $SDB = \langle s_1, s_2, ..., s_p \rangle$ having sequence identifiers (SIDs) $1, 2.., p$.

Example 1. A sequence database is shown in Fig. 1 (left). It contains four sequences having the SIDs 1, 2, 3 and 4. Each single letter represents an item. Items between curly brackets represent an itemset. The first sequence $\langle \{a, b\}, \{c\}, \{f\}, \{g\}, \{e\}\rangle$ contains five itemsets. It indicates that items a and b occurred at the same time, were followed by c, then f and lastly e.

ID	Sequences
seq1	⟨{a, b},{c},{f},{g},{e}⟩
seq2	⟨{a, d},{c},{b},{a, b, e, f}⟩
seq3	⟨{a},{b},{f},{e}⟩
seq4	⟨{b},{f, g, h}⟩

ID	Rule	Support	Confidence
r1	{a, b, c}→{e}	0.5	1.0
r2	{a}→{c, e, f}	0.5	0.66
r3	{a, b}→{e, f}	0.75	1.0
r4	{b}→{e, f}	0.75	0.75
r5	{a}→{e, f}	0.75	1.0
r6	{c}→{f}	0.5	1.0
r7	{a}→{b}	0.5	0.66

Fig. 1. A sequence database (left) and some sequential rules found (right)

Definition 2 (sequential rule). A sequential rule $X \to Y$ is a relationship between two unordered itemsets $X, Y \subseteq I$ such that $X \cap Y = \emptyset$ and $X, Y \neq \emptyset$. The interpretation of a rule $X \to Y$ is that if items of X occur in a sequence, items of Y will occur afterward in the same sequence.

Definition 3 (itemset/rule occurrence). Let $s : \langle I_1, I_2...I_n \rangle$ be a sequence. An itemset I occurs or is contained in s (written as $I \sqsubseteq s$) iff $I \subseteq \bigcup_{i=1}^{n} I_i$. A rule $r : X \to Y$ occurs or is contained in s (written as $r \sqsubseteq s$) iff there exists an integer k such that $1 \leq k < n$, $X \subseteq \bigcup_{i=1}^{k} I_i$ and $Y \subseteq \bigcup_{i=k+1}^{n} I_i$.

Example 2. The itemset $\{a, b, f\}$ is contained in sequence $\langle \{a\}, \{b\}, \{f\}, \{e\} \rangle$. The rule $\{a, b, c\} \to \{e, f, g\}$ occurs in $\langle \{a, b\}, \{c\}, \{f\}, \{g\}, \{e\} \rangle$, whereas the rule $\{a, b, f\} \to \{c\}$ does not, because item c does not occur after f.

Definition 4 (sequential rule size). A rule $X \to Y$ is said to be of size $k * m$ if $|X| = k$ and $|Y| = m$. Furthermore, a rule of size $f * g$ is said to be larger than another rule of size $h * i$ if $f > h$ and $g \geq i$, or alternatively if $f \geq h$ and $g > i$.

Example 3. The rules $r : \{a, b, c\} \to \{e, f, g\}$ and $s : \{a\} \to \{e, f\}$ are respectively of size $3 * 3$ and $1 * 2$. Thus, r is larger than s.

Definition 5 (support). The *support* of a rule r in a sequence database SDB is defined as $sup_{SDB}(r) = |\{s | s \in SDB \wedge r \sqsubseteq s\}| / |SDB|$.

Definition 6 (confidence). The *confidence* of a rule $r : X \to Y$ in a sequence database SDB is defined as $conf_{SDB}(r) = |\{s | s \in SDB \wedge r \sqsubseteq s\}| / |\{s | s \in SDB \wedge X \sqsubseteq s\}|$.

Definition 7 (sequential rule mining). Let $minsup, minconf \in [0, 1]$ be thresholds set by the user and SDB be a sequence database. A rule r is a *frequent sequential rule* iff $sup_{SDB}(r) \geq minsup$. A rule r is a *valid sequential rule* iff it is frequent and $conf_{SDB}(r) \geq minconf$. The *problem of mining sequential rules* from a sequence database is to discover all valid sequential rules [6].

Example 4. Fig 1 (right) shows 7 valid rules found in the database illustrated in Table 1 for $minsup = 0.5$ and $minconf = 0.5$. For instance, the rule $\{a, b, c\} \to \{e\}$ has a support of $2/4 = 0.5$ and a confidence of $2/2 = 1$. Because those values are respectively no less than $minsup$ and $minconf$, the rule is deemed valid.

3   The ERMiner Algorithm

In this section, we present the ERMiner algorithm. It relies on the novel concept of equivalence classes of sequential rules, defined as follows.

Definition 8 (rule equivalence classes). For a sequence database, let $\mathcal{R}$ be the set of all frequent sequential rules and I be the set of all items. A left equivalence class $LE_{W,i}$ is the set of frequent rules $LE_{W,i} = \{W \rightarrow Y | Y \subseteq I \wedge |Y| = i\}$ such that $W \subseteq I$ and i is an integer. Similarly, a *right equivalence class* $RE_{W,i}$ is the set of frequent rules $RE_{W,i} = \{X \rightarrow W | X \subseteq I \wedge |X| = i\}$, where $W \subseteq I$, and i is an integer.

Example 5. For $minsup = 2$ and our running example, $LE_{\{c\},1} = \{\{c\} \rightarrow \{f\}, \{c\} \rightarrow \{e\}\}$, $RE_{\{e,f\},1} = \{\{a\} \rightarrow \{e,f\}, \{b\} \rightarrow \{e,f\}, \{c\} \rightarrow \{e,f\}\}$ and $RE_{\{e,f\},2} = \{\{a,b\} \rightarrow \{e,f\}, \{a,c\} \rightarrow \{e,f\}, \{b,c\} \rightarrow \{e,f\}\}$.

Two operations called left and right merges are used by ERMiner to explore the search space of frequent sequential rules. They allows to directly generate an equivalence class using a smaller equivalence class.

Definition 9 (left/right merges). Let be a left equivalence class $LE_{W,i}$ and two rules $r : W \rightarrow X$ and $s : W \rightarrow Y$ such that $r, s \in LE_{W,i}$ and $|X \cap Y| = |X - 1|$, i.e. X and Y are identical except for a single item. A *left merge* of r, s is the process of merging r, s to obtain $W \rightarrow X \cup Y$. Similarly, let be a right equivalence class $RE_{W,i}$ and two rules $r : X \rightarrow W$ and $r : Y \rightarrow W$ such that $r, s \in RE_{W,i}$ and $|X \cap Y| = |X - 1|$. A *right merge* of r, s is the process of merging r, s to obtain the rule $X \cup Y \rightarrow W$.

Property 1 (generating a left equivalence class). Let be a left equivalence class $LE_{W,i}$. $LE_{W,i+1}$ can be obtained by performing all left merges on pairs of rules from $LE_{W,i}$. **Proof.** Let be any rule $r : W \rightarrow \{a_1, a_2, ...a_{i+1}\}$ in $LE_{W,i+1}$. By Definition 8, rules $W \rightarrow \{a_1, a_2, ...a_{i-1}, a_i\}$ and $W \rightarrow \{a_1, a_2, ...a_{i-1}, a_{i+1}\}$ are members of $LE_{W,i}$, and a left merge of those rules will generate r.$\square$

Property 2 (generating a right equivalence class). Let be a right equivalence class $RE_{W,i}$. $RE_{W,i+1}$ can be obtained by performing all right merges on pairs of rules from $RE_{W,i}$. **Proof.** The proof is similar to Property 1 and is therefore omitted.

To explore the search space of frequent sequential rules using the above merge operations, ERMiner first scans the database to build all equivalence classes for frequent rules of size $1 * 1$. Then, it recursively performs left/right merges starting from those equivalence classes to generate the other equivalence classes. To ensure that no rule is generated twice, the following ideas have been used.

First, an important observation is that a rule can be obtained by different combinations of left and right merges. For example, consider the rule $\{a, b\} \rightarrow \{c, d\}$. It can be obtained by performing left merges for $LE_{\{a\},1}$ and $LE_{\{b\},1}$

followed by right merges on $RE_{\{c,d\},1}$. But it can also be obtained by performing right merges on $RE_{\{c\},1}$ and $RE_{\{d\},1}$ followed by left merges using $LE_{\{a,b\},1}$. A simple solution to avoid this problem is to not allow performing a left merge after a right merge but to allow performing a right merge after a left merge. This solution is illustrated in Fig. 2.

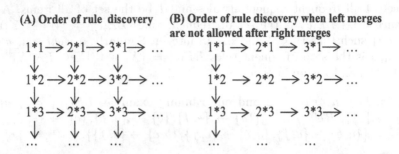

Fig. 2. The order of rule discovery by left/right merge operations

Second, another key observation is that a same rule may be obtained by merging different pairs of rules from the same equivalence class. For example, a rule $\{a, b, c\} \rightarrow \{e\}$ may be obtained by performing a left merge of $\{a, b\} \rightarrow \{e\}$ with $\{a, c\} \rightarrow \{e\}$ or with $\{b, c\} \rightarrow \{e\}$. To avoid generating the same rule twice, a simple solution is to impose a total order on items in rule antecedents (consequents) and to only perform a left merge (right merge) if the rule consequent (rule antecedent) shares all but the last item according to the total order. In the previous example, this means that $\{a, c\} \rightarrow \{e\}$ would not be merged with $\{b, c\} \rightarrow \{e\}$.

Using the above solutions, it can be easily seen that all rules are generated only once. However, to be efficient, a sequential rule mining algorithm should be able to prune the search space. This is done using the following properties for merge operations.

Property 3 (antimonotonicity with left/right merges). Let be a sequence database SDB and two frequent rules r, s. Let t be a rule obtained by a left or right merge of r, s. The support of t is lower or equal to the support of r and that of s. **Proof.** Since t contains exactly one more item than r and s, it can only appear in the same number sequences or less. $\square$

Property 4 (pruning). If the support of a rule is less than $minsup$, then it should not be merged with any other rules because all such rules are infrequent. **Proof.** This directly follows from Property 3.

Because there does not exist any similar pruning properties for confidence, it is necessary to explore the search space of frequent rules to get the valid ones.

Fig. 1 shows the main pseudocode of ERMiner, which integrates all the previous idea. ERMiner takes as input a sequence database SDB, and the $minsup$

and *minconf* thresholds. It first scans the database once to build all equivalence classes of rules of size $1 * 1$, i.e. containing a single item in the antecedent and a single item in the consequent. Then, to discover larger rules, left merges are performed with all left equivalence classes by calling the *leftSearch* procedure. Similarly, right merges are performed for all right equivalence classes by calling the *rightSearch* procedure. Note that the *rightSearch* procedure may generate some new left-equivalence classes because left merges are allowed after right merges. These equivalence classes are stored in a structure named *leftStore*. To process these equivalence classes, an additional loop is performed. Finally, the algorithm returns the set of rules found *rules*.

Algorithm 1. The ERMiner algorithm

---

input : *SDB*: a sequence database, *minsup* and *minconf*: the two
user-specified thresholds
output: the set of valid sequential rules

1 *leftStore* ← ∅ ;
2 *rules* ← ∅ ;
3 Scan *SDB* once to calculate *EQ*, the set of all equivalence classes of rules of size 1*1;
4 **foreach** *left equivalence class H ∈ EQ* **do**
5 | leftSearch (*H*, *rules*);
6 **end**
7 **foreach** *right equivalence class J ∈ EQ* **do**
8 | rightSearch (*J*, *rules*, *leftStore*);
9 **end**
10 **foreach** *left equivalence class K ∈ leftStore* **do**
11 | rightSearch (*K*);
12 **end**
13 **return** *rules*;

Fig. 2 shows the pseudocode of the *leftSearch* procedure. It takes as parameter an equivalence class *LE*. Then, for each rule *r* of that equivalence class, a left merge is performed with every other rules to generate a new equivalence class. Only frequent rules are kept. Furthermore, if a rule is valid, it is output. Then, *leftSearch* is recursively called to explore each new equivalence class generated that way. The *rightSearch* (see Fig. 3) is similar. The main difference is that new left equivalences are stored in the left store structure because their exploration is delayed, as previously explained in the main procedure of ERMiner.

Now, it is important to explain how the support and confidence of each rule is calculated by ERMiner (we had previously deliberately ommitted this explanation). Due to space limitation and because this calculation is done similarly as in the RuleGrowth [4] algorithm, we here only give the main idea. Initially, a database scan is performed to record the first and last occurrences of each item in each sequence where it appears. Thereafter, the support of each rule

Algorithm 2. The leftSearch procedure

input : LE: a left equivalence class, $rules$: the set of valid rules found until
now, $minsup$ and $minconf$: the two user-specified thresholds

```
1  foreach rule r ∈ LE do
2  |   LE' ← ∅ ;
3  |   foreach rule s ∈ LE such that r ≠ s and the pair r, s have not been
   |   processed do
4  |   |   Let c, d be the items respectively in r, s that do not appear in s, r ;
5  |   |   if countPruning(c, d) = false then
6  |   |   |   t ← leftMerge(r, s) ;
7  |   |   |   calculateSupport(t, r, s);
8  |   |   |   if sup(t) ≥ minsup then
9  |   |   |   |   calculateConfidence(t, r, s);
10 |   |   |   |   if conf(t) ≥ minconf then
11 |   |   |   |   |   rules ← rules ∪ {t};
12 |   |   |   |   end
13 |   |   |   |   LE' ← LE' ∪ {t};
14 |   |   |   end
15 |   |   end
16 |   end
17 |   leftSearch (LE', rules);
18 end
```

of size 1*1 is directly generated by comparing first and last occurrences, without scanning the database. Similarly, the first and last occurrences of each rule antecedent and consequent are updated for larger rules without scanning the database. This allows to calculate confidence and support efficiently (see [4] for more details about how this calculation can be done).

Besides, an optimization is to use a structure that we name the *Sparse Count Matrix* (SCM) (aka CMAP [7]). This structure is built during the first database scan and record in how many sequences each item appears with each other items. For example, Fig. 3 shows the structure built for the database of Fig. 1 (left), represented as a triangular matrix. Consider the second row. It indicates that item b appear with items b, c, d, e, f, g and h respectively in 2, 1, 3, 4, 2 and 1 sequences. The SCM structure is used for pruning the search space as follows (implemented as the *countPruning* function in Fig. 3 and 2). Let be a pair of rules r, s that is considered for a left or right merge and c, d be the items of r, s, that respectively do not appear in s, r. If the count of r, s is less than $minsup$ in the SCM, then the merge does not need to be performed and the support of the rule is not calculated.

Lastly, another important optimization is how to implement the left store structure for efficiently storing left equivalence classes of rules that are generated by right merges. In our implementation, we use a hashmap of hashmaps, where the first hash function is applied to the size of a rule and the second hash function

---

Algorithm 3. The rightSearch procedure

---

    input : RE: a right equivalence class, $rules$: the set of valid rules found until
            now, $minsup$ and $minconf$: the two user-specified thresholds,
            $leftStore$: the structure to store left-equivalence classes of rules
            generated by right-merges

1 **foreach** $rule\ r \in RE$ **do**
2     $RE' \leftarrow \emptyset$;
3   **foreach** $rule\ s \in RE$ such that $r \neq s$ and the pair r, s have not been
       processed **do**
4   Let c, d be the items respectively in r, s that do not appear in s, r ;
5   **if** $countPruning(c, d) = false$ **then**
6             $t \leftarrow rightMerge(r, s)$;
7             $calculateSupport(t, r, s)$;
8   **if** $sup(t) \geq minsup$ **then**
9                 $calculateConfidence(t, r, s)$;
10   **if** $conf(t) \geq minconf$ **then**
11                    $rules \leftarrow rules \cup \{t\}$;
12   **end**
13                $RE' \leftarrow RE' \cup \{t\}$;
14                $addToLeftStore(t)$
15   **end**
16   **end**
17   **end**
18   rightSearch $(RE', rules)$;
19 **end**

Item	a	b	c	d	e	f
b	3					
c	2	2				
d	1	1	1			
e	3	3	2	1		
f	3	4	2	1	3	
g	1	2	1	0	1	2
h	0	1	0	0	0	1

Fig. 3. The Sparse Count Matrix

is applied to the left itemset of the rule. This allows to quickly find to which equivalence class belongs a rule generated by a right merge.

4 Experimental Evaluation

We performed experiments to assess the performance of the proposed algorithm. Experiments were performed on a computer with a third generation Core i5 processor running Windows 7 and 5 GB of free RAM. We compared the performance of ERMiner with the state-of-the-art algorithms for sequential rule mining Rule-Growth [4]. All algorithms were implemented in Java.

All memory measurements were done using the Java API. Experiments were carried on five real-life datasets having varied characteristics and representing four different types of data (web click stream, sign language utterances and protein sequences). Those datasets are *Sign, Snake, FIFA, BMS* and *Kosarak10k*. Table 2 4 summarizes their characteristics. The source code of all algorithms and datasets used in our experiments can be downloaded from `http://goo.gl/aAegWH`.

Table 1. Dataset characteristics

dataset	sequence count	distinct item count	avg. seq. length (items)	type of data
Sign	730	267	51.99 (std = 12.3)	language utterances
Snake	163	20	60 (std = 0.59)	protein sequences
FIFA	20450	2990	34.74 (std = 24.08)	web click stream
BMS	59601	497	2.51 (std = 4.85)	web click stream
Kosarak10k	10000	10094	8.14 (std = 22)	web click stream

We ran all the algorithms on each dataset while decreasing the *minsup* threshold until algorithms became too long to execute, ran out of memory or a clear winner was observed. For these experiments, we fixed the *minconf* threshold to 0.75. However, note that results are similar for other values of the *minconf* parameter since the confidence is not used to prune the search space by the compared algorithms. For each dataset, we recorded the execution time, the percentage of candidate pruned by the SCM structure and the total size of SCMs.

Execution times. The comparison of execution times is shown in Fig. 4. It can be seen that ERMiner is faster than RuleGrowth on all datasets and that the performance gap increases for lower *minsup* values. ERMiner is up to about five times faster than RuleGrowth. This is because RuleGrowth has to perform costly database projection operations.

Memory overhead of using SCM. We have measured the overhead produced by using the SCM structure by ERMiner. The size of SCM is generally quite small (less than 35 MB). The reason is that we have implemented it as a sparse matrix (a hashmap of hashmaps) rather than a full matrix (a $n \times n$ array for n items). If a full matrix is used the size of SCM increased up to about 300 MB.

Overall memory usage. The maximum memory usage of RuleGrowth / ERMiner for the Snake, FIFA, Sign, BMS and Kosarak datasets were respectively 300 MB / 1950 MB, 478 MB / 2030 MB, 347 MB / 1881 MB, 1328 MB / 2193 MB and 669 MB / 1441 MB. We therefore notice that thhere is a trade-off between having faster execution times with ERMiner versus having lower memory consumption with RuleGrowth. The higher memory consumption for ERMiner is in great part due to the usage of the left store structure which requires maintaining several equivalence classes into memory at the same time.

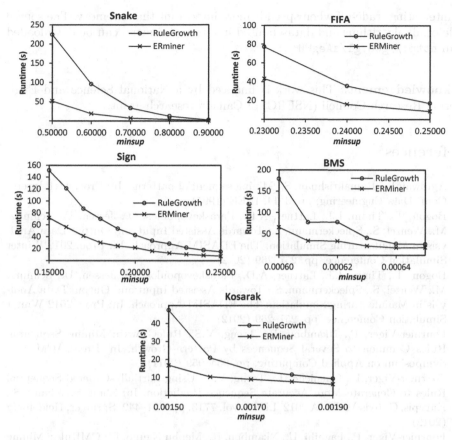

Fig. 4. Execution times

Effectiveness of candidate pruning. The percentage of candidate rules pruned by using the SCM data structure in ERMiner for the Snake, FIFA, Sign, BMS and Kosarak datasets were respectively 1 %, 0.2 %, 3.9 %, 3 % and 51 %. It can be concluded that pruning based on SCM is less effective for datasets containing dense or long sequences(e.g. Snake, FIFA, Sign) where each item co-occurs many times with every other items. It could therefore be desactivated on such datasets.

5   Conclusion

In this paper, we proposed a new sequential rule mining algorithm named ER-Miner (Equivalence class based sequential Rule Miner). It relies on the novel idea of searching using equivalence classes of rules having the same antecedent or consequent. Furthermore, it an includes a data structure named SCM (Sparse Count Matrix) to prune the search space. An extensive experimental study with five real-life datasets shows that ERMiner is up to five times faster than the state-of-the-art algorithm but comsumes more memory. It can therefore be seen as an interesting trade-off when speed is more important than memory. The source code of all algorithms and datasets used in our experiments can be downloaded from http://goo.gl/aAegWH.

Acknowledgement. This work is financed by a National Science and Engineering Research Council (NSERC) of Canada research grant.

References

1. Agrawal, R., Ramakrishnan, S.: Mining sequential patterns. In: Proc. 11th Intern. Conf. Data Engineering, pp. 3–14. IEEE (1995)
2. Bogon, T., Timm, I.J., Lattner, A.D., Paraskevopoulos, D., Jessen, U., Schmitz, M., Wenzel, S., Spieckermann, S.: Towards Assisted Input and Output Data Analysis in Manufacturing Simulation: The EDASIM Approach. In: Proc. 2012 Winter Simulation Conference, pp. 257–269 (2012)
3. Bogon, T., Timm, I.J., Lattner, A.D., Paraskevopoulos, D., Jessen, U., Schmitz, M., Wenzel, S., Spieckermann, S.: Towards Assisted Input and Output Data Analysis in Manufacturing Simulation: The EDASIM Approach. In: Proc. 2012 Winter Simulation Conference, pp. 257–269 (2012)
4. Fournier-Viger, P., Nkambou, R., Tseng, V.S.: RuleGrowth: Mining Sequential Rules Common to Several Sequences by Pattern-Growth. In: Proc. ACM 26th Symposium on Applied Computing, pp. 954–959 (2011)
5. Fournier-Viger, P., Gueniche, T., Tseng, V.S.: Using Partially-Ordered Sequential Rules to Generate More Accurate Sequence Prediction. In: Zhou, S., Zhang, S., Karypis, G. (eds.) ADMA 2012. LNCS, vol. 7713, pp. 431–442. Springer, Heidelberg (2012)
6. Fournier-Viger, P., Faghihi, U., Nkambou, R., Mephu Nguifo, E.: CMRules: Mining Sequential Rules Common to Several Sequences. Knowledge-based Systems 25(1), 63–76 (2012)
7. Fournier-Viger, P., Gomariz, A., Campos, M., Thomas, R.: Fast Vertical Mining of Sequential Patterns Using Co-occurrence Information. In: Tseng, V.S., Ho, T.B., Zhou, Z.-H., Chen, A.L.P., Kao, H.-Y. (eds.) PAKDD 2014, Part I. LNCS, vol. 8443, pp. 40–52. Springer, Heidelberg (2014)
8. Han, M., Wang, Z., Yuan, J.: Mining Constraint Based Sequential Patterns and Rules on Restaurant Recommendation System. Journal of Computational Information Systems 9(10), 3901–3908 (2013)
9. Kamsu-Foguem, B., Rigal, F., Mauget, F.: Mining association rules for the quality improvement of the production process. Expert Systems and Applications 40(4), 1034–1045 (2012)
10. Leneve, O., Berges, M., Noh, H.Y.: Exploring Sequential and Association Rule Mining for Pattern-based Energy Demand Characterization. In: Proc. 5th ACM Workshop on Embedded Systems For Energy-Efficient Buildings, pp. 1–2. ACM (2013)
11. Lo, D., Khoo, S.-C., Wong, L.: Non-redundant sequential rules - Theory and algorithm. Information Systems 34(4-5), 438–453 (2009)
12. Mabroukeh, N.R., Ezeife, C.I.: A taxonomy of sequential pattern mining algorithms. ACM Computing Surveys 43(1), 1–41 (2010)

13. Pham, T.T., Luo, J., Hong, T.P., Vo, B.: An efficient method for mining non-redundant sequential rules using attributed prefix-trees. Engineering Applications of Artificial Intelligence 32, 88–99 (2014)
14. Nayrolles, M., Moha, N., Valtchev, P.: Improving SOA antipatterns detection in Service Based Systems by mining execution traces. In: Proc. 20th IEEE Working Conference on Reverse Engineering, pp. 321–330 (2013)
15. Zhao, Y., Zhang, H., Cao, L., Zhang, C., Bohlscheid, H.: Mining both positive and negative impact-oriented sequential rules from transactional data. In: Theeramunkong, T., Kijsirikul, B., Cercone, N., Ho, T.-B. (eds.) PAKDD 2009. LNCS, vol. 5476, pp. 656–663. Springer, Heidelberg (2009)

Mining Longitudinal Epidemiological Data to Understand a Reversible Disorder

Tommy Hielscher[1], Myra Spiliopoulou[1], Henry Völzke[2], and Jens-Peter Kühn[2]

[1] Otto-von-Guericke University Magdeburg, Germany
[2] University Medicine Greifswald, Germany
{tommy.hielscher,myra}@iti.cs.uni-magdeburg.de,
{voelzke,kuehn}@uni-greifswald.de

Abstract. Medical diagnostics are based on epidemiological findings about reliable predictive factors. In this work, we investigate how sequences of historical recordings of routinely measured assessments can contribute to better class separation. We show that predictive quality improves when considering old recordings, and that factors that contribute inadequately to class separation become more predictive when we exploit historical recordings of them. We report on our results for factors associated with a multifactorial disorder, hepatic steatosis, but our findings apply to further multifactorial outcomes. [1] [2]

Keywords: medical data mining, longitudinal epidemiological studies, hepatic steatosis, classification.

1 Introduction

Diagnostic procedures for health incidents are based on epidemiological findings. Epidemiological studies encompass sociodemographic assessments and medical tests for randomly selected participants. In this study, we investigate how *longitudinal* epidemiological data can contribute to class separation w.r.t. a disease that has been reliably identified only in the most recent moment of the observation horizon. The outcome we study is hepatic steatosis, a liver disorder that indicates a risk of hepatic sequels (like cirrhosis) and extrahepatic ones (like cardiovascular diseases [1]), but our mining workflow applies to other multifactorial diseases.

Epidemiological advances have lead to the discovery of elaborate features, such as genetic markers, that are associated with clinical outcomes. Nonetheless, major importance is allotted to features that are easily and *routinely recorded*, like the "fatty liver index" [2], which is termed by Bedogni et al. as "simple and accurate". This leads to the question of how the history of recordings for a routinely measured feature can be exploited to increase its predictive power. We investigate this question on data of 578 participants from the Study of Health in Pomerania (SHIP) [3], denoted as SHIP·578 hereafter, for whom we obtained (in 2013) the Magnetic Resonance Imaging (MRI) results on fat accumulation in the liver.

---

[1] Part of this work was supported by the German Research Foundation project SP 572/11-1 "IMPRINT: Incremental Mining for Perennial Objects".
[2] Data made available through cooperation SHIP/2012/06/D "Predictors of Steatosis Hepatis".

H. Blockeel et al. (Eds.): IDA 2014, LNCS 8819, pp. 120–130, 2014.

Historical recordings for cohort participants are often incomplete, because some assessments become part of the protocol after the study has started, while others are discontinued. The assessments of the SHIP participants have been measured at three moments (SHIP-0, SHIP-1, SHIP-2), but liver sonography was omitted in SHIP-1, and liver MRI was done only in SHIP-2. Hence, our mining approach must deal with the additional challenge of learning from incomplete sequences of recordings.

Our contributions are as follows. First, we propose a mining workflow for longitudinal epidemiological data, in which sequences of recordings of each feature are exploited for classification. We address the challenges of incomplete sequences and of the absence of a recording for the target variable in all but the last moment. We omit the large diversity among the individual values per participant and moment by deriving *sequence profiles* from the sequences, and by using these profiles as new features. Finally, in the context of the specific disorder, we identify informative features and show that the sequence profiles can be more informative than the features they originate from.

The paper is organized as follows. In section 2, we discuss related work. In section 3, we first present our materials (the subcohort SHIP·578). Our mining methods, organized as a workflow, are provided in the following section 4. In section 5, we report on our results. In the last section, we reflect on our findings and propose further steps.

Our approach is a followup of our recent work [4]: we proposed a mining workflow for classification on the basis of participant similarity, considering only one moment of SHIP (SHIP{2}·578). Here, we use *all* available historical data, i.e. the complete SHIP·578. We employ an adjusted similarity function and also use a kNN classifier with new input, since we organize the historical recordings on each variable into sequences, and we derive sequence profiles for them, which we use as new features.

2 Related Work

Learning prediction models on epidemiological data is a promising approach to identify potential risk factors of diseases or disorders. For example, Oh et al. [5] use health records to study diabetic retinophaty and use multiple penalized logistic regression models, to help to account for the high dimensionality of the data and varying relevance of predictors. Although study assessments were conducted multiple times, each survey was constituted of different participants so that outcome sequences could not be considered. Fuzzy Association Rule models on epidemiological data to predict risk of future dengue incidence are proposed in [6]: they show potential alternatives to regression models.

Approaches that share similarities to ours are presented in [7] and [8]. Moskovitch and Shahar [7] mine temporal interval patterns on multivariate time-oriented data of differing temporal granularity, and use the frequent time interval relation patterns as features for classification. Berlingerio et al mine time annotated sequences considering sequentiality and elapsed time between events [8]. Both approaches intend to find time-dependent relations among multiple features, whereas we consider only relations between time-specific realizations of the same medical assessment. Rather than value abstraction (for continuous features), we conduct supervised clustering to detect informative groups of similar sequences.

The SHIP-data we built our workflow on were analyzed in [9], [10], [11], [12] and in our earlier work [4]. In [4] the viability of similarity-based classification and identification of important features was investigated. In [12], a mining workflow is proposed for SHIP{2}·578 . However, these studies did not consider past recordings of participants which we utilize in our workflow. On top of that we provide insights regarding their impact on class separation quality and on the identification of important features and subpopulations.

3   Materials

SHIP is a population-based project [3]: persons are chosen who reside in Pomerania (Northeast Germany) and are between 20 and 79 years old. SHIP participants undergo an examination program consisting of interviews, exercise tests, laboratory analysis, ultrasound examinations and whole-body magnetic resonance tomography (MRT). Three examination "'moments'" of the first SHIP cohort exist, SHIP-0 (1997-2001, n= 4308), SHIP-1 (2002-2006, n= 3300) and SHIP-2 (2008-2012, n= 2333). We have the SHIP-2 liver fat concentration only for 578 participants (mrt_liverfat_s2). The values of assessments at the three moments 0,1,2 are recorded in SHIP·578 as different features, e.g. som_bmi_s0, som_bmi_s1, som_bmi_s2 for the somatographic Body Mass Index (see also Figure 2, left upper part). In [12], the use of the original target variable with regression led to poor results, we therefore discretize the continuous target variable into a positive and negative class, with cut-off value choices as shown in [4] and [12] to formulate a classification problem. Like in both works, we also consider the partition of female participants ($Subset_w$) and male participants ($Subset_m$) separately, where $Subset_w$ contains 314 individuals with a relative negative-class frequency of 0.81 and $Subset_m$ contains 264 individuals with a relative negative-class frequency of 0.69.

4   Methods

Our mining workflow consists of three main steps as shown in Figure 1: (1) generating sequence features per assessment/variable; (2) identifying a subset of informative

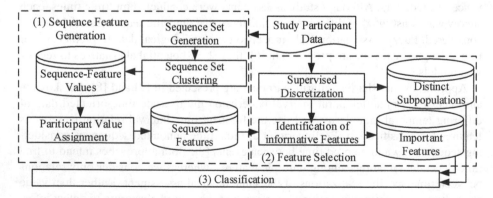

Fig. 1. Workflow for classification and risk factor identification on longitudinal data

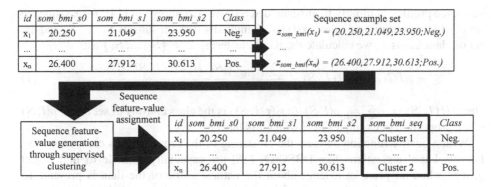

Fig. 2. Example workflow for the generation of the som_bmi sequence-feature

features; (3) similarity-based class separation of study participants.These steps are described in the following.

4.1 Sequence-Feature Generation

Workflow input is the data collected during all rounds of an epidemiological study. Such a study consists of medical examinations or measurements (e.g. Body Mass Index measurement, sonography examination etc.), hereafter referred to as assessment, which study participants (individuals) have to undergo multiple times. During each round a specific realization of an assessment is conducted and given as a single feature. First, we incorporate recordings of past epidemiological study rounds into our classification workflow. To do this, we generate new features, so called sequence-features, from assessments obtained at different times during the multiple study moments. As an example, consider the Body Mass Index (BMI) assessment as shown in Figure 2. For each individual we have the BMI in SHIP-0,1 and 2, som_bmi_s0, som_bmi_s1 and som_bmi_s2 (the post-fix $s\{0, 1, 2\}$ corresponds to the SHIP study moment). We then build a sequence per individual and assign each sequence the class of the corresponding individual (as shown in Figure 2, right upper part), cluster the set of sequence examples and generate a new sequence-feature. Finally each of the individuals gets assigned a nominal sequence-feature value, that is the cluster id correspondent to the cluster membership of their exhibited sequence.

Let $a_t \in F$ be assessment a, at study moment t, and $a_t(x)$ its value for individual $x \in X$. A feature sequence example is a tuple $z_a(x) = (s_a(x), y)$ where $s_a(x)$ is the value-vector $(a_{t_1}(x), a_{t_2}(x), ..., a_{t_m}(x))^T$ with $t_j < t_{j+1}$, $j = 1, ..., m$ and y is the label of x at the most recent study moment. For assessment a, we build a set of sequence examples $S(a) = \{z_a(x) | x \in X\}$. In the case of missing feature values we impute them by their subpopulation average.

After extracting each assessment's sequence set, we separately cluster them to sequence profiles. Equivalently to supervised clustering [13] we want to find groups of objects which are homogenous w.r.t. the class variable. We apply Information Gain on the data partitions aiming at high priority w.r.t. class, so that we can find predictive sequence profiles. Similarly to attribute discretization as shown in [14], for a sequence set S which is partitioned (clustered) into k mutually disjoint subsets $S_1, ..., S_k$ and the set of class labels L, we calculate the IG of clustering $Z = \{S_1, ..., S_k\}$ as:

$$IG(Z) = H(L, S) - \frac{\sum_{SUB \in S}(|SUB| \cdot H(L, SUB))}{|S|}, \tag{1}$$

Here, $H(L, S) = -\sum_{l \in L} p(l, S) \cdot log_2(p(l, S))$ is the class entropy of set S and $p(l, S)$ is the proportion of examples in S with class label l.

With a quality function defined, the clustering of the sequence sets can commence. Density-based algorithms like DBSCAN [15] are especially suitable to find object groups when no prior knowledge about the natural shape of the data is present. To evaluate different clusterings we conduct a grid-search of the $minPts$ and eps parameter space and afterwards choose the data partitioning with the highest Information Gain, using Euclidean Distance (for continuous features) or Overlap Metric (for nominals) as proximity measure. We execute multiple clusterings per assessment-type exploring all values $minPts \in \{4, 5, ..., 10\}$ and $eps \in \{0.1, 0.2, ..., 0.5\}$ in the case of continuous features and $eps \in \{0, 1, 1.5\}$ when dealing with nominal features. A continuous feature's range is normalized in advance to $[0, 1]$. The approach is parallelizable and its potentially high costs are mitigated by the low data dimensionality and number of instances. Through this clustering method, for each assessment a with corresponding sequence set $S(a)$, we build a nominal sequence-feature α. The values of α are the cluster-ids obtained during the best sequence set clustering w.r.t. Information Gain. Sequence-feature values $\alpha(x)$ for known individual x are the observed sequence cluster-ids, denoted as $cid(s_a(x))$, i.e. $\alpha(x) = cid(s_a(x))$. Unknown individuals sequence-features are assigned the cluster-id of their nearest labeled sequence vectors, that means each sequence-feature value of an unlabeled individual is given by the sequence-feature value of a known individual who exhibits the most similar sequence vector (as measured by a distance function). In addition to the existing features derived from the study assessments, generated sequence-features are then used to represent participants as feature-vectors.

4.2   Feature Selection

After representing individuals with existing features and generated sequence-features we aim to find feature subsets with high feature-class and low feature-feature interdependencies. To achieve this we use Correlation-based Feature Selection (CFS) (cf. [16]) to compute the *merit value* M_F of a set of features F. The merit is the ratio between average feature-class and feature-feature association of features in F. We compute the associations as described in (cf. [16]), calculating the standardized IG between each feature-feature and feature-class-variable pair and separately averaging over both types of associations. To apply CFS on a dataset I, we first use entropy-based MDL discretization on all continuous features (cf. [14]), to obtain a nominal representation consisting of feature-value ranges which induce distinct subpopulations with highly skewed class distributions. Then, we start with an empty set of features F and iteratively add to F the feature f that leads to the highest new merit-value $M_{F \cup f}$. The procedure ends when adding any feature decreases the merit.

4.3 Classification

Our classification approach is based on the similarities between individuals. Individuals are modeled as vectors in a d-dimensional feature space, we then identify their most similar individuals by computing their k Nearest Neighbors (kNN). Our approach assigns each individual x with unknown label to the majority class among the classes of the k nearest labeled neighbors of x using simple majority voting. Similarity between individuals is calculated through the mutual distance of their vector representations in the feature space. We use the Heterogeneous Euclidean Overlap Metric (HEOM) as base function: HEOM uses Euclidean distance for continuous features and the Overlap Metric (OM) for nominals [17].

Yet, base HEOM does not take the nature of epidemiological data into account. In epidemiological studies a plethora of features from different measurements and examinations are derived. Even after feature selection the importance of features can differ widely. Some features can contain more useful information for class separation than others and thus should have a higher impact when calculating participant similarity. To address this issues we adjust the HEOM distance as follows.

Let A be the set of features over dataset I and $a(x)$, $a(y)$ be the values of $a \in A$ for $x, y \in I$. Let further $QF()$ be a quality function with value $QF(a)$ for feature a, and let QF_{max} be the highest observed quality function value in the considered feature set. Then we denote d_{HEOM_QF} as the adjusted HEOM distance measure:

$$d_{HEOM\_QF}(x,y) = \sqrt{\sum_{a \in A} \left(\frac{QF(a)}{QF_{max}} \cdot \delta_{HEOM}(a(x), a(y)) \right)^2}, \qquad (2)$$

$$\delta_{HEOM}(a(x), a(y)) = \begin{cases} \delta_{OM}(a(x), a(y)) & \text{if } a \text{ is nominal} \\ \frac{a(x) - a(y)}{range(a)} & \text{if } a \text{ is continuous} \\ 1 & \text{otherwise} \end{cases}$$

$$\text{with } \delta_{OM}(a(x), a(y)) = \begin{cases} 0 & \text{if } a(x) = a(y) \\ 1 & \text{otherwise} \end{cases}$$

The unadjusted HEOM measure comes from the adjusted measure when $QF(a) = 1, \forall a \in A$. For our computations we set $QF = IG$, i.e. we use Information Gain as feature quality measure. For each participant $x \in I$, we then build the set $NN(k, x) \subseteq I$, which contains the k labeled participants most similar to x, such that: (i) the cardinality of $NN(k, x)$ is k, and (ii) for each $y \in NN(k, x)$ and for each $z \in I \setminus (NN(k, x) \cup x)$ it holds that $d_{HEOM\_QF}(x, y) \leq d_{HEOM\_QF}(x, z)$.

5   Results

In our experiments we study class separation performance on the disorder hepatic steatosis when using the presented workflow. Based on attribute quality measures, we show a subset of relevant features according to our study data including sequence-features which contain useful information not present in the existing set of regular features.

5.1 Variants under Evaluation

Prior to the evaluation we balance the class distribution of our dataset by applying random under-sampling on both $Subset_w$ and $Subset_m$, resulting in 118 ($Subset_w$) and 162 ($Subset_m$) examples. Beside the unknown distribution in a future medical scenario, this is also done to better show the impact on class separation of our workflow without classifier bias having a too strong effect on classification performance (highly dominant negative class in the original dataset). Classification results on the unbalanced dataset are provided in [4]. For classification, we consider a kNN classifier with neighborhood size $k = 7$, majority voting and HEOM as base proximity measure. We built multiple variants of this kNN classifier to show the impact of each step of our workflow as follows: we use the kNN classifier with unadjusted HEOM, using only the original SHIP-2 features. Then we use CFS on the set of SHIP-2 features and contrast the use of the unadjusted HEOM with the adjusted HEOM ($HEOM_{IG}$). Lastly we generate sequence-features (SFG), apply CFS on the combined set of SHIP-2 and sequence-features and use adjusted HEOM for similarity computation. All variants are separately applied on $Subset_w$ and $Subset_m$. The CFS algorithm, sequence-feature generation and HEOM adjustment exclusively use labeled information available from training data.

5.2 Findings on Classification Performance

In our experiments, we evaluate classification performance on sensitivity, specificity, accuracy and calculate the area under the receiver operating characteristic curve (AUC). As explained in subsection 5.1, we consider several classifier variants on the basis of different feature sets. For each variant, we perform five-fold cross-validation and average over the evaluation measure values and size of the feature set used for classification. Table 1 shows the results of each variant for $Subset_w$ and $Subset_m$.

Classification performance is generally high for $Subset_w$ participants. Here, the HEOM base-classifier achieves lowest performance values. This was expected because there may be features with redundant or irrelevant information. Each successive step within our workflow leads to better or same evaluation results compared to its direct predecessor. In comparison to the plain base classifier, the complete workflow leads to an average increase of 8.5%- sensitivity, 9.1%- specificity, 8.5%- accuracy and 2.8%-points in AUC, with absolute performance values in the vicinity of 0.9. Highest gains are achieved with feature selection while distance measure adjustment and sequence feature generation contribute approximately equally to performance enhancement. On average, best results are obtained with just 2.2 features in contrast to the original set of 67 features. Considering the classification performance and small feature count, we conclude that there exist a small subset of uncorrelated, highly informative features for $Subset_w$.

The classification results for $Subset_m$ differ strongly. No variant performs best for all evaluation measures, however $\{CFS, HEOM_{IG}\}$ and $\{SFG, CFS, HEOM_{IG}\}$ show the best accuracy and AUC results. All in all, average performance gain when using the complete workflow in comparison to the plain base classifier accumulates to 20%- sensitivity, -2.2%- specificity, 8.5%- accuracy and 12.3%-points AUC. Again, the amount of features used for classification is heavily reduced after applying CFS but,

Table 1. For each variant: Average evaluation measures: best values are printed in boldface

Variant	Avg Sensitivity	Avg Specificity	Avg Accuracy	Avg AUC	Avg Num of Features
$Subset_w$					
$HEOM$	0.847	0.796	0.822	0.889	67
$CFS, HEOM$	**0.932**	0.816	0.873	0.898	3.2
$CFS, HEOM_{IG}$	**0.932**	0.851	0.890	0.909	3.2
$SFG, CFS, HEOM_{IG}$	**0.932**	**0.887**	**0.907**	**0.917**	2.2
$Subset_m$					
$HEOM$	0.629	**0.797**	0.716	0.762	61
$CFS, HEOM$	0.840	0.775	0.808	0.862	7.4
$CFS, HEOM_{IG}$	**0.865**	0.787	**0.827**	0.867	7.4
$SFG, CFS, HEOM_{IG}$	0.829	0.775	0.801	**0.885**	8

in contrast to $Subset_w$, the variants associated with our workflow achieve lower performance and maintain bigger feature subsets. This signals that the association between features and target concept are weaker which in turn leads to more difficult classification for the subcohort of male participants.

5.3   Findings on Important Features

The evaluation results show that there exist small subsets of highly informative features which are relevant to class separation. We present these features, a short description and their Information Gain in Table 2 and Table 4. Note that these are the remaining features when using our workflow on the whole dataset. Because of varying training and test sets during cross validation, these can slightly differ in comparison to the resultant subsets of the workflow evaluation in Section 5.2. We observe that with the exception of the hrs_s_s2 feature, the selected $Subset_w$ feature set is a subset of the $Subset_m$ feature set. However, the information quality is much higher for $Subset_w$. Within both feature sets, the stea_seq sequence feature contains the most information regarding the class variable. $Subset_m$'s twice as big feature set additionally contains the ggt_s_seq sequence-feature. The relative frequency distributions of these features are presented in

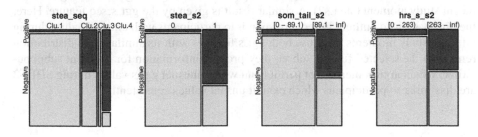

Fig. 3. Mosaic plots of most predictive features for $Subset_w$.

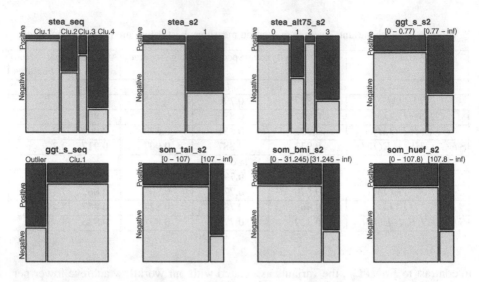

Fig. 4. Mosaic plots of most predictive features for $Subset_m$

the mosaic plots of Figure 3 and Figure 4. A plot's rectangle corresponds to a subpopulation of participants which exhibit a feature-value and class combination as specified by the plot axis. The area of a rectangle directly reflects the relative frequency of participants which have the associated feature-value and class combination. For example, by comparing the lower left rectangle with the lower right rectangle of $Subset_m$'s stea_s2 mosaic plot, we can see that number of negative-class participants with a stea_s2 value of 0 is more than twice as big as the number of negative-class participants with stea_s2 values of 1. The mosaic plots show that each feature in both subsets induce at least one subpopulation where a majority of participants belong to the negative class and one which is more balanced ($Subset_w$) or which contains a majority of positive-class pariticipants ($Subset_m$). Especially the newly generated sequence features show some interesting distributions. Of all features, the stea_seq feature with its cluster 1 and cluster 4 values induces the most skewed subpopulations regarding the class variable. Table 3 shows the regular stea feature values associated with each stea_seq cluster. It becomes apparent that either two sequential negative or positive sonography diagnostics (cluster 1 and 4) are much more informative in comparison to a single diagnosis during the most recent study moment (stea_s2). A similar result is given by the ggt_s_seq feature. Here, the highly skewed outlier cluster corresponds to participants with unusual values during all three study moments. Because both ggt_s features with its similar class distribution remain in the selected feature subset, they provide information for different subpopulations which in turn means that pariticipant whith unusual ggt_s values during SHIP-2 are dissimilar to participants which exhibit unsual values consistently.

Table 2. Most relevant features in $Subset_w$

Subset_w		
Feature	**IG**	**Description**
stea_seq	0.347	Sono. result cluster
stea_s2	0.304	Sono. result
som_tail_s2	0.193	Waist circumference (cm)
hrs_s_s2	0.147	Serum uric acid ($\mu mol/l$)

Table 4. Most relevant features in $Subset_m$

Subset_m		
Feature	**IG**	**Description**
stea_seq	0.321	Sono. result cluster
stea_s2	0.250	Sono. result
stea_alt75_s2	0.186	Sono. result
ggt_s_s2	0.149	Serum GGT ($\mu mol/sl$)
ggt_s_seq	0.127	Serum GGT ($\mu mol/sl$) cluster
som_tail_s2	0.118	Waist circumference (cm)
som_bmi_s2	0.117	Body Mass Index (kg/m^2)
som_huef_s2	0.095	Hip size (cm)

Table 3. Value mapping of the stea features

stea_seq	stea_s0	stea_s2
Cluster 1	0	0
Cluster 2	0	1
Cluster 3	1	0
Cluster 4	1	1

6 Conclusions

We have presented a workflow for the identification of important features and distinct subpopulations from epidemiological study data w.r.t. a multifactorial disorder. We investigated how past study recordings can contribute to class separation by first executing multiple clusterings of participants with similar feature value sequences and then building sequence-features on the basis of discovered clusters. In the case of hepatic steatosis, we further evaluated the classification performance impact of each workflow step and provided the set of most predictive features for the subcohorts of female and male study participants. Our contribution shows the viability of a data-driven approach to mine important features from longitudinal epidemiological data w.r.t. a target disease or disorder. The workflow provides a way to validate the predictive quality of mined features through a similarity-based participant classification. For hepatic steatosis, our workflow achieved good classification results on a balanced two-class problem with AUC values of 0.917 (female participants) and 0.885 (male participants). The automatically generated sequence-features incorporated historical (up to 10 years old) participant assessment knowledge and enhanced the discriminative ability of the regular feature set which reflected assessment outcomes from only the most recent study moment. Our analysis of the set of important features showed that some values of sequence-features induce highly distinct subpopulations which go unidentified when solely considering their regular counterparts. For example, the stea_seq sequence feature indicates highly skewed class distributions, and thus allows for better class separation, through the differentiation of participants with two consecutive positive or negative sonography results.

In our future work we want to validate our workflow for further multifactorial diseases and disorders like coronary heart diseases or diabetes. We also aim to incorporate additional time-series data like smoking- and drinking-behavior and identify new potential risk-factors. We further want to split the male and female subgroups into more homogenous subgroups that help to better predict the target variable.

References

1. Targher, G., Day, C.P., Bonora, E.: Risk of Cardiovascular Disease in Patients with Nonalcoholic Fatty Liver Disease. N. Eng. J. Med. 363(14), 1341–1350 (2010)
2. Bedogni, G., Bellentani, S., ..., Castiglione, A., Tiribelli, C.: The Fatty Liver Index: a simple and accurate predictor of hepatic steatosis in the general population. BMC Gastroenterology 6(33), 7 (2006)
3. Völzke, H., Alte, D., ..., Biffar, R., John, U., Hoffmann, W.: Cohort profile: the Study of Health In Pomerania. Int. J. of Epidemiology 40(2), 294–307 (2011)
4. Hielscher, T., Spiliopoulou, M., Völzke, H., Kühn, J.P.: Using participant similarity for the classification of epidemiological data on hepatic steatosis. In: Proc. of the 27th IEEE Int. Symposium on Computer-Based Medical Systems (CBMS 2014), Mount Sinai, NY, IEEE (accepted March 2014)
5. Oh, E., Yoo, T., Park, E.C.: Diabetic retinopathy risk prediction for fundus examination using sparse learning: a cross-sectional study. BMC Med. Inf. and Decision Making 13(1), 1–14 (2013)
6. Buczak, A., Koshute, P., Babin, S., Feighner, B., Lewis, S.: A data-driven epidemiological prediction method for dengue outbreaks using local and remote sensing data. BMC Med. Inf. and Decision Making 12(1), 1–20 (2012)
7. Moskovitch, R., Shahar, Y.: Medical temporal-knowledge discovery via temporal abstraction. In: AMIA Annu. Symp. Proc., vol. 2009, p. 452. American Medical Informatics Association (2009)
8. Berlingerio, M., Bonchi, F., Giannotti, F., Turini, F.: Mining clinical data with a temporal dimension: A case study. In: IEEE 2007 Int. Conf. on Bioinf. and Biomed. (BIBM 2007), pp. 429–436 (November 2007)
9. Völzke, H., Craesmeyer, C., Nauck, M., ..., John, U., Baumeister, S.E., Ittermann, T.: Association of Socioeconomic Status with Iodine Supply and Thyroid Disorders in Northeast Germany . Thyroid 23(3), 346–353 (2013)
10. Haring, R., Wallaschofski, H., Nauck, M., Dörr, M., Baumeister, S.E., Völzke, H.: Ultrasonographic hepatic steatosis increased prediction of mortality risk from elevated serum gamma-glutamyl transpeptidase levels. Hepatology 50, 1403–14011 (2009)
11. Baumeister, S.E., Völzke, H., Marschall, P., ..., Schmidt, C., Flessa, S., Alte, D.: Impact of fatty liver disease on health care utilization and costs in a general population: A 5-year observation. Gastroenterology 134(1), 85–94 (2008)
12. Niemann, U., Völzke, H., Kühn, J.P., Spiliopoulou, M.: Learning and inspecting classification rules from longitudinal epidemiological data to identify predictive features on hepatic steatosis. Expert Systems with Applications (accepted February 2014)
13. Eick, C., Zeidat, N., Zhao, Z.: Supervised clustering - algorithms and benefits. In: 16th IEEE Int. Conf. on Tools with Artif. Int (ICTAI 2004), pp. 774–776. IEEE Computer Society, Boca Raton (2004)
14. Fayyad, U.M., Irani, K.B.: Multi-interval discretization of continuous-valued attributes for classification learning. In: Int. Joint Conf. on Artif. Inf (IJCAI 1993), pp. 1022–1029 (1993)
15. Ester, M., Kriegel, H.P., Sander, J., Xu, X.: A density-based algorithm for discovering clusters in large spatial databases with noise. In: Proc. of the 2nd Int. Conf. on Knowledge Discovery and Data Mining (KDD 1996), pp. 226–231. AAAI Press (1996)
16. Hall, M.A.: Correlation-based feature selection for discrete and numeric class machine learning. In: Proc. of 17th Int. Conf. on Machine Learning, pp. 359–366. Morgan Kaufmann, San Francisco (2000)
17. Wilson, D.R., Martinez, T.R.: Improved heterogeneous distance functions. J. Artif. Int. Res. 6(1), 1–34 (1997)

The BioKET Biodiversity Data Warehouse: Data and Knowledge Integration and Extraction

Somsack Inthasone, Nicolas Pasquier,
Andrea G.B. Tettamanzi, and Célia da Costa Pereira

Univ. Nice Sophia Antipolis, CNRS, I3S, UMR 7271,
06903 Sophia Antipolis, France
{somsacki,pasquier}@i3s.unice.fr,
{andrea.tettamanzi,celia.pereira}@unice.fr

Abstract. Biodiversity datasets are generally stored in different formats. This makes it difficult for biologists to combine and integrate them to retrieve useful information for the purpose of, for example, efficiently classify specimens. In this paper, we present BioKET, a data warehouse which is a consolidation of heterogeneous data sources stored in different formats. For the time being, the scopus of BioKET is botanical. We had, among others things, to list all the existing botanical ontologies and relate terms in BioKET with terms in these ontologies. We demonstrate the usefulness of such a resource by applying FIST, a combined biclustering and conceptual association rule extraction method on a dataset extracted from BioKET to analyze the risk status of plants endemic to Laos. Besides, BioKET may be interfaced with other resources, like GeoCAT, to provide a powerful analysis tool for biodiversity data.

Keywords: Biodiversity, Information Technology, Ontologies, Knowledge Integration, Data Mining.

1 Introduction

Biological diversity, or biodiversity, refers to the natural variety and diversity of living organisms [26]. Biodiversity is assessed by considering the diversity of ecosystems, species, populations and genes in their geographical locations and their evolution over time. Biodiversity is of paramount importance for a healthy environment and society, as it ensures the availability of natural resources and the sustainability of ecosystems [6,10,13,16,22,25]. The effects of biodiversity loss on the environment, caused by habitat loss and fragmentation, pollution, climate change, invasive alien species, human population, and over-exploitation can affect all life forms and lead to serious consequences [9]. Understanding biodiversity is an essential prerequisite for sustainable development.

For many years, biodiversity datasets have been stored in different formats, ranging from highly structured (databases) to plain text files, containing plant descriptions (vocabularies and terms). Numerous data and knowledge repositories containing biodiversity and environmental information are available on

H. Blockeel et al. (Eds.): IDA 2014, LNCS 8819, pp. 131–142, 2014.

the Internet as on-line and off-line resources nowadays. Data repositories store large amounts of information depicting facts on concrete objects related to a specific domain of application, e.g., results of environmental studies or inventories of species in a geographic location. This makes it difficult for botanists or zoologists to combine and integrate them to retrieve useful information for the purpose of identifying and describing new species.

The ever increasing availability of data relevant to biodiversity makes the idea of applying data mining techniques to the study of biodiversity tempting [12].

Data mining, also known as knowledge discovery from data (KDD), is a set of concepts, methods and tools for the rapid and efficient discovery of previously unknown information, represented as knowledge patterns and models, hidden inside massive information repositories [11].

One important obstacle to the application of data mining techniques to the study of biodiversity is that the data that might be used to this aim are somewhat scattered and heterogeneous [24]. Different datasets cover different aspects of the problem or focus on some geographical areas only. None of them is complete and there is no standard format.

To overcome these limitation, we have designed and implemented BioKET, a data warehouse whose purpose is to consolidate a maximum of data sources on biodiversity in a logically organized, coherent, and comprehensive resource that can be used by the scientific community as a basis for data-intensive studies.

The main contribution of this paper is to provide a detailed account of how the BioKET data warehouse has been designed and populated, by consolidating and integrating multiple and heterogeneous sources of data. The reader should not underestimate the methodological challenges and the practical problems that had to be overcome in order to achieve that result. As all data mining practitioners agree, pre-processing, which includes data cleaning, integration, and transformation is the most time-consuming and critical phase of the data mining process [14,15] illustrated in Figure 1.

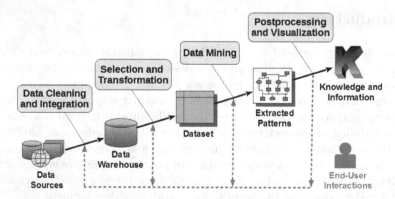

Fig. 1. Main phases of a data mining process

We demonstrate the use of such resource by applying FIST, a combined bi-clustering and conceptual association rule extraction method already described in the literature [17], on a dataset extracted from it to analyze the risk status of plants endemic to Laos.

2   The BioKET Data Warehouse

The BioKET data warehouse is the consolidation of four main data sources:

- BIOTIK [2] (Western Ghats of India and National University of Laos), which contains 652 species records;
- the herbarium from the BRAHMS repository [3] (National University of Laos, Faculty of Forestry), with 7548 species records;
- the herbarium from the NAPIS repository [18] (Lao Ministry of Public Health, Institute of Traditional Medicine), with 747 species records;
- the IUCN Red List Data [27], with 71570 species records.

These data sources are stored in different formats: BIOTIK and IUCN Red List are in HTML, while the two others use, respectively, the dBase and Paradox file formats. Integrating such diverse data sources required performing the following tasks:

1. Listing all botanical and plant ontologies available on the Internet.
2. Selecting relevant information (phenotypic/plant traits/features/characteristics).
3. Relating terms in our database with terms in these ontologies.
4. Searching for thesauri/glossaries/taxonomies of terms for plants available on the Internet.
5. Relating terms in our database with terms in these thesauri.
6. Relating terms in Plant Ontology (PO) (which seems to be the most complete ontology in Botany) with terms/definitions (e.g., Latin terms) in these thesauri.

The first step was to extract data from sources and store them in a standard file format (such as an Excel spreadsheet), by using database management tools. Then, data cleaning was performed by using advanced Excel functions. The next step was to generate and link Google Maps Geocoding Service with the BIOTIK, BRAHMS, and NAPIS data by using VBA script (GoogleGeoLocation Function). The last step was to import the data thus obtained into the BioKET database, under MySQL.

A key factor for the integration and the enrichment of the data was the use of ontologies. Formal ontologies are a key for the semantic interoperability and integration of data and knowledge from different sources. An ontology may be regarded as "a kind of controlled vocabulary of well-defined terms with specified relationships between those terms, capable of interpretation by both humans and computers" [28]. From a practical point of view, an ontology defines a set of concepts and relations relevant to a domain of interest, along with axioms stating their properties. An ontology thus includes a taxonomy of concepts, a formally defined vocabulary (called a terminology), and other artifacts that help structure a knowledge base. A knowledge base that uses the terms defined in an ontology becomes usable by and interoperable with any other system that has access to that ontology and is equipped by a logic reasoner for it [19].

It was thus important to construct a map among all the concepts in all the data sources and all the considered ontologies. It is worth noting that (i) some concepts are not equally represented in all the sources, (ii) some are represented in some sources and not in others and (iii) other concepts are not represent at all. The mapping process works as follows: the textual descriptors of plants are segmented into small chunks, which are then matched with the labels of concepts in the target ontology. For instance, from the descriptor "evergreen tree up to 8 m", we can in infer that "evergreen" is related to "shedability", "up to 8m" is related to "height" and "tree" is related to "plant type ". In the process, new concepts may be generated (e.g., from the textual descriptor "branches ascending or horizontal", where "branch", "branch ascending", and "branch horizontal" match concepts in the ontology, a new concept "branch ascending or horizontal", subsumed by "branch" and subsuming the latter two is generated). The plant record can thus be automatically enriched with a large number of "implicit" fields, inferred from the ontology. We have designed a relational data base of concepts that make it possible to relate concepts represented in different ways.

The result of this integration process — the BioKET data warehouse — is schematized in Figure 2. BioKET contains 77 relationship entities and a total of 80,517 records.

As pointed out by many researchers (see, e.g., [1]), to conserve organisms, whether plant or animal, one important step to take is to identify rare and endangered organisms in a given geographical area or country. The integration of geographical information from Geographical Information Systems (GIS) with

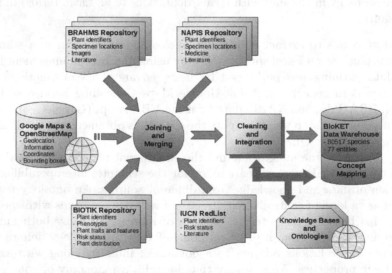

Fig. 2. An Overview of the BioKET Data Integration Process

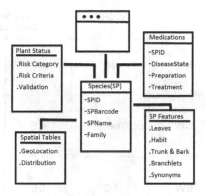

Fig. 3. Overview of the BioKET Datawarehouse Entity-Relationship Model

species data, and its use in data mining studies is the object of the biodiversity informatics project of the W. P. Fraser Herbarium (SASK) [20]. The participants in such project developed an integrated bio-geography GIS model, using Google Maps API, based on data mining concepts to map and explore flora data. This allows data to be explored on a map and analyzed in several ways to reveal patterns showing relationships and trends that are not discernible in other representations of information.

The BioKET data warehouse integrates geographical information and 8,947 species out of the 80,517 total species have descriptions of specimen location and risk status that may differ depending on the area considered. This information is described at different levels of precision, from continent to specific places such as cities or villages. For example, *Cratoxylum formosum* grows up in Myanmar, South China, Thailand, Indochina, and Laos (Khammouan) [2]. This species is also reported in the Lower Risk/Least Concern category by IUCN Red List data [27]. The integration of geolocation information allows to explore species properties in different areas using the GeoCAT (Geospatial Conservation Assessment Tool) platform [8]. GeoCAT is based on Google Maps to explore geographical information if coordinates, i.e., latitudes and longitudes, are provided. We already linked Google Maps with the terms of geographical information of each source (BIOTIK, BRAHMS, and NAPIS) and extracted their coordinates into BioKET database system. Google Maps does not support coordinates of directions (South, North, East, West, etc.) like "South China", but Google Bounding Box (BBox) coordinates are provided. We propose to improve this issue by calculating the coordinates for each direction (Figure 4) from Google BBox coordinates.

In the geolocation domain, the BBox of an area on Earth is defined by two points corresponding to the minimal and maximal longitudes and latitudes of the area [4]. Figure 4 shows the 13 partitions of an area: the 9 elementary partitions and the North, South, East and West partitions that result of merging the 3 corresponding elementary partitions, e.g., NW, NC and NE for North. This multilevel partitioning allows to represent location related properties of species, such as risk status or abundance for instance, at different area covering levels.

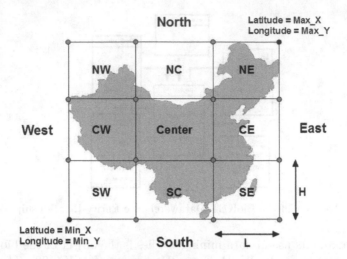

Fig. 4. Thirteen partitions for an area on Earth

Table 1. Partition Bounding Box Computations

Area	Min_Long	Min_Lat	Min_Long	Max_Lat
South	Min_Y	Min_X	$Min\_Y + H$	Max_X
North	$Min\_Y + 2H$	Min_X	Max_Y	Max_X
West	Min_Y	Min_X	Max_Y	$Max\_X + L$
East	Min_Y	$Min\_X + 2L$	Max_Y	Max_X
SW	Min_Y	Min_X	$Min\_Y + H$	$Min\_X + L$
SC	Min_Y	$Min\_X + L$	$Min\_Y + H$	$Min\_X + 2L$
SE	Min_Y	$Min\_X + 2L$	$Min\_Y + H$	Max_X
CW	$Min\_Y + H$	Min_X	$Min\_Y + 2H$	$Min\_X + L$
Center	$Min\_Y + H$	$Min\_X + L$	$Min\_Y + 2H$	$Min\_X + 2L$
CE	$Min\_Y + H$	$Min\_X + 2L$	$Min\_Y + 2H$	Max_X
NW	$Min\_Y + 2H$	Min_X	Max_Y	$Min\_X + L$
NC	$Min\_Y + 2H$	$Min\_X + L$	Max_Y	$Min\_X + 2L$
NC	$Min\_Y + 2H$	$Min\_X + 2L$	Max_Y	Max_X

Formulas to calculate the BBox of each partition are given in Table 1. These computations use the L and H values computed from the minimal (Min_X, Min_Y) and maximal (Max_X, Max_Y) longitude and latitude coordinates of the BBox of the partitioned area as follows:

$$L = \frac{(Max\_X - Min\_X)}{3}, \quad H = \frac{(Max\_Y - Min\_Y)}{3}.$$

This computation of partitions can be applied to all objects defined by a geolocation bounding box, from continents to cities. For example, using the BBox of China, that is {73.4994137, 18.1535216, 134.7728100, 53.5609740}, the BBox of Southern China will be computed as {73.4994137, 18.1535216, 93.9238791, 53.5609740}.

3   BioKET Experimental Analysis

For experimental purpose, we constructed a dataset containing information on the 652 species extracted from the Biotik repository. This information is represented as 1834 binary attributes describing morphological and environmental properties (characteristics of part of the plant, size, habitat, exudation, etc.) and risk status of species. Extracting knowledge patterns can then provide support to relate increases and decreases in risk status to environmental factors impacting specific species (climate change, pollution, etc.). They can also help taxonomists to analyze the different types of species in an ecosystem, e.g., associating species with specific features and risk categories, and their viability, or growth rate, in some particular areas.

3.1   Conceptual Bicluster Extraction

This dataset was analyzed using the FIST approach which is based on the frequent closed itemsets framework. FIST extracts minimal covers of conceptual association rules and biclusters jointly.

Conceptual biclusters of the form $\{I_1\ V_1\}$ associate to a maximal set of instances I_1, a maximal set of variable values V_1 that are common to all instances. In other words, a bicluster is a sub-matrix associating a subset of rows and a subset of columns such that all these rows have a similar value for each of these columns. Conceptual biclusters are partially ordered according to the inclusion relation and form a lattice. This hierarchical organization allows to explore groups of instances (species) and properties (characteristics) at different levels of abstraction: the highest biclusters in the lattice regroup a large number of properties shared by small groups of instances; the lowest biclusters regroup small set of properties that are common to large group of instances.

Conceptual association rules are rules with the form $\{V_1 \longrightarrow V_2, I_1, \textit{support},$ $\textit{confidence, lift}\}$ where V_1 and V_2 are sets of variable values (properties) and I_1 is the set of instances (species) supporting the rule. Statistical measures computed for each rule are:

- $\textit{support} = P(V_1 \cup V_2)$ (or $\textit{count}(V_1 \cup V_2) = |I_1|$ if given as an absolute number) evaluates the scope, or weight, of the rule in the dataset. It corresponds to the proportion of instances containing V_1 and V_2 among all instances.
- $\textit{confidence} = \frac{P(V_1 \cup V_2)}{P(V_1)}$ evaluates the precision of the rule. It corresponds to the proportion of instances containing V_2 among those containing V_1. Rules with $\textit{confidence} = 1$, that have no counter-example in the dataset, are called $\textit{exact rules}$. Rules with $\textit{confidence} < 1$ are called $\textit{approximate}$ rules.
- $\textit{lift} = \frac{P(V_1 \cup V_2)}{P(V_1)P(V_2)}$ corresponds to the correlation between occurrences of V_1 and V_2:
  - $\textit{lift} > 1$ means there is positive correlation between V_1 and V_2,
  - $\textit{lift} = 1$ means V_1 and V_2 are independent,
  - $\textit{lift} < 1$ means there is a negative correlation between V_1 and V_2.

Extraction parameters are the *minsupport* threshold, which corresponds to the minimal number of supporting instances required for a rule to be considered valid, and the *minconfidence* threshold, which corresponds to the minimal *confidence* required in order to consider a rule valid.

Experiments were conducted on a Dell PowerEdge R710 server with 2 Intel Xeon X5675 processors at 3.06 GHz, each possessing 6 cores, 12 MB cache memory, 24 GB of DDR3 RAM at 1333 MHz and 2 Hot Plug SAS hard disks of 600 GB at 15000 rounds/min with RAID 0 running under the 64 bits CentOS Linux operating system.

The numbers of patterns extracted, i.e., generators, biclusters, and rules, are shown in Figure 5. For this experiment, the *minsupport* threshold was varied between 50% (326 species) and 0.5% (3 species). The *minconfidence* threshold was varied between 50% and 1%. It should be noted that the vertical axes are on a logarithmic scale.

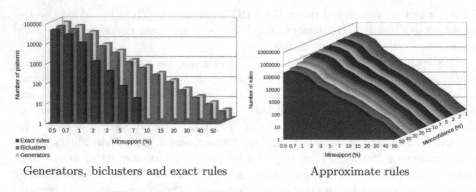

Generators, biclusters and exact rules   Approximate rules

Fig. 5. Number of Patterns Generated by FIST

3.2 Extracted Pattern Evaluation

In this section, we present some interesting conceptual association rules obtained from FIST. We would like to stress that these results mainly depend on the data collected within BioKET, which, as far as we know, is the only data warehouse consolidating different biodiversity information sources. These rules make it possible to estimate the risk status of a plant species according to IUCN RedList categories (*Lower Risk, Endangered, Least Concern, Vulnerable, Critically Endangered, Rare, Data Deficient, Rare & Threatened, Possibly Extinct*) with respect to their characteristics and *vice-versa*. For this experiment, the *minsupport* threshold was set to 1%, which corresponds to 6 species in the dataset, and the *minconfidence* threshold was set to 50%.

One of the obtained rules with the highest lift (11.75) is

INFL:pedicels up to 3 mm long, BBT:Twigs terete, INFL:axillary $\Rightarrow$ RS:Lower Risk.

(1)

According to this rule, of the six species with pedicels up to 3 mm long, twigs terete, and axillary inflorescence, 66,67% belong to the *lower risk* category. The identified species are *Cratoxylum cochinchinense, Cratoxylum formosum, Engelhardtia serrata, Engelhardtia spicata, Irvingia malayana,* and *Knema globularia.* This result is corroborated, for example, by information from Singapore flore.[1]

The following rule states, with 83.33% confidence, that a plant species classified as *critically endangered* has simple leaves:

$$\text{RS:Rare} \Rightarrow \text{LEAVES:Leaves simple.} \tag{2}$$

This rule is corroborated, for example, by [7], which describes *Gaultheria paucinervia,* a new species restricted to the eastern slopes of Mt. Kinabalu in Sabah State, Borneo, Malaysia, which has been confused with *Gaultheria borneensis Stapf,* but differs in its more erect habit and larger stature, longer nonappressed leaf trichomes, purple (vs. white) fruiting calyx, and lower elevation range, among other features. *Gaultheria paucinervia* has not yet been assessed for the IUCN Red List, (but is in the Catalogue of Life: *Gaultheria paucinervia P.W. Fritsch & C.M. Bush* apparently). Besides, by taking into account the features in the geographical data source, the FIST algorithm finds the rule

$$\text{RS:Rare, GEO:Western Ghats} \Rightarrow \text{LEAVES:Leaves simple,} \tag{3}$$

which identifies species *Bentinckia condapanna, Drypetes malabarica, Glycosmis macrocarpa, Holigarna grahamii, Lasianthus jackianus, Pittosporum dasycaulon,* and *Vepris bilocularis,* all found in the Western Ghats.

The following rule states, with 79.59% confidence, that a plant species classified as *Vulnerable* has simple leaves:

$$\text{RS:Vulnerable} \Rightarrow \text{LEAVES:Leaves simple.} \tag{4}$$

This result is corroborated, for example, by [23,29]. In [23], the author discusses the applicability of the Accelerated Pioneer-Climax Series (APCS) method for restoring forests to degraded areas in Southern Vietnam using many local species such as *Hopea odorata* directly concernend by the above rule and which has been identified as *vulnerable* in the IUCN red list. Wickneswari [29], instead, proposes a document which can help the readers to understand the entire life cycle of *Hopea odorata Roxb* in Malaysia, Vietnam, Cambodia, and Thailand.

The following rule, whose lift is 1.189 and whose support is 4.14%, states, with 55.1% confidence, that a plant species classified as *vulnerable* has both *glabrous* and *simple* leaves:

$$\text{RS:Vulnerable} \Rightarrow \text{LEAVES:glabrous, LEAVES:Leaves simple.} \tag{5}$$

Indeed, [21], proposing a deep and comprehensive botanical study of two rock outcrops in India, corroborates this rule.

---

[1] URL: http://florasingapura.com/Home.php. The aim of this site is to to bridge the gap between the terse technical descriptions of plants found in various botanical text books and what is observed in the Singapore forests.

Another interesting rule with a support of 3.37% and a lift of 1.07, states, with 59.46% confidence, that a plant species classified as having a *lower risk* has alternate leaves:

RS:Lower Risk ⇒ LEAVES:alternate.   (6)

This result is corroborated, for example, by results obtained by Craenel [5]. Species concerned include *Aglaia elliptica, Aphanamixis polystachya,* and *Prunus arborea.* As seen for Rule 3, the integration of geolocation information with multiple heterogeneous biological data can show common properties related to species with a specific risk status and/or in a specific area. For instance, the following rule with a lift of 4.26 states that 88.9% of species having a *lower risk* in the Indochina geographic region (i.e., 8 species) have leaves with entire margin:

RS:Lower Risk, GEO:Indochina ⇒ LEAVES:Margin entire.   (7)

Another example of such rules is the following, showing that 88.2% of *endangered* species in Western Ghats have alternate leaves:

RS:Endangered, GEO:Western Ghats ⇒ LEAVES:Alternate.   (8)

This rule, whose lift is 1.96, concerns 15 species. Such patterns can help comparisons between different geographical areas, at different levels of abstraction. For instance, considering the Malaysia geographic region, a part of Indochina, only 61.5% of species having a *lower risk* have leaves with entire margin as stated by the following rule, whose lift is 2.95 and which concerns 8 species:

RS:Lower Risk, GEO:Malaysia ⇒ LEAVES:Margin entire.   (9)

If we consider the Agasthyamalai area, lying at the extreme southern end of the Western Ghats mountain range along the western side of Southern India, we can see from the following rule that only 50% of *endangered* species in this area have alternate leaves, whereas the percentage is of 88.2% in the whole Western Ghats:

RS:Endangered, GEO:Agasthyamalai ⇒ LEAVES:Alternate.   (10)

This rule, which has a lift of 4.27, concerns 10 species.

All the above rules have been constructed from the consolidation of data from the four data sources presented above. Although some of the species are not yet included in the IUCN red list, combining information from the three other data sources allowed us to infer their risk status using the rules constructed by FIST. This is the case, e.g., for the species related to Rule 3, with the sole exception of *Bentinckia condapanna*, whose risk category is explicitly in IUCN. Indeed, *Glycosmis macrocarpa*'s taxon has not yet been assessed for the IUCN Red List, but is listed in the Catalogue of Life as *Glycosmis macrocarpa Wight.* The same holds for *Drypetes malabarica* (in the Catalogue of Life as *Drypetes malabarica (Bedd.) Airy Shaw*), *Lasianthus jackianus* (in the Catalogue of Life as *Lasianthus jackianus Wight*), *Pittosporum dasycaulon* (in the Catalogue of Life as *Pittosporum dasycaulon Miq*), and *Vepris bilocularis* (in the Catalogue of Life as *Vepris bilocularis (Wight & Arn.) Engl.*).

4   Conclusion

We presented BioKET, a data warehouse obtained by consolidation of a number of heterogeneous data sources on biodiversity. As far as we know, this is the first data warehouse containing that amount of heterogeneous data which can be used for conducting data-intensive studies about biodiversity. For the moment, the scopus of BioKET is botanical, but we plan to integrate other types of data.

We have demonstrated the usefulness of BioKET by applying FIST, an existing conceptual biclustering method, on a dataset extracted from BioKET to analyze the risk status of plants endemic to Laos. The evaluation of the extracted patterns against the botanical literature shows that meaningful knowledge can be infered from BioKET.

References

1. Benniamin, A., Irudayaraj, V., Manickam, V.S.: How to identify rare and endangered ferns and fern allies. Ethnobotanical Leaflets 12, 108–117 (2008)
2. Biodiversity informatics and co-operation in taxonomy for interactive shared knowledge base (BIOTIK), http://www.biotik.org (accessed September 2011)
3. Botanical research and herbarium management system (BRAHMS), http://herbaria.plants.ox.ac.uk/bol/ (accessed January 2013)
4. http://wiki.openstreetmap.org/wiki/Bounding_box (Accessed April 2014)
5. De Craenel, L.R., Wanntorp, L.: Floral development and anatomy of salvadoraceae. Ecological Applications 104(5), 913–923 (2009)
6. Eldredge, N.: Life on Earth: An Encyclopedia of Biodiversity, Ecology, and Evolution, Life on Earth, vol. 1. ABC-CLIO (2002)
7. Fritsch, P.W., Bush, C.M.: A new species of gaultheria (ericaceae) from mount kinabalu, borneo, malaysia. Novon: A Journal for Botanical Nomenclature 21(3), 338–342 (2011), http://dx.doi.org/10.1371/journal.pone.0005725
8. Geocat: Geospatial conservation assessment tool, http://geocat.kew.org/ (accessed April 2014)
9. Global biodiversity outlook 3, http://www.cbd.int/gbo3 (accessed January 2013)
10. Grillo, O., Venora, G. (eds.): Biological Diversity and Sustainable Resources Use. InTech (2011)
11. Han, J., Kamber, M., Pei, J.: Data Mining: Concepts and Techniques, 3rd edn. Morgan Kaufmann Publishers Inc., San Francisco (2011)
12. Hochachka, W.M., Caruana, R., Fink, D., Munson, A., Riedewald, M., Sorokina, D., Kellings, S.: Data-mining discovery of pattern and process in ecological systems. The Journal of Wildlife Management 71(7), 2427–2437 (2007)
13. Institute, W.R.: Ecosystems and human well-being: Biodiversity synthesis. Millennium Ecosystem Assessment (2005)
14. Marbán, O., Mariscal, G., Segovia, J.: A data mining & knowledge discovery process model. In: Data Mining and Knowledge Discovery in Real Life Applications, InTech, Vienna (2009)
15. Mariscal, G., Marbán, O., Fernández, C.: A survey of data mining and knowledge discovery process models and methodologies. The Knowledge Engineering Review 25(2), 137–166 (2010), http://journals.cambridge.org/article_S0269888910000032

16. Midgley, G.: Biodiversity and ecosystem function. Science 335(6065), 174–175 (2012), http://www.sciencemag.org/content/335/6065/174.short
17. Mondal, K.C., Pasquier, N., Mukhopadhyay, A., Maulik, U., Bandyopadhyay, S.: A new approach for association rule mining and bi-clustering using formal concept analysis. In: MLDM 2012, pp. 86–101 (2012)
18. Natural products information system (NAPIS), http://whitepointsystems.com (accessed February 2013)
19. Obrst, L.: Ontologies for semantically interoperable systems. In: CIKM 2003, pp. 366–369 (2003), http://doi.acm.org/10.1145/956863.956932
20. Peters, C., Peters, D., Cota-Sánchez, J.: Data mining and mapping of herbarium specimens using geographic information systems: A look at the biodiversity informatics project of the W. P. Fraser Herbarium, SASK (2009), http://www.herbarium.usask.ca/research/Data%20Mining,%20CBA%202009.pdf
21. Rahangdale, S.S., Rahangdale, S.R.: Plant species composition on two rock outcrops from the northern western ghats, maharashtra, india. Journal of Threatened Taxa 6(4), 5593–5612 (2014)
22. Shah, A.: Why Is Biodiversity Important? Who Cares? Global Issues (April 2011), http://www.globalissues.org/article/170/why-is-biodiversity-important-who-cares
23. So, N.V.: The potential of local tree species to accelerate natural forest succession on marginal grasslands in southern vietnam, http://www.forru.org/extra/forru/PDF_Files/frfwcpdf/part2/p28
24. Spehn, E.M., Korner, C. (eds.): Data Mining for Global Trends in Mountain Biodiversity. CRC Press (2009)
25. Talent, J.: Earth and Life: Global Biodiversity, Extinction Intervals and Biogeographic Perturbations Through Time. International Year of Planet Earth. Springer (2012)
26. The convention on biological diversity (CBD), http://www.cbd.int (accessed September 2013)
27. The IUCN Red List of Threatened Species, http://www.iucnredlist.org/ (accessed January 2014)
28. Whetzel, P., Noy, N., Shah, N., Alexander, P., Nyulas, C., Tudorache, T., Musen, M.: What are ontologies (accessed March 2013), http://www.bioontology.org/learning-about-ontologies
29. Wickneswari, R.: Hopea odorata roxb, http://www.apforgen.org/apfCD/Information

Using Time-Sensitive Rooted PageRank
to Detect Hierarchical Social Relationships

Mohammad Jaber[1], Panagiotis Papapetrou[2], Sven Helmer[3],
and Peter T. Wood[1]

[1] Department of Comp. Sci. and Info. Systems, Birkbeck, University of London, UK
[2] Department of Computer and Systems Sciences, Stockholm University, Sweden
[3] Faculty of Computer Science, Free University of Bozen-Bolzano, Bolzano, Italy

Abstract. We study the problem of detecting hierarchical ties in a
social network by exploiting the interaction patterns between the actors
(members) involved in the network. Motivated by earlier work using a
rank-based approach, i.e., Rooted-PageRank, we introduce a novel time-
sensitive method, called T-RPR, that captures and exploits the dynamics
and evolution of the interaction patterns in the network in order to identify the underlying hierarchical ties. Experiments on two real datasets
demonstrate the performance of T-RPR in terms of recall and show its
superiority over a recent competitor method.

1   Introduction

Interactions between groups of people and the patterns of these interactions are
typically affected by the underlying social relations between the people. In social
networks such social relationships are usually implicit. Nonetheless, analysing
patterns of social interactions between the members of a social network can help
to detect these implicit social relations. For example, consider a social network
where the members declare explicitly some type of social relationship with others, such as x is a *colleague* of y. Now, suppose that we also have available
the communication patterns between x and y, e.g., how often they exchange
e-mails in a month. Using this information we may be able to infer additional
relationships between these two members, such as x is the *manager* of y.

In this paper, we study the problem of detecting implicit *hierarchical* relationships in a social network by exploiting the interactions between the members of
the network. We mainly focus on two key features that play a central role in
our problem: (a) the structure of the interaction network, and (b) the evolution
of the interactions, or in other words, the "dynamics" of the interactions over
time. Given the interactions, we are interested in finding for each member of the
social network their parent in the hierarchy, which we call their *superior*.

Figure 1 illustrates the problem we are addressing. The input graph (a) is the
interaction network, where nodes represent actors and edges represent interactions between the actors. By analysing this network, we can infer a hierarchical
relationship network (b), where nodes represent the same group of actors and
edges represent the hierarchical relationships detected between them.

H. Blockeel et al. (Eds.): IDA 2014, LNCS 8819, pp. 143–154, 2014.

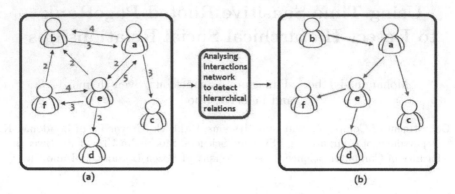

Fig. 1. (a) An interaction graph, where each arc is weighted by the total number of interactions. (b) Inferred hierarchical relationships between actors in (a).

Most related work has focussed on either the relationship structure of the network or the interactions between the nodes, but not on both. In our previous work [9], we proposed two methods for detecting hierarchical ties between a group of members in a social network. The first one, *RPR*, exploits the interaction graph of the network members and employs the Rooted-PageRank algorithm [17], whereas the second, *Time-F*, studies the interaction patterns between the network members over time.

In this paper, we propose a novel method, *Time-sensitive Rooted-PageRank* (*T-RPR*), to capture the interaction dynamics of the underlying network structure. The method proves to be more effective in detecting hierarchical ties, especially when the period over which the interactions occur is long enough.

The **contributions** of this paper include: (1) a novel time-sensitive method (T-RPR) which builds upon Rooted-PageRank and captures how the structure of the interaction network changes over time, (2) two approaches for aggregating scores from each of the time slots over which T-RPR is run, one based on a simple weighted average and the other based on voting, (3) an extensive experimental evaluation of the performance of these methods in terms of recall on two large real datasets, the Enron e-mail network and a co-authorship network. Our experiments show that T-RPR achieves considerably better results than the competitor RPR: in the Enron network, T-RPR detects up to 58% of manager-subordinate relationships, compared to only 29% by RPR, while in the co-author network it detects about 65.5% of PhD advisor-advisee relationships, a significant improvement over the 39.5% achieved by RPR.

2   Related Work

Few researchers have focussed on finding implicit ties in social networks. In our previous paper [9], we started an investigation into how the time dimension of interactions between actors could improve the detection of hierarchical ties. We defined two methods, Time-F and FiRe, which are based on predefined time functions. However, both Time-F and FiRe are ineffective in detecting hierarchical ties when there are relatively few or no interactions between the actors connected by a hierarchical relationship, a problem we address in the current paper.

Gupte at al. [6] propose an algorithm to find the best hierarchy in a directed network. They define a global measure of hierarchy which is computed by analysing the direction of interaction edges. They do not consider the temporal dimension of the interactions nor do they infer the superior of each actor, as we do. Buke et al. [3] focus on child-parent relationships at many life stages and how communication varies with the age of child, geographic distance and gender. In contrast to our approach, they model the language used between users to generate text features. Backstrom et al. [1] developed a new measure of tie strength which they termed "dispersion" to infer romantic and spouse relationships. However, dispersion does not seem relevant to our problem of detecting hierarchical relations.

On the other hand, many methods have been developed in social-network analysis to assess the importance of individuals in implicitly- or explicitly-defined social networks. Measures of importance in social networks include in-degree, degree centrality, closeness centrality, betweenness centrality and eigenvector centrality [8,10,13,14,16].

The PageRank [2] and HITS [11] algorithms have been used and adapted to address a range of problems. For example, Xiong et al. [18] evaluate a user's influence based on PageRank. They considere three factors: the number of the user's friends, the quality of their friends and the community label, i.e. the similarity between the user and the community. Fiala et al. [5] employ and adapt PageRank to analyse both co-authorship and citation graphs to rank authors by their influence. Their results are improved in [4] by introducing time-aware modifications in which citations between researchers are weighted according to a number of factors, such as the number of common publications and whether or not they were published before a citation was made.

In the same context, Yan and Ding [19] used weighted PageRank to discover author impact on a community. In the area of search engines, Li et al. [12] investigated how time-based features improve the results of retrieving relevant research publications. They consider both the structure of the citation network and the date of publication, giving older papers lower weight.

Predicting future link formation in networks has attracted much research. Huang and Lin [7] implemented an approach that considers the temporal evolution of link occurrences within a social network to predict link occurrence probabilities at a particular time in the future. Sun [15] proposed a meta path-based model to predict the future co-author relationships in a bibliographic network. Different types of objects (e.g., venues, topic) and the links among them were analysed. However, our approach differs from these two studies in that it detects *hierarchical* social ties.

3   Problem Setting

Let V denote the set of *actors* (members) of a social network. We consider two types of graphs defined over V: the *interaction graph* and the *hierarchy graph*.

Definition 1 (Interaction Graph). An *interaction graph* is defined as $G_I = (V, E^c, W)$, where E^c is the set of edges (directed or undirected) representing the interactions between the actors in V and W is a vector of edge weights, where $w_{uv} \in W$ corresponds to the weight of the edge connecting nodes u and v.

We note that G_I can be modeled both as a directed or undirected graph, as well as weighted or unweighted, depending on the nature of the interactions and the application domain at hand.

Definition 2 (Hierarchy Graph). A *hierarchy graph* is a directed graph defined as $G_H = (V, E^s)$, where $E^s \subseteq V \times V$ is a set of of edges representing the hierarchical relationship between the actors in V. Each edge $(u, v) \in E^s$ indicates that actor $u \in V$ is the direct superior of actor $v \in V$ in the hierarchy.

For example, in the context of an e-mail network among a group of employees a hierarchy graph may represent the set of manager-subordinate relationships, where $(u, v) \in E^s$ indicates that u is the manager of v. Based on the above definitions, the problem studied in this paper can be formulated as follows:

Problem 1. *Given a set of actors V and their corresponding interaction graph G_I, infer the hierarchy graph G_H of V.*

In Figure 1 we can see an example of the problem we want to solve. Given the interaction graph (a) of the five actors a, b, c, d, e, f, we want to infer their corresponding hierarchy graph (b).

4   Static Rooted-PageRank (S-RPR)

In our previous paper [9], we proposed an approach that employs Rooted Page-Rank (RPR) to detect hierarchical ties between a group of actors who interact over a time period. The approach relies on the fact that RPR scores reflect the importance of nodes relative to the root node. For each actor $x \in V$ (root node) we rank each other actor $y \in V \setminus \{x\}$ according to the score $RS_x(y)$ obtained by RPR, which reflects the chance of y being the superior of x. In the ideal case, the actual superior of x should have the highest score and be ranked first. The main feature of this approach is that it considers the static structure of the interaction graph over the whole time period of the interaction.

After running RPR for all actors, a ranking list $L(x) = [y_1, y_2, \ldots, y_{|V|-1}]$, $y_i \in V$, is produced for each $x \in V$, such that $RS_x(y_i) \geq RS_x(y_{i+1})$, $1 \leq i \leq n-1$. Finally, the hierarchy graph G_H is inferred from $L(x)$ by assigning to each node $x \in V$ one of the candidate managers that ranked high in $L(x)$, e.g., within the top-K places, for some K.

5 Time-Sensitive Rooted PageRank (T-RPR)

We investigate whether "time matters" in detecting hierarchical social relationships; in other words, whether significant improvements in detecting hierarchical ties can be obtained by taking into account the temporal aspects of the interactions. We adapt Rooted PageRank (RPR), as described in the previous section, and introduce *Time-Sensitive Rooted Pagerank (T-RPR)*, which captures how the ranking scores of the interactions change over time. The proposed method consists of three parts: time segmentation, ranking, and rank aggregation.

5.1 Time Segmentation

We consider the interaction graph G_I of V. As opposed to S-RPR, now G_I is not static. Let $T = [t_1, t_m]$ be the time period of interactions in G_I, starting at time t_1 and ending at time t_m. First, T is divided into n equal-sized non-overlapping time slots $\{T_1, T_2, ..., T_n\}$, with $T_j = [t_{jk}, t_{jl}]$, $\forall j \in [1, n]$, such that $t_{jl} - t_{jk} = d$, $\forall j$, where $d \in \mathbb{Z}^+$ is the size of the time segments. Observe that a time slot can be any time unit (e.g., day, fortnight, month, or year) depending on the application. Next, we define an interaction graph for each time slot.

Definition 3 (Time-Interaction Graph). A *time-interaction* graph is defined as $G_I^k = (V_k, E_k^c, W)$, where $V_k \subseteq V$ is the set of actors who interacted with at least one other actor within time slot T_k, $E_k^c \subseteq V_k \times V_k$ is the set of edges (directed or undirected) corresponding to the interactions between the set of actors V_k which took place within T_k, and W is the vector of edge weights.

Finally, a set of time-interaction graphs $\mathcal{G}_I = \{G_I^1, ..., G_I^n\}$ is produced for the n time slots. The next task is to rank the nodes in each graph.

5.2 Segment-Based Ranking

For each time-slot T_k and each actor $x \in V$, we run RPR on the corresponding time-interaction graph $G_I^k = (V_k, E_k, W)$. Let $score_{x,k}(v_i)$ denote the RPR score of actor v_i when x is used as root on G_I^k. This results in a list of actors sorted in descending order with respect to their RPR scores at time slot k:

$$L(x)_k = [v_1, v_2, .., v_N] \,,$$

where $N = |V_k| - 1$, $V_k \backslash \{x\} = \{v_1, v_2, .., v_N\}$, and $score_{x,k}(v_i) \geq score_{x,k}(v_{i+1})$ for $i = 1, ..., N - 1$.

The rankings obtained over the n time-slots are aggregated for each root actor x and all remaining actors $v_i \in V$, resulting in an aggregate score $aggScore_x(v_i)$. Finally, the aggregate scores are sorted in descending order resulting in the following aggregate list of actor ranks:

$$L(x) = [v_1, v_2, .., v_M] \,,$$

where $M = |V| - 1$, $V/\{x\} = \{v_1, v_2, .., v_M\}$, and $aggScore_x(v_i) \geq aggScore_x$ (v_{i+1}), for $i = 1, ...M - 1$. More details on the aggregation techniques are given below.

Finally, as in S-RPR, the hierarchy graph G_H is inferred from $L(x)$ by assigning to each node $x \in V$ one of the candidate managers that ranked high in $L(x)$, e.g., within the top-K places, for some K.

5.3   Rank Aggregation

We explored two rank aggregation techniques, one based on averaging and one based on voting.

Average-based Time-sensitive RPR (AT-RPR). In this approach, the ranking in $L(x)$ is based on a weighted average of the individual RPR scores over all time-slots. We define a set of weights $\Omega = \{\omega_1, \ldots, \omega_n\}$, where ω_k is the weight assigned to time slot T_k. Each actor $y \in L(x)$ is ranked according to the obtained scores over all time-slots:

$$aggScore_x(y) = 1/n \cdot \sum_{k=1}^{n} \omega_k \cdot score_{x,k}(y) \ . \tag{1}$$

Assigning the values in Ω is application-dependent. For example, if the interactions between actors and their superiors are distributed regularly over the whole period T, then all weights can be equal. On the other hand, the interactions between actors and their superiors may be more intensive in earlier or later time-slots. An example of the former case is when detecting PhD advisor-advisee relationships in a co-author network; higher weights are given to scores in early time-slots when the advisees are expected to publish more papers with their advisors, decreasing in later time-slots.

Vote-based Time-sensitive RPR (VT-RPR). An alternative approach is to assign candidate actors with votes at each time-slot T_k based on their rank in that slot. The final rank of an actor is determined according to the total number of votes they win over all time-slots.

More precisely, given $L(x)_k$ at slot T_k, a vote is assigned to actor $y \in V \setminus \{x\}$, if y appears among the first c actors in $L(x)_k$. We call c the *vote-based cut-off*. Let $pos(L(x)_k, y)$ denote the position of y in $L(x)_k$. The total number of votes obtained by each candidate y is then defined as:

$$aggScore_x(y) = \sum_{k=1}^{n} \omega_k \cdot vote_{x,k,c}(y) \ , \tag{2}$$

where:
$$vote_{x,k,c}(y) = \begin{cases} 1 & \text{if } pos(L(x)_k, y) \leq c \\ 0 & \text{otherwise} \end{cases}$$

and ω_k is the weight of time slot T_k, which is set depending on the application.

5.4   Example

Let us consider the example shown in Figure 1. Assume that we want to detect the superior of actor a, namely, actor b. We will apply S-RPR and T-RPR, and compare the findings. We emphasise that the RPR scores used in the example are made up; however, we wish to illustrate how ranks are aggregated in our approach rather than how RPR scores are computed.

S-RPR. We set a to be the root and run RPR over the interaction graph, which produces $L(a)$. Specifically, the list contains the actors in the following order:

$$[(e, 0.30), (f, 0.25), (b, 0.20), (d, 0.18), (c, 0.07)] .$$

We observe that the position of actor b in $L(a)$ is 3 (out of 5).

T-RPR. Suppose that T consists of four equal-sized time-slots. We generate four time-interaction graphs, $G_I^1(V_1, E_1^c), G_I^2(V_2, E_2^c), G_I^3(V_3, E_3^c), G_I^4(V_4, E_4^c)$, one for each time-slot, as shown in Table 1.

Table 1. The set of time-interaction graphs: we list the set of vertices V_k and edges E_k^c for each time slot T_k

	V_k	E_k^c
T_1	a, b, e, f	$(a, b), (e, b), (e, f), (f, e)$
T_2	a, b, c, d, e, f	$(a, b), (a, c), (b, a), (e, d), (f, e),$
T_3	a, b, e, f	$(a, b), (a, e), (e, a), (e, b), (e, f), (f, b), (f, e)$
T_4	a, c, e	$(a, c), (a, e)$

Next, to detect the superior of a, we run RPR with a as the root for each time-slot T_k using G_I^k for $k = 1, \ldots, 4$. A ranked list $L(a)_k$ is produced for each T_k, as shown in Table 2.

Table 2. Rank lists produced by T-RPR over all time-slots with root a

L(a)_1			L(a)_2			L(a)_3			L(a)_4		
rank	actor	$score_{a,1}$	rank	actor	$score_{a,2}$	rank	actor	$score_{a,3}$	rank	actor	$score_{a,4}$
1	b	1.00	2	b	0.50	1	b	0.50	2	c	0.50
5	c	0.00	2	c	0.50	2	e	0.30	2	e	0.50
5	d	0.00	5	d	0.00	3	f	0.20	5	b	0.00
5	e	0.00	5	e	0.00	5	c	0.00	5	d	0.00
5	f	0.00	5	f	0.00	5	d	0.00	5	f	0.00

Finally, the lists are aggregated using Eq. (1) (average-based approach with weights $\omega_k = 1$) and Eq. (2) (vote-based approach with weights $\omega_k = 1$ and cut-off $c = 2$). The final aggregated lists for each aggregation approach are shown in Tables 3 and 4. For example, the score for e in Table 3 is 0.20 because its average score is $(0.30 + 0.50)/4$, while its score in Table 4 is 2 because e appears in ranks 1 or 2 in 2 time slots (3 and 4). We observe that in both cases $T\text{-}RPR$ places actor b at position 1, as opposed to position 3 ($S\text{-}RPR$).

Table 3. Final ranked list using the average-based approach

position	actor	$score_a$
1	b	**0.50**
2	c	0.25
3	e	0.20
4	f	0.05
5	d	0.00

Table 4. Final ranked list using the vote-based approach

position	actor	$score_a$
1	b	3
3	c	2
3	e	2
5	d	0
5	f	0

6 Results and Analysis

We evaluated the methods in terms of recall on two datasets: the Enron email dataset and a co-authorship network, both of which are available online[1].

The Enron dataset includes more than 255000 emails exchanged among 87474 email addresses between January 2000 and November 2001. However, only 155 of these email addresses belong to Enron employees. Each email in the dataset has a sender, subject, timestamp, body, and a set of recipients. The dataset also contains the hierarchical manager-subordinate relationship between employees. The co-author dataset includes more than 1 million authors who contributed to about 80000 papers in total between 1967 and 2011. Each paper has a title, date, conference where it was published and a list of co-authors. The dataset includes hierarchical relationships between PhD advisors and their advisees.

To evaluate the performance of the two methods, we compute for each subordinate/advisee x the *rank* of their correct superior/advisor x^* in $L(x)$:

$$rank(x, x^*) = |\{y : score_x(y) \geq score_x(x^*)\}| \, , \forall y \in L(x) \, . \tag{3}$$

Hence, the rank of the manager x^* of x is the number of actors in $L(x)$ who have an RPR score greater than or equal to the score of x^* (see Table 2).

Finally, given a threshold K, we can define the *overall rank* $\rho(K)$ of V as the percentage of actors with rank at most K over all hierarchical relations that exist in G_H:

$$\rho(K) = \frac{|\{x : rank(x, x^*) \leq K\}|}{|E^s|} \cdot 100 \tag{4}$$

Enron. We excluded all email addresses of people who were not Enron employees. In cases where an employee used more than one email address, we chose one

---

[1] http://arnetminer.org/socialtie/

randomly. We explored two versions of the interaction graph: *directed*, where a directed edge exists from employee u to v if u sent at least one email to v; *undirected*, where any interaction (sent or received email) between u and v is represented as an edge between them. So, in both cases, all edge weights are 1.

For the T-RPR approach, each time-slot represents 1 month, giving 24 time-slots in total. In addition, since we expect to have regular interaction between a subordinate and their manager over the whole time period, each weight ω_k ($k = 1, \ldots, n$) in the aggregation functions given in Eq. (1) and (2) was set to 1.

The experimental results of the performance benchmark of AT-RPR and VT-RPR against S-RPR are shown in Figure 2(a) for the directed case and in Figure 2(b) for the undirected case. S-RPR performs considerably better for the directed graph. This becomes clear when we consider the number of managers ranked first in a subordinate's ranked list. About 30% of the managers are ranked first when using a directed graph compared to only 9.7% for the undirected graph. However, this picture changes when we consider the time dimension in T-RPR. Both AT-RPR and VT-RPR give better results on the undirected graph and especially when using vote-based aggregation. We consider the best value for the voting cut-off c for each case, i.e., 3 for the directed and 4 for the undirected graph. This finding suggests that the volume of email matters more than direction for detecting hierarchical ties in an employer-employee setting. A possible explanation is that employees may have similar communication patterns with respect to the fraction of sent vs. received emails when they communicate with other employees and also when they communicate with their boss. However, the volume of the email traffic as a whole can be a more distinctive feature of the underlying hierarchical tie.

In Tables 5 and 6, we can see how different values of the voting cut-off c affect the performance percentages of VT-RPR. Using these tables in combination with Figure 2 we can make several observations. Firstly, for the undirected graph, we observe that VT-RPR with the voting cut-off at 4 is preferable to both S-RPR and AT-RPR with significant improvement in detecting managers who rank

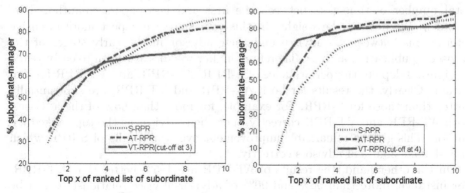

Fig. 2. Results using S-RPR and T-RPR (both aggregation strategies) on Enron using (a) the directed interaction graph and (b) the undirected interaction graph

in the top *three* of their subordinate's lists. For example, in 58.9% of manager-subordinate relations, managers come first in the ranked lists compared to 39.8% and 9.7% detected by AT-RPR and S-RPR respectively. In addition, for the directed graph, VT-RPR is still better than both AT-RPR and S-RPR, which perform similarly. For AT-RPR and S-RPR, about 30–33% of managers are ranked first in their subordinate's lists. VT-RPR performs substantially better, detecting over 48% of managers in the top position ($K = 1$).

Table 5. VT-RPR results on Enron dataset using vote cut-off $c = 1$–6 with *directed* interaction graph

c	ρ				
	$K = 1$	$K = 2$	$K = 3$	$K = 4$	$K = 5$
1	39.72	52.05	56.84	60.27	62.32
2	43.83	55.47	62.32	67.80	68.49
3	**48.63**	**56.84**	**62.32**	**66.43**	**67.80**
4	44.52	58.90	63.01	67.12	67.80
5	42.46	58.21	63.69	65.75	67.80
6	40.41	59.58	63.69	65.75	67.80

Table 6. VT-RPR results on Enron dataset using vote cut-off $c = 1$–6 with *undirected* interaction graph

c	ρ				
	$K = 1$	$K = 2$	$K = 3$	$K = 4$	$K = 5$
1	45.20	60.95	67.80	68.49	68.49
2	54.10	67.12	75.34	78.08	79.45
3	57.53	72.60	76.02	78.08	79.45
4	**58.90**	**73.28**	**76.71**	**78.76**	**80.13**
5	58.21	73.28	76.02	77.39	80.13
6	54.79	72.60	75.34	78.08	80.82

Co-author. For the purposes of our study, we excluded all single-author papers as well as papers without a publication date. Due to the symmetric nature of the co-author relationship, the interaction graph representing this dataset is undirected. Once again, each edge weight was set to 1. For the T-RPR approach, we defined 45 time-slots, one per publication year. Moreover, the weights used by both aggregation methods were defined for each $aggScore_x(y)$, as $\omega_k = 1 - \frac{N_{first}}{N_{all}}$, where N_{first} is the number of time-slots (years) between time-slot k and the slot in which the first paper co-authored by x and y appeared, and N_{all} is the total number of time-slots between the first and last papers co-authored by x and y. We defined the weights in this way since we expect more intensive interactions between an advisee and their advisor in the early stages of the advisee's publication activity. Therefore, higher weights are given to early years.

Figure 3 depicts the performance of S-RPR, AT-RPR, and VT-RPR for Co-author. Clearly, the results for both AT-RPR and VT-RPR, are substantially better than those for S-RPR. For example, for more than 65% of the advisees, both AT-RPR and VT-RPR correctly infer their advisor as the top-ranked co-author. This gives a remarkable improvement over the results of S-RPR which only detects 39.6% of advisors correctly.

On the other hand, the results of AT-RPR are 4–5% better than VT-RPR. For instance, more than 95% and 90% of advisor-advisee relationships can be detected within the top 7 authors by AT-RPR and VT-RPR respectively. Table 7 shows that the best results for VT-RPR are with voting cut-off $c = 1$.

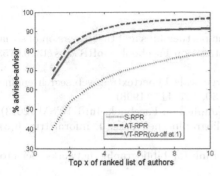

Fig. 3. Results for S-RPR, T-RPR (both aggregations) on Co-author

Table 7. VT-RPR results for Co-author using vote cut-off $c = 1\text{-}5$

	ρ				
c	$K = 1$	$K = 2$	$K = 3$	$K = 4$	$K = 5$
1	**65.52**	**79.25**	**85.12**	**87.55**	**89.50**
2	60.80	78.54	84.64	88.03	89.93
3	56.84	75.58	82.30	85.78	87.88
4	53.55	72.72	79.97	83.83	86.17
5	50.69	69.48	77.25	81.78	84.21

Main Findings. For both the Enron and co-author datasets, the time-sensitive methods AT-RPR and VT-RPR are significantly better than S-RPR. This demonstrates that time matters when detecting hierarchical relationships in social networks. However, AT-RPR and VT-RPR perform differently on each dataset, with VT-RPR being more effective in detecting subordinate-manager relationships in the Enron data and AT-RPR being slightly better in detecting advisee-advisor relationships in the co-author network.

One interpretation of these results is that, when the interactions between actors and their superiors extend over many time-slots, then VT-RPR is more appropriate. An example of this is the Enron dataset, where the interactions occur over 24 time-slots. On the other hand, when the interactions with the superior are intensive within a few time-slots, AT-RPR is preferable to VT-RPR. This is the case for the co-author dataset where usually an advisee publishes papers with their advisor within only 4–5 time-slots while the advisee is completing their PhD. When compared to our previous work [9], our new time-sensitive methods prove to be effective in detecting hierarchical ties even when there are no, or relatively few, interactions between an actor and their superior.

7   Conclusion

We introduced T-RPR, a method for detecting hierarchical ties in an interaction graph. We investigated the impact of the temporal dimension in the ranking process and adapted Rooted-PageRank to capture the dynamics of the interactions over time between the actors in the network. We explored two variants for aggregating the rankings produced at each time slot. Experiments on two real datasets showed the superiority of T-RPR against our previous static approach, S-RPR, hence providing reasonable empirical justification for our claim that *"time matters"* in detecting hierarchical ties.

References

1. Backstrom, L., Kleinberg, J.M.: Romantic partnerships and the dispersion of social ties: A network analysis of relationship status on Facebook. CoRR, abs/1310.6753 (2013)
2. Brin, S., Page, L.: The anatomy of a large-scale hypertextual web search engine. Computer Networks and ISDN Systems 30, 107–117 (1998)
3. Burke, M., Adamic, L., Marciniak, K.: Families on Facebook. In: ICWSM (2013)
4. Fiala, D.: Time-aware pagerank for bibliographic networks. J. Informetrics 6(3), 370–388 (2012)
5. Fiala, D., Rousselot, F., Jezek, K.: Pagerank for bibliographic networks. Scientometrics 76(1), 135–158 (2008)
6. Gupte, M., Shankar, P., Li, J., Muthukrishnan, S., Iftode, L.: Finding hierarchy in directed online social networks. In: WWW, pp. 557–566. ACM (2011)
7. Huang, Z., Lin, D.K.J.: The time-series link prediction problem with applications in communication surveillance. INFORMS Journal on Computing 21(2), 286–303 (2009)
8. Hubbell, C.H.: An input-output approach to clique identification. Sociometry 28(4), 377–399 (1965)
9. Jaber, M., Wood, P.T., Papapetrou, P., Helmer, S.: Inferring offline hierarchical ties from online social networks. In: WWW Companion 2014, pp. 1261–1266 (2014)
10. Katona, Z., Zubcsek, P.P., Sarvary, M.: Network effects and personal influences: the diffusion of an online social network. J. Marketing Research 48 (2011)
11. Kleinberg, J.M.: Authoritative sources in a hyperlinked environment. J. ACM 46(5), 604–632 (1999)
12. Li, X., Liu, B., Yu, P.S.: Time sensitive ranking with application to publication search. In: ICDM, pp. 893–898 (2008)
13. Newman, M.E.J.: Scientific collaboration networks. II. shortest paths, weighted networks, and centrality. Physical Review E 64 (2001)
14. Newman, M.E.J.: The mathematics of networks. The New Palgrave Encyclopedia of Economics (2007)
15. Sun, Y., Barber, R., Gupta, M., et al.: Co-author relationship prediction in heterogeneous bibliographic networks. In: ASONAM, pp. 121–128 (2011)
16. Zlatic, V., Bianconi, G., et al.: On the rich-club effect in dense and weighted networks. EPJ B-Condensed Matter and Complex Systems 67(3), 271–275 (2009)
17. White, S., Smyth, P.: Algorithms for estimating relative importance in networks. In: KDD, pp. 266–275 (2003)
18. Xiong, Z., Jiang, W., Wang, G.: Evaluating user community influence in online social networks. In: IEEE TrustCom, pp. 640–647 (2012)
19. Yan, E., Ding, Y.: Discovering author impact: A pagerank perspective. CoRR, abs/1012.4870 (2010)

Modeling Daily Profiles of Solar Global Radiation Using Statistical and Data Mining Techniques

Pedro F. Jiménez-Pérez and Llanos Mora-López

Dpto. Lenguajes y C.Computación. ETSI Informática.
Universidad de Málaga. Campus de Teatinos. 29071 Málaga, Spain
{pjimenez,llanos}@lcc.uma.es

Abstract. Solar radiation forecasting is important for multiple fields, including solar energy power plants connected to grid. To address the need for solar radiation hourly forecasts this paper proposes the use of statistical and data mining techniques that allow different solar radiation hourly profiles for different days to be found and established. A new method is proposed for forecasting solar radiation hourly profiles using daily clearness index. The proposed method was checked using data recorded in Malaga. The obtained results show that it is possible to forecast hourly solar global radiation for a day with an energy error around 10% which means a significant improvement on previously reported errors.

Keywords: k-means, clearness index, forecasting hourly solar radiation.

1 Introduction

The number of solar energy plants has increased significantly in recent years, mainly due to the following factors: the need to use energy sources that contribute to reduce carbon emissions, support policies being established to introduce this type of systems, improved efficiency of these systems and the significant reduction in the price of all the components that make up those systems. As the number of these systems rises, there is an increasingly greater need for systems to be developed that enable these energy sources to be integrated with the traditional generation system. Therefore, predicting energy production by these plants has become a requirement on competitive electricity markets. In fact, accurate solar radiation forecasting is critical to large-scale adoption of photovoltaic energy.

Electrical energy production planning is usually made on an hourly basis to adjust production to consumption. Solar power plants depend on meteorological factors (basically solar irradiance) to produce energy. Knowing how much energy these systems will produce is necessary in order to ensure correct integration into the electrical system. For instance, since 1998, the Spanish electricity market

H. Blockeel et al. (Eds.): IDA 2014, LNCS 8819, pp. 155–166, 2014.
© Springer International Publishing Switzerland 2014

has moved from a centralized operational approach to a competitive one. It encourages the deployment of solar plants with a financial penalty for incorrect prediction of solar yields for the next day on an hourly basis.

Estimating the energy generated by solar plants is difficult mainly due to its dependence on meteorological variables, such as solar radiation and temperature, [1], [2]. In fact, photovoltaic production prediction is mainly based on global solar irradiation forecasts. The behavior of this variable can change quite dramatically on different days, and even on consecutive days.

Short-term prediction of solar radiation can be addressed by other process forecasting. In general, a wide type of statistical and data mining techniques have been developed for process forecasting. Statistical time series methods are based on the assumption that the data have an internal structure that can be identified by using simple and partial autocorrelation, [3], [4], [5]. Time series forecasting methods detect and explore such a structure. In particular, ARMA (autoregressive moving average), ARIMA (autoregressive integrated moving average) models have been widely used; for instance, [6], [7], [8], [9], [10] and [11] propose different methods for modeling hourly and daily series of clearness index (parameter related to solar global radiation). These models are particularly useful for long-term characterization and prediction of the clearness index as they pick up the statistical and sequential properties of series. However, these methods have not been used for short-term prediction of clearness index as the error in the prediction of isolated values (next value in a series) is too large.

Data mining techniques have been also used for process forecasting. These approaches have been proposed to overcome the limitations of statistical methods as they do not require any assumptions to be made, particularly with respect to the linearity of the series. Some of the data mining models developed for forecasting global solar radiation can be found in [12], [13], [14], [15], [16]. Short-term forecasting models have been also proposed. For instance, a multilayer perceptron (MLP) for predicting 24 hours irradiance values for the next day is proposed in [17]. MLP input values are mean daily values for solar irradiance and air temperature; the correlation coefficient obtained is about 98-99% for sunny days and 94-96% for cloudy days; however, it is around 32% in energy terms. In [18] short-term forecasting with continuous time sequences was performed and the mean square errors of the proposed models range from 0.04 to 0.4 depending on the value of clearness index. An artificial neural network (ANN) model was used to estimate the solar radiation parameters for seven cities from the Mediterranean region of Anatolia in Turkey in [19]. The maximum RMSE was found to be 6.9% for Mersin citation and the best value obtained was 3.58% for Isparta. Artificial neural network and ARIMA models are proposed in [20]; the errors range from 30 to 40% in energy terms. Similar models are used in [21]; the obtained errors for predicting hourly values for a day range from 23 to 28%.

This paper proposes the use of a clustering data mining technique that allows different solar radiation hourly profiles for different days to be found to address the need to forecast hourly solar radiation values. The main objective is to develop a method that is able to forecast accurately solar radiation on an hourly basis for a day. The main idea is to establish different hourly profiles for different types of days, that is, separate days into different clusters and then, for each profile, apply the information of its cluster to get accurate predictions. The rest of the paper is organized as follows. The materials and methods used are described in the second section. The third section describes the data used. The results of the proposed methodology are detailed in the fourth section. Finally, the last section concludes.

2   Materials and Methods

Hourly irradiance values distribution over a day seems to have a clear dependency on the daily clearness index, [22]. In order to describe how solar irradiance is distributed in each hour along the day, cumulative distribution probability functions (c.p.d.f.) of the recorded values in 20 minutes intervals will first be estimated in order to determine how many different functions have been observed. K-means will use all these curves to cluster these functions into groups, which satisfy that all c.p.d.f.'s in each group can be considered homogeneously equal using the Kolmogorov-Smirnov two sample test. For each group, tests will be conducted to establish which parameters explain curves in that group. Finally, a model capable of simulating the hourly profile of the group will be proposed.

The proposed models will be evaluated using the total energy error estimated using the predicted hourly energy and the actual hourly energy.

2.1   Solar Radiation and Clearness Index

Solar radiation received in the Earth has seasonality and daily variability. The clearness index is commonly used to remove these trends, instead of using solar radiation data directly. The instantaneous clearness index is defined as the ratio of the horizontal global irradiance to the irradiance available from the atmosphere or extraterrestrial irradiance. In a similar way, the clearness for different time periods (such as hourly or daily) is defined as the ratio of the horizontal global irradiation received for a period of time to the irradiation received for the same period out of the atmosphere, according to the expression 1:

$$K_t = \frac{G_t}{G_{t,0}} \tag{1}$$

where G_t is the solar global radiation recorded for time t and $G_{t,0}$ is the extraterrestrial solar global radiation for this period. We have used two different periods, 20 minutes and 1 hour, both from sunrise to sunset.

Solar radiation received on Earth and solar radiation received out of the atmosphere depend on the distance to the Sun and the Earth's relative position to the Sun, that is defined by three angles: the solar elevation, the azimuth and the hourly angle. Extraterrestrial solar radiation is estimated using the Earth-Sun distance, the declination, the latitude of the place and the hour angle considered. The extraterrestrial irradiance (instantaneous radiation), G_0 received on an horizontal surface is obtained using the expression:

$$G_0 = I_{sc}E_0 \cos\theta_z = I_{sc}E_0(\sin\delta\sin\phi + \cos\delta\cos\phi\cos\omega_s)(Wm^{-2}) \qquad (2)$$

where I_{sc} is the solar constant, E_0 is the eccentricity factor, δ is the declination angle, ϕ is the latitude, ω_s is the hour angle. The expressions for estimating E_0, δ and ω_s can be found in [23]. The expression for obtaining the extraterrestrial solar radiation for a period of time is obtained integrating expression 2 for this period. For the hourly solar extraterrestrial global radiation, the expression is Eq.2 being ω_s the value of this angle for the center of the hour.

2.2   Cumulative Probability Distribution Functions

The cumulative probability distribution function (c.p.d.f.), $F_X(\cdot)$, of independent and identically distributed observations, $\{X_i\}_{i=1}^n$, with the same distribution as a random variable X, is calculated using the expression:

$$F_X(t) = \Pr(X \leq t) \qquad (3)$$

assuming $F_X(\cdot)$ is continuous.

These functions were estimated from the clearness index values calculated using the solar radiation recorded every 20 minutes for each day. We used these functions for each day instead of directly using the clearness index values because of the different length of hours with data (due of the different length of days during the year).

2.3   K-means

We propose using the data mining technique known as k-means clustering to establish the number of different clearness index c.p.d.f.'s. Clustering techniques have already been used in different areas such as text mining, statistical learning and pattern recognition, [24], [25], [26]. It is based on analyzing one or more attributes (variables) to identify a cluster of correlating results. The distance from the sample to its cluster is used to measure the similarity between each c.p.d.f. and the centroid of each cluster as all the variables are numerical. We have used squared Euclidean distances, as it is defined in [24].

$$d_p((\mathbf{X}_i, \mathbf{X}_j) = (\sum_{k=1}^d |x_{i,k} - x_{j,k}|^p)^{1/p} = ||\mathbf{X}_i - \mathbf{X}_j||_p \qquad (4)$$

where $\mathbf{X}_i$ and $\mathbf{X}_j$ are the variables included for characterizing observation i and j respectively; they are the c.p.d.f. vector for each day in our study . These vectors are the inputs for the k-means algorithm. The hypothesis is that K sets will be obtained, with each one representing a different type of day.

2.4 Kolmogorov-Smirnov Two Sample Test

We propose to use the Kolmogorov-Smirnov two sample test to analyze whether two c.p.d.f.'s are homogeneously equal. This test will be used to check whether each c.p.d.f.'s of each group obtained with k-means is equal to the centroid of its group. The final number of groups used will be selected according to the results of this test. The "test of homogeneity between two samples" is defined as follows. Let the cumulative probability distribution function (c.p.d.f.) of X as $F_X(\cdot)$ and the c.p.d.f. of Y as $F_Y(\cdot)$, i.e. $F_X(t) = \Pr(X \leq t)$, $F_Y(t) = \Pr(Y \leq t)$, according to 2.2. Both $F_X(\cdot)$ and $F_Y(\cdot)$ are assumed to be continuous. Suppose that we want to test the null hypothesis

$$H_0 : F_X(\cdot) = F_Y(\cdot),$$

versus the general alternative hypothesis

$$H_a : F_X(\cdot) \neq F_Y(\cdot),$$

making no parametric assumption about the shape of these c.p.d.f.'s.

The test can be performed using the Kolmogorov-Smirnov statistic that compares the empirical c.p.d.f.'s obtained with each sample.

Specifically, if for any real number t we define $\hat{F}_X(t) \equiv n^{-1} \sum_{i=1}^{n} \mathbf{I}(X_i \leq t)$ and $\hat{F}_Y(t) \equiv m^{-1} \sum_{i=1}^{m} \mathbf{I}(Y_i \leq t)$, where $\mathbf{I}(A)$ is the indicator function of event A, which takes the value 1 if A is true or 0 otherwise, then the Kolmogorov-Smirnov statistic is

$$D_{n,m} \equiv \left(\frac{nm}{n+m} \right)^{1/2} \sup_{t \in \mathbb{R}} \left| \hat{F}_X(t) - \hat{F}_Y(t) \right|.$$

The null hypothesis is rejected with significance level α if $D_{n,m} > c_\alpha$, where c_α is a critical value that only depends on α (for details, see e.g. [27]).

2.5 Metrics for Evaluating the Proposed Methodology

The metric for homogeneity in clusters obtained with k-means (Kolmogorov-Smirnov statitistic) is labeled in the direct value comparison category [28] which tests whether the model output shows similar characteristics as a whole to the set of comparison data, but does not directly compare observed and modeled data points.

The metrics for evaluating the hourly estimated modeled profiles are:

- The standard deviation of daily profile hourly values for each cluster.
- The difference between the total hourly solar radiation estimated, $\hat{G}_h$, and the total hourly solar radiation received for the whole period of data normalized to the total hourly solar radiation received, G_h, according to the expression:

$$Error_{rad} = \frac{\sum_{i=1}^{m} |G_{h,i} - \hat{G}_{h,i}|}{\sum_{i=1}^{m} G_{h,i}} 100(\%) \tag{5}$$

The error in the estimation of total solar radiation was calculated as this information is very useful for solar plants connected to the grid. The data used is the difference between the values of total predicted radiation (directly used to predict the energy produced) and the total actual radiation received, because of the penalization applied for this difference.

3 Proposed Model for Simulating Hourly Profiles of Solar Radiation

We propose to use the k-means and Kolmogorov-Smirnov two sample test to cluster the observations and establish the number of different solar radiation hourly profiles. The hourly profile type of each cluster is obtained as the hourly mean values estimated using all the observations in the cluster.

The proposed model for simulating solar radiation hourly profiles uses the daily clearness index value and the solar radiation hourly profiles obtained for each cluster as inputs. The daily clearness index can be estimated using different previously proposed models, such as [8], [9].

The procedure for forecasting hourly global radiation is described in 1.

Input : K_d Daily clearness index; Clearness index hourly profiles.
Using K_d select the cluster. Estimate solar global radiation hourly values using the clearness index hourly profile for the selected cluster and the extraterrestrial solar global radiation hourly values (Eq.2).
Output: Hourly solar radiation values for day d

Algorithm 1. Procedure for obtaining solar global radiation hourly values

4 Data

Data used were recorded at the Photovoltaic Systems Laboratory of the University of Malaga (latitude is 36º42'54" N, Longitude 4º28'39 W, 45 meters elevation). Available data ranges from 2010/11/1 to 2012/10/31 in order to have two complete years.

The available data are recorded every minute, in total 1440 records per day. First, data are preprocessed to remove undesired values, such as night hour values or out of range values, measurements error, etc. Sunset and sunrise angles were used remove the night hours. Moreover, measurements corresponding to sun elevation angles less than $5º$ were removed due to observed distortions in these measurements (values greater that solar extraterrestrial radiation or values in the range of the error of the measurement device). Extraterrestrial solar horizontal radiation and clearness index are calculated and stored within the whole data set.

5   Results

Clearness index values were estimated for each of the measured solar global radiation values using Eq. 1. After calculating these values we calculated the cumulative probability distribution function for the values of each day where a vector of dimension 100 was obtained as the accuracy used for estimating clearness index is 0.01. These vectors are the inputs for k-means clustering. The goal of using a clustering method is to cluster functions in groups with similar c.p.d.f's. The number of clusters need to be fixed in advance in order to use k-means. We checked four different number of clusters, from 4 to 7, taking into account the expected different c.p.d.f.'s. For each execution of k-means we obtained the centroid of each cluster. Using these centroids we checked the equality of c.p.d.f's of the cluster in each test using the Kolmogorov-Smirnov two sample test. Table 1 shows the results obtained for the different numbers of checked clusters.

Table 1. Results of Kolmogorov-Sminorv two sample test for different number of clusters

Number of clusters	c.p.d.f's for which $D_{n,m} > c_\alpha$	% c.p.d.f.'s	% of clusters $D_{n,m} > c_\alpha$
4	29	4.4	100
5	26	3.9	80
6	21	3.1	33
7	19	2.9	29

Taking these results into account, we selected a total of 6 clusters instead of 7 clusters as only 2 of 6 clusters in both cases have some c.p.d.f.'s (approximately 3%) that are significantly different from theirs centroids and therefore 6 clusters are enough to capture the different c.p.d.f.'s observed. The days included in each cluster are from different months in all cases which means that the clustering method allows us to capture the different c.p.d.f.'s observed along the year and not the season of the year.

After clustering all the observations, the relationship between the daily clearness index corresponding to each c.p.d.f. and its cluster was analyzed. Figure 1 shows these values.

As can be observed, the daily clearness index is related to the cluster to which the c.p.d.f. for the day in question belongs. These results can be used to decide the cluster to which one day belongs. We have checked that for c.p.d.f.'s that can belong to two different clusters (see Fig.1) the $D_{n,m}$ between these each c.p.d.f. and the centroid of each possible cluster is always lower than the critical value

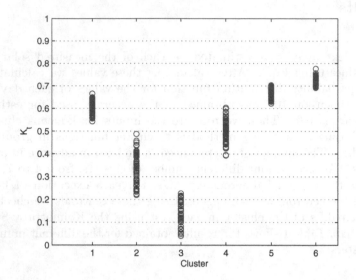

Fig. 1. Daily clearness index, K_t, vs cluster to which the day belongs

c_α. We propose to assign the cluster depending on the daily clearness index value using the expression 6.

$$Number\ of\ cluster = \begin{cases} 3 & \text{if } K_d \leq 0.22 \\ 2 & \text{if } 0.22 < K_d \leq 0.42 \\ 4 & \text{if } 0.42 < K_d \leq 0.55 \\ 1 & \text{if } 0.55 < K_d \leq 0.62 \\ 5 & \text{if } 0.62 < K_d \leq 0.7 \\ 6 & \text{if } K_d > 0.7 \end{cases} \qquad (6)$$

The observed relationship between daily clearness index and cluster suggests that k-means could produce clusters with days with a similar hourly solar radiation profile. Moreover, the relationship between the solar radiation distribution during a day and the daily clearness index value has been pointed out in [22]. However, solely one hourly profile cannot be used for all days due to the observed differences in different clusters. Using these two facts, we propose to use all the days of each cluster to estimate the solar radiation hourly profile for that cluster. Therefore, the hourly clearness index mean value and its standard deviation have been calculated for each cluster. Figures 2, 3 and 4 show these values. As can be observed, the standard deviation for most clusters and hours is not large and the results show that these values change with solar time, particularly for the hours at the start and end of the day. These results agree with those previously obtained in [10]. Moreover, these values decrease significantly for the hours with more radiation (central hours of day). This fact is the reason for our proposing the use of these mean values as the hourly profile model for each cluster.

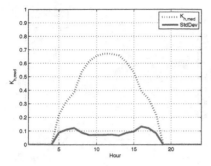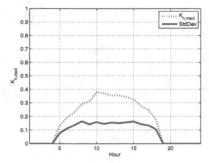

Fig. 2. Mean hourly clearness index and hourly standard deviation for clusters 1 (left) and 2 (right)

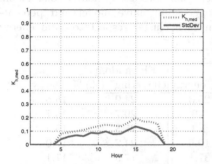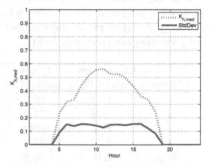

Fig. 3. Mean hourly clearness index and hourly standard deviation for clusters 3 (left) and 4 (right)

Following the procedure described in 1, the estimated hourly radiation values for all the recorded data were estimated. Eq.5 was used to estimate the energy error for each cluster to check the accuracy of these predictions. A naive persistent model that assumes that the hourly profile for a day is the same as the profile for the previous day was also used in order to evaluate whether the proposed model improves this naive model. Table 2 shows the obtained results.

As can be observed, the proposed method is able to estimate the daily profiles of hourly global radiation with an error lower or equal to 5% for the 57% of energy received; this total increases to 84% with an error less than 11%. The highest error occurs for only 1.3% of energy received. The total error for all the clusters is 10.5% that is less than both naive model errors, 20.6%, and errors reported in previous works that range between 20 and 40%, as set out in the Introduction section. These results indicate that it is possible to use the obtained daily profiles of hourly solar radiation distribution to forecast the hourly values of this variable.

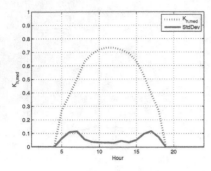 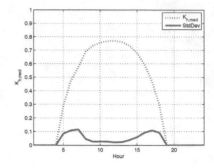

Fig. 4. Mean hourly clearness index and hourly standard deviation for clusters 5 (left) and 6 (right)

Table 2. Energy error (EE)(%) for each cluster when forecasting hourly solar global radiation with proposed model and with a persistent naive model and percentage of energy received in all days included in each cluster respect to total energy received.(*clusters are randomly built)

Cluster	EE proposed model (%)	EE naive model (%)	% energy	EE random clustering* (%)	% energy
1	10.5	16.5	26.3	24.9	15.5
2	36.8	71.0	4.9	29.7	15.7
3	49.1	193.8	1.3	25.3	15.4
4	25.0	35.9	10.5	24.2	17.0
5	5.0	10.8	37.1	17.9	17.1
6	4.0	12.3	19.9	22.6	19.3
All clusters	10.5	20.6		24.0	

6   Conclusions

We propose a methodology to characterize and model the solar global radiation hourly profiles observed for different types of days. The cumulative probability distribution functions of data and k-means clustering data mining technique were used for these tasks. Moreover, the Kolmogorov-Smirnov two sample test was used for analyzing the c.p.d.f.'s of each clusters. The 97% of c.p.d.f.'s are equal to the centroid of its cluster.

The observed relationship between daily clearness index and cluster means that we can state a one hourly radiation profile can be defined for each cluster.

The proposed method has been checked for data recorded in Malaga. The energy error is lower or equal to 5% for 57% of the energy received and lower than 11% for 84% of the energy. The obtained results show that hourly solar global radiation for a day can be forecasted with an energy error around 10% which is a significant improvement over previously reported errors that range from 20% to 40%.

As future research, the authors will conduct new experiments with data from different locations to prove the validity and universality of the proposed model.

Acknowledgments. This research has been supported by the Spanish Consejería de Economía, Innovación y Ciencia of the Junta de Andalucía under projects P10-TIC-6441 and P11-RNM-07115.

References

1. Luque, A., Hegedus, S.: Handbook of photovoltaic science and engineering. John Wiley & Sons Ltd., Berlin (2002)
2. Chang, T.P.: Output energy of a photovoltaic module mounted on a single-axis tracking system. Applied Energy 86, 2071–2078 (2009)
3. Box, G.E.P., Jenkins, G.M.: Time Series Analysis forecasting and control. Prentice Hall (1976)
4. De Gooijer, J.G., Hyndman, R.J.: 25 years of iif time series forecasting: A selective review. Monash Econometrics and Business Statistics Working Papers 12/05, Monash University, Department of Econometrics and Business Statistics (2005)
5. Brockwell, P.J., Davis, R.A.: Introduction to Time Series and Forecasting. Springer Texts in Statistics (2002)
6. Brinkworth, B.J.: Autocorrelation and stochastic modelling of insolation sequences. Solar Energy 19, 343–347 (1997)
7. Bartoli, B., Coluaai, B., Cuomo, V., Francesca, M., Serio, C.: Autocorrelation of daily global solar radiation. Il nuovo cimento 40, 113–122 (1983)
8. Aguiar, R., Collares-Pereira, M., Conde, J.P.: Simple procedure for generating sequences of daily radiation values using a library of markov transition matrices. Solar Energy 4(3), 269–279 (1988)
9. Graham, V.A., Hollands, K.G.T., Unny, T.E.A.: A time series model for kt with application to global synthetic weather generation. Solar Energy 40, 83–92 (1988)
10. Aguiar, R.J., Collares-Pereira, M.: TAG: A time dependent autoregressive gaussian model for generating synthetic hourly radiation. Solar Energy 49(3), 167–174 (1992)
11. Mora-López, L., Sidrach de Cardona, M.: Multiplicative arma models to generate hourly series of global irradiation. Solar Energy 63, 283–291 (1998)
12. Perez, R., et al.: Forecasting solar radiation preliminary evaluation of an approach based upon the national forecast database. Solar Energy 81(6), 809–812 (2007)
13. Mora-López, L., Mora, J., Sidrach de Cardona, M., Morales-Bueno, R.: Modelling time series of climatic parameters with probabilistic finite automata. Environmental modelling and software 20(6), 753–760 (2005)
14. Viorel, B.: Modeling Solar Radiation at the Earths Surface. Recent Advances. Springer (2008)
15. Guarnieri, R.A., Pereira, E.B., Chou, S.C.: Solar radiation forecast using articial neural networks in south brazil. In: 8 ICSHMO, INPE, Foz do Iguau, Brasil, April 24-28, pp. 1777–1785 (2008)
16. Heinemann, D., Lorenz, E., Girodo, M.: Forecasting of solar radiation, solar energy resource management for electricity generation from local level to global scale. Nova, Hauppauge (2005)

17. Mellit, A., Pavan, A.M.: A 24-h forecast of solar irradiance using artificial neural network: Application for performance prediction of a grid-connected {PV} plant at trieste, italy. Solar Energy 84(5), 807–821 (2010)

18. Mora-López, L., Martínez-Marchena, I., Piliougine, M., Sidrach-de-Cardona, M.: Binding statistical and machine learning models for short-term forecasting of global solar radiation. In: Gama, J., Bradley, E., Hollmén, J. (eds.) IDA 2011. LNCS, vol. 7014, pp. 294–305. Springer, Heidelberg (2011)

19. Koca, A., Oztop, H.F., Varol, Y., Koca, G.O.: Estimation of solar radiation using artificial neural networks with different input parameters for mediterranean region of anatolia in turkey. Expert Systems with Applications 38(7), 8756–8762 (2011)

20. Reikard, G.: Predicting solar radiation at high resolutions: A comparison of time series forecast. Solar Energy 83, 342–349 (2009)

21. Voyant, C., Paoli, C., Muselli, M., Nivet, M.-L.: Multi-horizon solar radiation forecasting for mediterranean locations using time series models. Renewable and Sustainable Energy Reviews 28, 44–52 (2013)

22. Bendt, P., Collares-Pereira, M., Rabl, A.: The frequency distribution of daily insolation values. Solar Energy 27, 1–5 (1981)

23. Iqbal, M.: An introduction to solar radiation. Academic Press Inc., New York (1983)

24. Jain, A., Murty, M., Flynn, P.: Data clustering: A review. ACM Computing Surveys 31(3), 264–323 (1999)

25. Duda, R., Hart, P., Stork, D.: Pattern classification. John Wiley & Sons (2001)

26. Hastie, T., Tibshirani, R., Friedman, J.: The elements of statistical learning: Data mining, inference and prediction. Springer (2001)

27. Rohatgi, V.K., Saleh, A.K.M.E.: An Introduction to Probability and Statistics, 2nd edn. Wiley-Interscience (2001)

28. Bennett, N.D., Croke, B.F.W., Guariso, G., Guillaume, J.H.A., Hamilton, S.H., Jakeman, A.J., Marsili-Libelli, S., Newham, L.T.H., Norton, J.P., Perrin, C., Pierce, S.A., Robson, B., Seppelt, R., Voinov, A.A., Fath, B.D., Andreassian, V.: Characterising performance of environmental models. Environmental Modelling & Software 40, 1–20 (2013)

Identification of Bilingual Segments
for Translation Generation

Kavitha Karimbi Mahesh[1,3], Luís Gomes[1,2], and José Gabriel P. Lopes[1,2]

[1] CITI (NOVA LINCS), Faculdade de Ciências e Tecnologia,
Universidade Nova de Lisboa, Quinta da Torre. 2829-516 Caparica, Portugal
k.mahesh@fct.unl.pt   luismsgomes@gmail.com   gpl@fct.unl.pt
[2] ISTRION BOX-Translation & Revision, Lda., Parkurbis, Covilhã 6200-865
Portugal
[3] Department of Computer Applications, St. Joseph Engineering College
Vamanjoor, Mangalore, 575 028, India
kavitham@sjec.ac.in

Abstract. We present an approach that uses known translation forms
in a validated bilingual lexicon and identifies bilingual stem and suffix
segments. By applying the longest sequence common to pair of orthographically similar translations we initially induce the bilingual suffix
transformations (replacement rules). Redundant analyses are discarded
by examining the distribution of stem pairs and associated transformations. Set of bilingual suffixes conflating various translation forms are
grouped. Stem pairs sharing similar transformations are subsequently
clustered which serves as a basis for the generative approach. The pri-
mary motivation behind this work is to eventually improve the lexicon
coverage by utilising the correct bilingual entries in suggesting translations for OOV words. In the preliminary results, we report generation
results, wherein, 90% of the generated translations are correct. This was
achieved when both the bilingual segments (bilingual stem and bilingual
suffix) in the bilingual pair being analysed are known to have occurred
in the training data set.

Keywords: Translation lexicon coverage, Cluster analysis, Bilingual morphology, Translation generation.

1 Introduction

Given a bilingual lexicon of translations, the approach returns the sets of probable bilingual stems and bilingual suffixes along with their frequencies. Also,
the clusters representing a set of transformations (suffix replacement rules) and
the associated set of stem pairs sharing those transformations as observed in the
training set is returned. The main objective of the work is to suggest new translations through a generative approach by productively combining the bilingual
stems and suffixes belonging to the same cluster. Further, by identifying bilingual
morphemes in new translations, probable translations could be suggested.

H. Blockeel et al. (Eds.): IDA 2014, LNCS 8819, pp. 167–178, 2014.

The bilingual translation lexicon used in this study is acquired from an aligned parallel corpora[1][1] using the extraction methods such as [2], [3], [4], [5]. First, a bilingual lexicon is used to align parallel texts [1] and then extract new[2] term-pairs from the aligned texts [2]. In order to continually improve the alignments, the extracted translations are validated and the correct ones are added to the bilingual lexicon, marked as 'accepted' which are used in subsequent alignment and translation extractions. Incorrect ones are as well added to the lexicon marked as 'rejected'. Thus, a cycle of iterations over parallel text alignment, term translation extraction and validation is carried out. The English (EN) - Portuguese (PT) unigram translations used in this work for learning the translation segments is acquired with the aforementioned approach, and are affirmed as 'accepted' by the human validators. The verification step is crucial for keeping alignment and extraction errors from being propagated back into subsequent alignment and extraction iterations, which would cause the system to degenerate.

The lexicon thus acquired is not complete as it does not contain all possible translation pairs. Table 1 shows the accepted translations extracted for each of the *word forms* corresponding to *ensure*. The translations seem exhaustive with respect to the first 2 columns. However, certain missing translation forms are, *garantam* for *ensure*, *garantiu*, *garantiram*, *garantidos*, *permitidas*, *permitido*, *permitidos*, *permitidas*, *permitidas permitiu*, *permitiram* for *ensured*, which can also be considered as possible translations. All the translation forms in the 3rd column are missing. This is because, the extraction techniques cannot handle what is not in a parallel corpora, unless we care about automatically learning and generalising word and multi-word structures. Moreover, they are not able to extract everything.

Term (EN)	Term (PT)			
ensure	assegurar	zelar	garantir, garantem	permitir
	asseguram			permitam
	assegurem			permitem
ensures	assegura		garanta	permite
	assegure		garante	permita
ensured	asseguradas,		garantidas, garantido	
	assegurados,			
	assegurado,			
	assegurou			
	asseguraram		garantidos	
ensuring	assegurando		garantindo	permitindo

Table 1. Translation Patterns in the Extracted Lexicon

On a whole, although it is evident that the existing lexicon is reasonably extensive, acknowledging its incompleteness with respect to the vocabularies in

---

[1] A collection of pair of texts that are translations of each other.

[2] By *new* we mean that they were not in the bilingual lexicon that was used for aligning the parallel texts.

either or both languages involved, accommodating most of the possible patterns demand learning the translation structure. In this paper, we only focus on generating word-to-word translations, by treating a translation lexicon itself as a parallel corpus. Having a hugely high degree of certainty associated with each bilingual pair asserting its correctness, we use it to learn and generalise translation patterns, infer new patterns and hence generate those OOV translation pairs that were not explicitly present in the training corpus used for the lexicon acquisition.

From Table 1, we observe that each of the terms in EN share the same set of suffixes -e, -es, - ed and -ing and stem ensur. Including their corresponding translations, we see that a term in EN ending with -ed is translated to a term ending with -adas, - ados, -ado, -ou, -aram. Likewise, the translations for declared[3] follow similar pattern and the translations end with -ada, -adas, -ado, -ados, -aram, -ava, -ou. Equivalently they share the stems assegur and declar[4]. Knowing that ensured and declared share suffixes -ed and by considering the intersection of suffixes corresponding to their translation endings in PT, we may say that a term with suffix -ed in EN might be translated to terms with suffixes -adas, -ados, -ado, -ados, -aram, -ou in PT. This simple knowledge allows us to generate new translation pairs based on the similarities observed from the known examples. But, in 4th column of Table 1, the translation endings such as, -ido, -idas, -idos corresponding to -ed are related and may be used to generate permitido, permitidas, permitidos thereby partially completing the translations for ensured in column 5^5.

Moreover, clusters of suffixes in English may translate as different clusters of suffixes in Portuguese and hence it is necessary to identify, for a specific case the best selection. Referring Table 1, we may see that suffix -e in English maps to -am, -em, -ar (inflections for root assegur), -ir, -em (inflections for root garant) and -am, -em, -ir (inflections for root permit). Suffixes -am, -em are shared by all the three verb forms, while the suffix -ar discriminates the verbs in -ar group from the verbs in -ir group. Equivalent phenomena occurs for the other suffixes. Generally, we may see that verbs belonging to -e, -es, - ed, -ing in English could be mapped to verbs belonging to one of the three Portuguese conjugation classes -ar, -ir or -er with the classes being discriminated by the ending of their infinitive forms. It is to be noted that, by chance, the suffix -ou corresponding to -ed in column 1 of table 1, is also a discriminator for Portuguese verbs belonging to -ar group.

2 Related Work

The fact that 'words consist of high-frequency strings (affixes) attached to low-frequency strings (stems)' has motivated several researches ranging from text

---

[3] Declared ⇔ declarada, declaradas, declarado, declarados, declararam, declarava, declarou.

[4] Stem can be determined as longest common sequence of characters.

[5] This follows as verbs garantir and permitir belong to the 2nd verb inflection class, unlike assegur which belongs to the 1st verb inflection class.

analysis for acquisition of morphology, to learning suffixes and suffixation operations for improving word coverage and for allowing word generation. Certain approaches are partially supervised [6]. Unsupervised, Minimum Description Length based models such as [7], [8] focus on finding a better compressed representation for lexicon of words. Other unsupervised approaches address language specific issues such as data and resource sparseness [9], agglutination [10]. In each of these works, morphological segmentation is induced considering monolingual data. Lexical inference or morphological processing techniques have been established to be interesting in suggesting translations for OOV words that are variations of known forms. Below, we discuss a few of them.

Predicting translation for unknown words based on inductive learning mechanism is discussed in [11]. Common and different parts of strings between known words and their translations represent the example strings, referred as Piece of Word (PW) and Pair of Piece of Word (PPW). The bilingual pairs of these extracted example strings maintained as a Pair of Piece of Word (PPW) dictionary form the basis of the prediction process.

In [12], morphological processing is used to learn translations for unknown German compounds from the translation of their parts. The splitting options are guided by parallel texts in such a way that all the parts[6] should have occurred as whole word translations in the training corpus. The guidance from parallel corpus relies on two[7] translation lexicons with correspondences learnt using toolkit Giza. The two lexicons are jointly used to guide the splitting process. The approach records 99.1% accuracy with an improvement of 0.039 BLEU in German-English noun phrase translation task.

The hierarchical back-off model discussed in [13] for translating unseen forms stand out to act on highly inflectional languages such as German and Finnish, particularly under the scenario of limited training data. Morphological decompositions mainly include alternative layers of *stemming* and *compound splitting*, requiring that "a more specific form (a form closer to the full word form) is chosen before a more general form (a form that has undergone morphological processing)". Unlike the splitting techniques reported in [12], the method involves investigating all possible ways of segmentation with the only constraint that each part has a minimum length of 3 characters. As in [12], acceptance of the segmentation is subjected to the appearance of subparts as individual items in the training data vocabulary. The approach relies on translation probabilities derived from stemmed or split versions of the word in its phrasal context. Experiments with varied amount of training data reveal its appropriateness under limited training data conditions and adaptability to highly inflected languages.

Gispert et al., [14] show that translations for unseen verb forms can be generated by generalising them using verb forms seen in training data. Verbs are

---

[6] To avoid prefixes and suffixes from splitting off, the parts are restricted only to content words, thus excluding the prepositions or determiners.

[7] The first lexicon is learnt from original versions of parallel texts and the second from the parallel corpus with split German and unchanged English text versions in order to learn specific translations for compound parts.

identified using rules incorporating word forms, POS-tags and word lemmas and are classified to the lemma of their head verb, such that they belong to only one class with such a classification done for each language separately. To translate an unseen verb form, the verb is classified into the lemma of its head word and all the tuples representing translation of that class of verbs (in training data) are identified. New target verb form is generated by replacing the personal pronoun in the seen form with the personal pronoun in the expression to be translated. The suggested translation is weighed based on the frequency of its occurrence in the training data. In case of any ambiguity in generalisation of verb forms, the approach over generates all possible forms, leaving the target language model to decide on the best translation alternative.

The need for dealing language-specific problems while translating from English to a morphologically rich languages by identifying morphological relationships that are not captured by current SMT models, the possibilities of handling these independently from translation model by morphology derivation are discussed in [15]. Proper derivations into the text are introduced by simplifying morphological information (or parts of it), followed by a morphology generation by means of a classification model which makes use of a set of relevant features for each simplified morphology word and its context. The study reveals that the main source of potential improvement lies in verb form morphology as this morphological category is seen to exhibit more derivation in Romance languages.

Discovery of abstract morphemes by simultaneous morphology learning for multiple languages is discussed in [16]. The discriminative log-linear model discussed in [17] uses overlapping contextual features to boost the segmentation decisions. Morphological analysis and generation in [18] is achieved through the suffix stripping and joining method. The approach relies on a bilingual dictionary consisting of the root/stem of the words with its grammatical category.

In this paper, we focus on learning the structure of translations, treating the bilingual lexicon as a parallel resource. We take advantage of the bilingual data to deal with the ambiguities and complexities in decompositions by focusing on the frequent forms occurring in translations rather than words in one language. As in [11], the approach identifies common and different bilingual segments occurring in several translation examples and employs them in generating new translations. We restrict the bilingual segments only to two parts, interpreting the first part as the bilingual stem and the second part as bilingual suffix. A pair of bilingual suffixes attached to the same bilingual stem indicate the suffix replacement option and hence motivates translation generation. To enable generalisation, clusters of bilingual stems sharing same transformations are identified.

The remaining sections are organised as follows. The learning phase for identifying bilingual segments is discussed in Section 3. The approach used in generating new translations is discussed in Section 4. Results are presented and analysed in section 5. Conclusions drawn based on the results and the scope for future work are discussed in section 6.

3   Learning the Bilingual Segments

Given a lexicon of bilingual entries (word-to-word translations) extracted from the aligned parallel corpora[8], we first look for orthographically similar translations. Translations are considered similar if they begin with the same substring[9].

Based on the longest sequence common to pair of similar translations, we note the bilingual stems and the pair of bilingual suffixes attached it. For example, with the translation forms *ensuring* ⇔ *assegurando* and *ensured* ⇔ *assegurou*, we obtain the bilingual stem *ensur* ⇔ *assegur* with a pair of bilingual suffixes, (*ing* ⇔ *ando*, *ed* ⇔ *ou*).

To determine their validity, the induced bilingual segments are analysed with respect to their occurrences as bilingual stems and bilingual suffixes. We require that the following conditions are satisfied with respect to the bilingual segments:

- Each candidate bilingual stem should attach to at least two unique morphological extensions (pair of bilingual suffixes).
  For the example discussed above, the bilingual stem (*ensur* ⇔ *assegur*) is retained if it appears as a stem with at least another pair of bilingual suffixes, say, (*ing* ⇔ *ando*, *ed* ⇔ *ou*) induced from of the translations *ensuring* ⇔ *assegurando* and *ensured* ⇔ *assegurou* with the decomposition as below:
  [*ensur* ⇔ *assegur*] + [*ing* ⇔ *ando*] and [*ensur* ⇔ *assegur*] + [*ed* ⇔ *ou*],
- Similarly, each pair of bilingual suffixes should have been attached to at least two unique bilingual stems.
  For example, (*ing* ⇔ *ando*, *ed* ⇔ *ou*) is considered, if it appears with another bilingual stem such as (*declar* ⇔ *declar*) induced from translations *declaring* ⇔ *declarando* and *declared* ⇔ *declarou* with the following decompositions: [*declar* ⇔ *declar*] + [*ing* ⇔ *ando*] and [*declar* ⇔ *declar*] + [*ed* ⇔ *ou*].

3.1   Filtering

For each of the bilingual stems obtained, we gather all the bilingual suffixes associated with that bilingual stem. For example, the candidate bilingual suffixes that associate with the candidate bilingual stems ('ensur', 'assegur') obtained from the segmentation of the bilingual pairs *ensure* ⇔ *assegurem*, *ensured* ⇔ *assegurou* and *ensuring* ⇔ *assegurando* would be as follows:
('ensur', 'assegur'):('e', 'em'), ('ing', 'ando'), ('ed', 'ou').

Each such grouping indicate the suffix pair replacement rules that enable one translation form to be obtained using the other. For instance, from the above grouping, it follows that replacing the suffix *e* with *ed* and the suffix *em* with *ou* in the bilingual pair *ensure* ⇔ *assegurem*, yields *ensured* ⇔ *assegurou*.

---

[8] DGT-TM - https://open-data.europa.eu/en/data/dataset/dgt-translation-memory
  Europarl - http://www.statmt.org/europarl/
  OPUS (EUconst, EMEA) - http://opus.lingfil.uu.se/
[9] same with respect to the first and the second language, where the minimum substring length is 3 characters

A few among the identified groups are redundant. The bilingual stems and the associated bilingual suffixes listed below exemplify such redundancies.
('ensur', 'assegur') : ('e', 'ar'), ('ed', 'ado'), ('ed', 'ados'), ('ed', 'ada'), ('ed', 'adas'), ('ing', 'ando'), ('es', 'e'), ('es', 'a'), ('e', 'am'), ('e', 'em'), ('ed', 'aram'), ('ed', 'ou').
('ensure', 'assegur') : ('', 'ar'), ('d', 'ado'), ('d', 'ados'), ('d', 'ada'), ('d', 'adas'), ('s', 'e'), ('s', 'a'), ('', 'am'), ('', 'em'), ('d', 'aram')
('ensur', 'assegura') : ('e', 'r'), ('ed', 'do'), ('ed', 'dos'), ('ed', 'da'), ('ed', 'das'), ('ing', 'ndo'), ('es', ''), ('e', 'm'), ('ed', 'ram')
('ensure', 'assegura') : ('', 'r'), ('d', 'do'), ('d', 'dos'), ('d', 'da'), ('d', 'das'), ('s', ''), ('', 'm'), ('d', 'ram')

Table 2. Occurrence frequencies of induced bilingual stems with respect to the translations in the bilingual lexicon

Stem pair	Frequency	Stem pair	Frequency
accord, acord	3	'abandon', 'abandon'	17
accord, acorde	2	'abandon', 'abandona'	2

We consider discarding the redundant groups, where the bilingual stems vary by single character in the boundary. We retain the bilingual stems that allow higher number of transformations. This is done by counting the number of unique translations in the lexicon that begin with each of the bilingual stems. To handle multiple such instances we prefer shorter bilingual stems over longer pairs. In the examples listed above, the first group is retained.

3.2 Clustering

A set of bilingual stems that share same suffix transformations form a cluster. The bilingual stems identified in the previous step characterised by suffix pairs (features) are clustered using the clustering tool, CLUTO[10]. The toolkit provides three different classes of clustering algorithms such as, partition, agglomerative and graph-partitioning, to enable the clustering of low and high dimensional data sets. The partition and agglomerative clustering, is driven by total of seven different criterion functions that are described and analysed in [19].

In the experiments presented here, partition approach was adapted for clustering. To prepare the data for clustering, the *doc2mat*[11] tool is used, which provides the necessary conversion of data into matrix form. We experimented with 10, 15, 20, 50 and 100 way clustering and the best results were obtained with 50 clusters. The clustering results are further analysed manually to remove outliers (bilingual suffixes) from each cluster and to identify the sub-clusters from among the clustered results. Next, we generate new translations by direct

---

[10] http://glaros.dtc.umn.edu/gkhome/views/cluto
[11] http://glaros.dtc.umn.edu/gkhome/files/fs/sw/cluto/doc2mat.html

concatenation of stem pair and suffix pair belonging to a same cluster. These newly generated pairs are validated manually and are included as training data for the subsequent iteration.

4 Generating New Translations

The output of the learning phase includes known list of bilingual stems, bilingual suffixes along with their observed frequencies in the training data set. We further have information about which set of bilingual suffixes attach to which set of bilingual stems. The underlying approach for suggesting new translations relies on these clusters of bilingual stems and bilingual suffixes identified in the learning phase. Now, for each of the new bilingual pairs to be analysed, we consider all possible splits restricting the first part (bilingual stem) to a minimum of 3 characters. We examine the following possibilities in generating new translations:

- If both the bilingual stems and the bilingual suffixes are known, we check whether they belong to the same cluster. If so, each of the new translations are suggested by concatenating the stem pair with the associated suffix pairs.
- If only the bilingual suffixes are known, we generate by replacing the identified bilingual suffix with other bilingual suffixes that have been recorded to co-occur with the identified bilingual suffix. To avoid over-generations from multiple matches, we restrict to generations where the longer bilingual suffixes are preferred over shorter matches or the bilingual suffixes with the higher frequencies are preferred over bilingual suffixes with lower frequencies.
- If only the stem pairs are known, we generate by concatenating the identified stem pair with all the suffix pairs that attach to the identified stem pair.

5 Results and Discussion

5.1 Clustering Results

Table 3 provides an overview of the data sets used in training and the corresponding clustering statistics. A list of the most common bilingual suffixes identified for the currently existing lexicon entries is shown in Table 4.

Table 3. Overview of the training data and generation statistics with different training sets

Training Data	Unique Bilingual Stems	Unique Bilingual Suffixes	Generated pairs	Correct Generations	Incorrect Generations
36K	6,644	224	4,279	3,862	306
210K	24,223	232	14,530	2,283/2,334	20/2,334

Table 4. Highly frequent Bilingual Suffixes identified for EN-PT bilingual bases with different training sets (Frequencies in the lexicon are considered)

Training Set1		Training Set2	
Suffix Pair	Frequency	Suffix Pair	Frequency
(", 'o')	4,644	(", 'o')	15,006
(", 'a')	2,866	(", 'a')	9,887
('e', 'o')	1,685	(", 'as')	5,840
(", 'os')	1,362	(", 'os')	5,697
(", 'as')	1,339	('ed', 'ado')	4,760
('e', 'a')	1,297	('ed', 'ados')	4,221
('ed', 'ado')	1,001	('ed', 'ada')	4,193
('ed', 'ada')	868	('e', 'o')	4,159
('ed', 'ados')	814	('ed', 'adas')	4,051

We present below, a few randomly chosen clusters along with the discriminating features (bilingual suffixes) and a few example bilingual stems under each of the verb, noun and adjective classes.

Verb-ar Cluster: ('e', 'ar'), ('e', 'arem'), ('e', 'am'), ('e', 'em'), ('es', 'e'), ('es', 'a'), ('ed', 'ada'), ('ed', 'adas'), ('ed', 'ado'), ('ed', 'ados'), ('ed', 'aram'), ('ed', 'ou'), ('ing', 'ando'), ('ing', 'ar')
Example Bilingual Stems: *toggl* ⇔ *comut, argu* ⇔ *afirm, shuffl* ⇔ *baralh*

Verb-er Cluster: (", 'er'), (", 'erem'), (", 'am'), (", 'em'), ('s', 'e'), ('s', 'a'), ('ed', 'ida'), ('ed', 'idas'), ('ed', 'ido'), ('ed', 'idos'), ('ed', 'eram'), ('ed', 'eu'), ('ing', 'endo'), ('ing', 'er')
Example Bilingual Stems: *spend* ⇔ *dispend, reply* ⇔ *respond, answer* ⇔ *respond*

Verb-ir Cluster: (", 'ir'), (", 'irem'), (", 'am'), (", 'em'), ('s', 'e'), ('s', 'a'),('ed', 'ida'), ('ed', 'idas'), ('ed', 'ido'), ('ed', 'idos'), ('ed', 'iram'), ('ed', 'iu'), ('ing', 'indo'), ('ing', 'ir')
Example Bilingual Stems: *expand* ⇔ *expand, acclaim* ⇔ *aplaud, reopen* ⇔ *reabr*

Adjective-ent Cluster: ('ent', 'ente'), ('ent', 'entes')
Example Bilingual Stems: *bival* ⇔ *bival, adjac* ⇔ *adjac, coher* ⇔ *consist, coher* ⇔ *coer, circumfer* ⇔ *circunfer*

Adjective-al Cluster: ('al', 'ais'), ('al', 'al')
Example Bilingual Stems: *categori* ⇔ *categori, cervic* ⇔ *cervic, coast* ⇔ *litor*

Noun-ence Cluster: ('ence', 'ência'), ('ences', ências')
Example Bilingual Stems: *compet* ⇔ *compet, recurr* ⇔ *reocorr, jurisprud* ⇔ *jurisprud, opul* ⇔ *opul*

Noun-ist Cluster: ('ist', 'ista'), ('ists', 'istas')
Example Bilingual Stems: *alchem* ⇔ *alquim, bapt* ⇔ *bapt, column* ⇔ *colun*

5.2   Generation Results

With a training data of approximately 210K bilingual pairs, about 15K new translations were generated. Among the 2,334 validated entries, 2283 were accepted, 27 were inadequate (*accept-*) indicating incomplete/inadequate translations and 20 were rejected (*reject*). Table 5 shows the statistics for the generated translations (correct, accepted) in the parallel corpora, where the co-occurrence frequency is less than 10. Among the generated entries, 9034 bilingual pairs did not occur in the parallel corpora even once. When both the bilingual stem and the bilingual suffix in the bilingual pair to be analysed are known, 90% of the generated translations were correct, with the first data set (Table 3).

Table 5. Co-occurrence frequency for the generated translations in the parallel corpora

Co-occurrence Frequency	# of generated bilingual pairs	Co-occurrence Frequency	# of generated bilingual pairs
9	45	4	148
8	62	3	207
7	64	2	324
6	80	1	489
5	102	-	-

5.3 Error Analysis

Analysing the newly generated translations, we observe that certain translations are incomplete (examples labelled as *accept-*), and some are incorrect (examples labelled as *reject*). Translation candidates such as *'intend ⇔ pretendem'* are inadequate as the correct translations require *'intend'* to be followed by *'to'*. Similarly, *'include'* should be translated either by *'contam-se'* or *'se contam'* and so *'include ⇔ contam'* is classified as *accept-*.

Other generated entries, such as, *'collector ⇔ coleccionadores'*, *'advisor ⇔ consultores'*, *'rector ⇔ reitores'*, *'elector ⇔ eleitores'* are instances wherein the noun acts as an adjective that is translated either by adding *'de'* before the plural noun translation in PT, eventually with an article after *'de'* as in *'de os'*. Again, the bilingual pair generated, *'wholesales ⇔ grossistas'* misses the noun *'vendas'* as in *'vendas grossistas'*. The English noun is compounded in this case.

Generation errors labelled as *'reject'* in Table 6, are a consequence of incorrect generalisations. Verbs in PT ending in *'uir'* form past participle forms adding *'u'*. *'wants'* is an irregular verb that is translated either by *'quer'* or *'queira'* or *'quizer'*.

6   Conclusion

In this paper, we have discussed an approach for identifying bilingual segments for translation generation. The motivation for the work reported in this paper is the fact that extraction techniques cannot handle what is not in a parallel corpora and they cannot extract everything. Above all, the way in which translations are extracted and evaluated does not guarantee that most of the possible

Table 6. Generated Translations

Accept-	Reject
languages ⇔ linguísticas	rights ⇔ adequados
instructor ⇔ instrutores	replaced ⇔ substituida / -idas / -idos / -ido
ambassador ⇔ embaixadores	several ⇔ vário
include ⇔ contam	wants ⇔ quere
emerged ⇔ resultados	electrical ⇔ electrica

translation pairs not found in parallel corpora might be automatically suggested for a translation engine or as bilingual entries.

90% of the generated translations were correct when both the stem and suffix pairs in the bilingual pair to be analysed are known. However, the approach fails to handle irregular forms and hence needs to be addressed in the future. While the current work focused on EN-PT, bilingual segments for other language pairs, added with limited training data conditions, needs to be further studied. Also, the improvement brought by the generated translations to the quality of alignment and extraction needs to be assessed.

Acknowledgements. K. M. Kavitha and Luís Gomes gratefully acknowledge the Research Fellowship by FCT/MCTES with Ref. nos., SFRH/BD/64371/2009 and SFRH/BD/65059/2009, respectively. The authors would like to acknowledge ISTRION project (Ref. PTDC/EIA-EIA/114521/2009) funded by FCT/MCTES that provided other means for the research carried out, thank CITI, FCT/UNL for providing partial financial assistance to participate in IDA 2014, and ISTRION BOX - Translation & Revision, Lda., for the data and valuable consultation received while preparing this manuscript.

References

1. Gomes, L., Pereira Lopes, J.G.: Parallel texts alignment. In: New Trends in Artificial Intelligence, 14th Portuguese Conference in Artificial Intelligence, EPIA 2009, pp. 513–524 (2009)
2. Aires, J., Pereira Lopes, J.G., Gomes, L.: Phrase translation extraction from aligned parallel corpora using suffix arrays and related structures. In: Progress in Artificial Intelligence, pp. 587–597 (2009)
3. Brown, P.F., Pietra, V.J.D., Pietra, S.A.D., Mercer, R.L.: The mathematics of statistical machine translation: Parameter estimation. Computational linguistics 19(2), 263–311 (1993)
4. Lardilleux, A., Lepage, Y.: Sampling-based multilingual alignment. In: Proceedings of Recent Advances in Natural Language Processing, pp. 214–218 (2009)
5. Gomes, L., Pereira Lopes, J.G.: Measuring spelling similarity for cognate identification. In: Antunes, L., Pinto, H.S. (eds.) EPIA 2011. LNCS, vol. 7026, pp. 624–633. Springer, Heidelberg (2011)
6. Déjean, H.: Morphemes as necessary concept for structures discovery from untagged corpora. In: Proceedings of the Joint Conferences on New Methods in Language Processing and Computational Natural Language Learning, pp. 295–298. ACL (1998)

7. Goldsmith, J.: Unsupervised learning of the morphology of a natural language. Computational linguistics 27(2), 153–198 (2001)
8. Creutz, M., Lagus, K.: Unsupervised discovery of morphemes. In: Proceedings of the ACL-02 Workshop on Morphological and Phonological Learning, vol. 6, pp. 21–30. ACL (2002)
9. Hammarström, H., Borin, L.: Unsupervised learning of morphology. Computational Linguistics 37(2), 309–350 (2011)
10. Monson, C., Carbonell, J., Lavie, A., Levin, L.: ParaMor and morpho challenge 2008. In: Peters, C., et al. (eds.) CLEF 2008. LNCS, vol. 5706, pp. 967–974. Springer, Heidelberg (2009)
11. Momouchi, H.S.K.A.Y., Tochinai, K.: Prediction method of word for translation of unknown word. In: Proceedings of the IASTED International Conference, Artificial Intelligence and Soft Computing, Banff, Canada, July 27-August 1, p. 228. Acta Pr. (1997)
12. Koehn, P., Knight, K.: Empirical methods for compound splitting. In: Proceedings of the Tenth Conference on European Chapter of the Association for Computational Linguistics, vol. 1, pp. 187–193 (2003)
13. Yang, M., Kirchhoff, K.: Phrase-based backoff models for machine translation of highly inflected languages. In: Proceedings of EACL, pp. 41–48 (2006)
14. de Gispert, A., Mariño, J.B., Crego, J.M.: Improving statistical machine translation by classifying and generalizing inflected verb forms. In: Proceedings of 9th European Conference on Speech Communication and Technology, Lisboa, Portugal, pp. 3193–3196 (2005)
15. de Gispert, A., Marino, J.B.: On the impact of morphology in english to spanish statistical mt. Speech Communication 50(11-12), 1034–1046 (2008)
16. Snyder, B., Barzilay, R.: Unsupervised multilingual learning for morphological segmentation, pp. 737–745. ACL (2008)
17. Poon, H., Cherry, C., Toutanova, K.: Unsupervised morphological segmentation with log-linear models. In: Proceedings of Human Language Technologies: The 2009 Annual Conference of the North American Chapter of the Association for Computational Linguistics, pp. 209–217 (2009)
18. Jisha, P.J., Rajeev, R.R.: Morphological analyser and morphological generator for malayalam-tamil machine translation. International Journal of Computer Applications 13(8), 15–18 (2011)
19. Zhao, Y., Karypis, G.: Evaluation of hierarchical clustering algorithms for document datasets. In: Proceedings of the Eleventh International Conference on Information and Knowledge Management, pp. 515–524. ACM (2002)

Model-Based Time Series Classification

Alexios Kotsifakos[1] and Panagiotis Papapetrou[2]

[1] Department of Computer Science and Engineering, University of Texas at Arlington, USA
[2] Department of Computer and Systems Sciences, Stockholm University, Sweden

Abstract. We propose MTSC, a filter-and-refine framework for time series Nearest Neighbor (NN) classification. Training time series belonging to certain classes are first modeled through Hidden Markov Models (HMMs). Given an unlabeled query, and at the filter step, we identify the top K models that have most likely produced the query. At the refine step, a distance measure is applied between the query and all training time series of the top K models. The query is then assigned with the class of the NN. In our experiments, we first evaluated the NN classification error rate of HMMs compared to three state-of-the-art distance measures on 45 time series datasets of the UCR archive, and showed that modeling time series with HMMs achieves lower error rates in 30 datasets and equal error rates in 4. Secondly, we compared MTSC with Cross Validation defined over the three measures on 33 datasets, and we observed that MTSC is at least as good as the competitor method in 23 datasets, while achieving competitive speedups, showing its effectiveness and efficiency.

1 Introduction

Time series data have become ubiquitous during the last decades. Sequences of numerical measurements are produced at regular or irregular time intervals in vast amounts in almost every application domain, such as stock markets, medicine, and sensor networks. Large databases of time series can be exploited so as to extract knowledge on what has happened in the past or to recognize what is happening in the present. More specifically, the task at hand is *classification*. Given an *unlabeled* time series, i.e., of which the *category/class* is *not* known, we wish to assign to it the *most* appropriate class, which corresponds to the class of the *Nearest Neighbor (NN)* time series. Consequently, there is a need for *searching* the time series database.

One way of implementing such search is, given a distance measure, to perform *whole sequence matching* between each time series in the database and the query, and finally select the closest time series, i.e., the one with the smallest distance to the query. Thus, it can be easily understood that the selection of the distance measure to be used for comparing time series is critical, as it essentially decides whether a time series is a good match for the query or not, influencing the NN classification accuracy (percentage of time series correctly classified).

Several distance measures have been proposed, which are often computed in a dynamic programming (DP) manner [6]. The most widely known and used measure is Dynamic Time Warping (DTW) [22]. Many variants of DTW have also been proposed, such as cDTW [30], EDR [8], and ERP [9]. All of these measures have the attractive

H. Blockeel et al. (Eds.): IDA 2014, LNCS 8819, pp. 179–191, 2014.

characteristic that they are robust to misalignments along the temporal axis. Moreover, they provide very good classification accuracy results [34].

Searching large time series databases with any of these DP-based distance measures can be computationally expensive due to their quadratic complexity with respect to the time series length, though speedups can be achieved for cDTW by applying a lower-bounding technique [16]. An alternative approach would be to first represent each class of the time series database with a *model*, such as a *Hidden Markov Model* (*HMM*) [5], and then perform searching based on the constructed models. HMMs are widely known and have been used in a variety of domains, such as speech recognition [29], and music retrieval [28]. In this paper, we deal with both effectiveness (accuracy) and efficiency (runtime) and we propose a novel approach, named MTSC (shorthand for Model-based Time Series Classification). Given sets of time series of certain classes, MTSC first models their underlying structure through the training of one HMM per class. An HMM is capable of identifying the relationships between the observations within the time series [18]. At runtime, given a query time series of unknown class, MTSC finds the top K models that have most likely produced the query. Then, it refines the search by applying an appropriate distance measure between the query and all the training time series that compose the K selected models. What remains to be answered is what distance measure to use during the refine step. Intuitively, given a collection of distance measures, we can choose the one providing the highest classification accuracy on the training set.

The **main contributions** of this paper include: 1) A novel way of representing time series of a specific class via an HMM, and a comparative evaluation of this representation against three distance measures (DTW, ERP, and MSM) in terms of classification accuracy on the training sets of 45 datasets [17]. The evaluation shows that HMMs can attain significantly higher accuracy in 18 datasets, relatively higher accuracy in 12, and equal accuracy in 4; hence better or equal accuracy in 34 datasets. 2) MTSC: an model-based framework for effective and efficient time series NN classification. The framework works in a filter-and-refine manner, by exploiting the novel model-based representation of time series belonging to the same class. 3) An extensive experimental evaluation on NN classification accuracy between MTSC and the Cross Validation method defined over DTW, ERP, and MSM, on 33 datasets. We observed that MTSC is at least as good as Cross Validation in 23 datasets, while achieving competitive speedups, showing *both* its effectiveness and efficiency.

2   Related Work

A plethora of distance/similarity methods for time series have been proposed during the last decades. As shown by Wang et al. [34], Dynamic Time Warping (DTW) [22] is not only a widely used distance measure for computing distances between time series, but also provides very good classification accuracy results. Hence, several lower bounds have been proposed to speed up its expensive computation [2,24]. Variants of DTW include constrained DTW (cDTW) [30], Edit Distance on Real sequence (EDR) [9], and Edit distance with Real Penalty (ERP) [8]. A common characteristic of these methods is that they allow aligning elements "warped" in time. ERP and EDR satisfy the triangle inequality, and ERP is found to be more robust to noise than EDR. Another very recently proposed measure is called Move-Split-Merge (MSM) [31], which applies three types of operations (Move, Split, Merge) to transform one time series to another. MSM is metric and invariant to the choice of origin as opposed to ERP. Longest Common SubSequence (LCSS) [32] finds the maximum number of elements being common in the compared sequences allowing for gaps during the alignment. Other distance or similarity measures include DISSIM [12], TWED [26], and SMBGT [19,21].

Several techniques for time series representation have been proposed in the literature that capture global or local structural characteristics, e.g., SpaDe [10] and SAX [25]. Moreover, and *Shapelets* [35] focus on determining discriminant time series subsequence patterns, and have been used for classification. For a thorough comparison of different representations and measures please refer to Wang et al. [34], which demonstrates that there is little difference among them. Speeding up similarity search in large databases based on different time series summarizations, such as DFT [1] has also attracted the attention of researchers. Some recent approaches for faster whole sequence matching are embedding-based, which use a set of reference objects to transform similarity matching in a new vector space instead of the original one [3].

Although the aforementioned approaches are very promising, they focus on representing each time series by taking advantage of its structure. However, in this work we try to represent "groups" of time series that belong to the same class, which is orthogonal to the previously proposed techniques.

We focus on HMMs, which model the underlying structure of sequences determining the relationships between their observations. HMMs have been applied to speech recognition [29] and music retrieval [20]. Although training may be computationally expensive, once they are constructed they can be highly applicable to time series, as shown in this work through the classification task. Our approach differs from selecting the best model in Markovian Processes [15], since at each state we neither perform an action nor give a reward. Furthermore, model-based kernel for time series analysis requires significant amount of time [7]. Conditional Random Fields (CRFs) [23,33] can be used for modeling temporal patterns. Nonetheless, we do not target in finding and modeling patterns, rather to represent groups of "homogeneous" sequences by identifying the relationships among their observations. In addition, non-parametric techniques have also been proposed within the field of functional data analysis [11,14], though our approach is orthogonal to those. Finally, HMM-based approaches have also been used for the problem of time series clustering [13,27]; however, our focus here is classification and hence the proposed approach is customised for this task.

3   MTSC: Model-Based Time Series Classification

Let $X = (x_1,\ldots,x_{|X|})$ and $Y = (y_1,\ldots,y_{|Y|})$ be two time series of length $|X|$ and $|Y|$, respectively, where $x_i, y_j \in \mathbb{R}$, $\forall (i = 1,\ldots,|X|; j = 1,\ldots,|Y|)$. Since we are interested in labeled time series, we denote the class of X as c_X. The pair (X,c_X) will correspond to X along with its label. Given a distance measure $dist_x$, the distance between X and Y is defined as a function $d_{dist_x}(X,Y)$. The problem we would like to solve is given next.

Nearest Neighbor Classification. Given a collection of N training time series $\mathscr{D} = \{(X_1,c_{X_1}),\ldots,(X_N,c_{X_N})\}$, a distance measure $dist_x$, and an unlabeled query time series

Q, find the class c_Q of Q as follows:

$$c_Q = \{c_{X_j} \mid \arg\min_j (d_{dist_x}(Q, X_j)), \forall (j = 1, \ldots, N)\}$$

Next, we provide an overview of HMMs, how we can represent classes of time series with appropriate training of HMMs, and describe the MTSC framework, which takes advantage of the trained models for NN classification.

3.1 Hidden Markov Models

An HMM is a doubly stochastic process containing a finite set of states [29]. Formally, it is defined by M distinct states, L values that can be observed at each state (for time series any real number can be observed), the set $T = \{t_{uv}\}$ of transition probabilities, where $t_{uv} = P[s_t = v | s_{t-1} = u]$ with $1 \le u, v \le M$ and s_t being the state at time t (first order Markov chain), the set $E = \{e_v(k)\}$ of the probabilities of values at state v, where $e_v(k) = P[o_t = k | s_t = v]$ with o_t being the observed/emitted value at time t, and the set $\Pi = \{\pi_v\}$ of prior probabilities, where $\pi_v = P[s_1 = v], 1 \le v \le M$.

When a database consists of sets of time series belonging to certain classes, HMMs can be used to model the different classes after being trained on their respective time series. This lies on the fact that a trained HMM can reflect the probabilistic relations of the values within the sequences, and consequently represent their common structure. Thus, HMMs can be highly applicable for retrieval or classification [28]. Given a query Q, we can look for the model that maximizes the likelihood of having generated Q. With this direction in mind, the time series matching problem is transformed to probabilistic-based matching.

3.2 Training HMMs

Assume that we have a dataset $\mathscr{D}$ with z classes $C_1, \ldots, C_z$. Let $\mathscr{C}_i$ be the set of training time series that belong to class C_i, with $i = 1, \ldots, z$. The size of $\mathscr{C}_i$ is denoted as $|\mathscr{C}_i| = n_i$. The training phase of an HMM for each $\mathscr{C}_i$ is split to two phases, which are performed offline: a) initialization and b) iterative refinement.

Initialization Step. For each time series X_j $(j = 1, \ldots, n_i)$ of $\mathscr{C}_i$ $(i \in [1, z])$ we compute the average distance of all other time series $X_k \in \mathscr{C}_i$ $(j \ne k)$ to X_j, which we denote as $a^j_{dist_x}$, i.e.,

$$a^j_{dist_x} = \frac{1}{n_i - 1} \sum_{\forall X_k \in \mathscr{C}_i, X_j \ne X_k} d_{dist_x}(X_j, X_k).$$

For the above computation we choose DTW to be the distance measure, i.e., $dist_x =$ DTW, since it has been shown to be one of the most competitive measures for time series matching [34]. In addition, we keep track of the *warping path* of all the pair-wise time series alignments involved in this process.

Next, we identify the *medoid* of $\mathscr{C}_i$, denoted as $X_{\mu_{C_i}}$, which is the time series with the minimum average distance to the rest of the training set, where

$$\mu_{C_i} = \arg\min_j (a^j_{dist_x}), \forall (j = 1, \ldots, n_i).$$

The medoid $X_{\mu_{C_i}}$ is broken into M equal-sized *segments*, where each segment $m \in [1, M]$ corresponds to one HMM state. Using these segments and the stored warping paths we can determine the observed values of each state. Specifically, for each state (that corresponds to a segment m) the observed values include all elements of $X_{\mu_{C_i}}$ in m, along with the elements of all time series in $\mathscr{C}_i$ that have been aligned to elements of m; the latter can be retrieved from the stored warping paths.

A common case in HMMs is to have for each state a Gaussian distribution for E, which is defined by the mean and standard deviation of the stored elements. To compute T, since we consider each segment as a state, at time t when an observation is emitted we can either stay at the same state or move forward to the next state. Let $|s_t|$ denote the total number of elements at time t of state s_t. The probability of jumping to the next state is $p = n_i / |s_t|$. This is quite straightforward: consider only one segment and one time series X_j ($j = 1, \ldots, n_i$), and suppose that X_j contributes y elements to that segment. Since only the last element in the segment can lead to a transition from s_t to s_{t+1}, the transition probability is $1/y$. Considering now that all n_i time series contribute to the segment, the probability of a state transition is $p = n_i / |s_t|$. Finally, the probability of staying at the same state is $(1 - p) = (|s_t| - n_i)/|s_t|$, while for the last state it is 1. In total, the complexity of this step is $O((n_i^2 max_{j \in [1, n_i]} |X_j|)^2)$.

Iterative Refinement Step. In this step, we refine the z HMMs constructed during initialization. For a specific class C_i ($i \in [1, z]$), for each $X_j \in \mathscr{C}_i$, we compute the Viterbi algorithm [29] to find its best state sequence. Specifically, let us denote as δ the $M \times |X_j|$ probability matrix. Each X_j always starts from state 1 (thus $\pi_1 = 1$), and in the initialization phase the log-likelihood of its first element is computed according to the Gaussian distribution. The remaining elements of the first column of δ are set to $-\infty$. Since to get an observation we have either stayed at the same state or have performed a transition from the previous state, in the recursion phase we consider only the values of δ representing the probabilities of the previous element for the previous and the current state. For cell (u, v) these values are $\delta(u, v - 1)$ and $\delta(u - 1, v - 1)$, which were computed in the initialization step. Hence, we first find the most probable transition by computing $m = max(\delta(u, v - 1) + \log(t_{uu}), \delta(u - 1, v - 1) + \log(t_{u-1u}))$, and then $\delta(u, v) = m + \log(e_u(k))$. Finally, we backtrack from $\delta(M, |X_j|)$ and store the elements of X_j falling within each state. Having done this step for all $X_j \in \mathscr{C}_i$, the mean and standard deviation for E of each state, and also T are updated. The complexity of the aforementioned procedure is $O(M \sum_{j=1}^{n_i} |X_j|)$.

The refinement step is performed for the z HMMs and is repeated until a stopping criterion is met, e.g., the classification accuracy on the N training time series composing $\mathscr{D}$ cannot be further improved (Section 4.1). The final outcome of this step is a set of z HMMs, denoted as $\mathscr{H} = \{H_1, \ldots, H_z\}$, where each $H_i \in \mathscr{H}$ defines a probability distribution for $\mathscr{C}_i$, which essentially describes the likelihood of observing any time series of class C_i.

3.3 Filter-and-Refine Framework

Given $\mathscr{D}$, the MTSC framework consists of three steps: offline, filter, and refine.

Offline Step. First, we construct $\mathcal{H}$ as described in Section 3.2. To make our framework more generic, assume that we have available a set of l distance measures $\{dist_1, \ldots, dist_l\}$. For each $dist_x$ ($x \in [1, l]$) we compute the NN classification accuracy on the N training time series using leave-one-out cross validation.

Filter Step. Since we have created a "new" probabilistic space for time series similarity matching, we should define a way of measuring how "good" each HMM model $H_i \in \mathcal{H}$ is for a query Q. This can be achieved by applying the Forward algorithm [29], which computes the likelihood of Q having been produced by H_i. Thus, the "goodness" of H_i is the likelihood estimate given by the algorithm. The complexity of the Forward algorithm is $O(|Q|M^2)$. Note that this step involves only z computations of the Forward algorithm and it is significantly fast, given that in practice z is rarely higher than 50. This is based on the fact that in the 45 datasets of the UCR archive [17], which cover real application domains, z is at most 50. After computing the likelihood of each H_i for Q, we identify the K models with the highest likelihood of producing Q.

Refine Step. Next, the training time series that comprise each of the top K selected models are evaluated with Q. This evaluation is performed using the distance measure that achieved the highest classification accuracy on the training set during the offline step. Finally, Q is assigned with the class of the closest time series. The complexity of this step is $O(K'comp(dist_x))$, where K' is the total number of training time series corresponding to the K selected models, and $comp(dist_x)$ is the complexity of computing the distance between two time series using $dist_x$ (selected in the offline step).

It has to be mentioned that the smaller the K the faster our approach is. However, since each HMM is a very compact representation of all time series of a certain class, reducing the number of models selected at the filter step may greatly reduce accuracy, as the time series that will be evaluated at the refine step may not include those of the correct class. On the contrary, as K increases towards z, more training time series will be evaluated, resulting in the brute-force approach evaluating Q with all time series when $K = z$, which is certainly undesirable. Hence, a good value for K is needed to achieve a good tradeoff between effectiveness and efficiency. The choice of distance measure also influences accuracy, but it is beyond our scope to select the "best" measure for *each Q*.

4   Experiments

In this section, we present the setup and the experimental evaluation for HMM-based representation and MTSC.

4.1   Experimental Setup

We experimented on the 45 time series datasets available from the UCR archive [17]. A summary of the dataset statistics is included in Table 1. We first evaluated the performance of HMMs (Section 3.2) against DTW, ERP, and MSM. Secondly, we compared MTSC with Cross Validation using the same three measures.

Although any distance measure can be used to perform NN classification, an exhaustive consideration of all of distance measures is beyond the scope of this paper. The rationales behind selecting these measures are the following: (1) DTW is extensively

Table 1. NN classification error rates attained by MSM, DTW, ERP, and HMMs on the training set of 45 datasets from the UCR repository. The table shows for each dataset: the number of training and test objects, the length of each time series in the dataset, the number of classes, the value of parameter c used by MSM on that dataset that yielded the lowest error rate on the training set (when two or three values are given, the one in italics was randomly chosen), the number of states as a percentage of the time series length and the number of iterations for which the HMMs achieved the lowest error rate on the training set. Numbers in bold indicate the smallest error rate.

ID	Dataset	train error rate (%) MSM	DTW	ERP	HMMs	train size	test size	length $|X|$	class num. z	parameter c (MSM)	state perc.	iter. num.
1	Synthetic	1.33	1.00	0.67	**0.33**	300	300	60	6	0.1	0.4	3
2	CBF	**0.00**	**0.00**	**0.00**	**0.00**	30	900	128	3	0.1	0.5	2
3	FaceAll	**1.07**	6.79	2.50	1.25	560	1,690	131	14	1	0.5	11
4	OSU	19.50	33.00	30.50	**17.00**	200	242	427	6	0.1	0.3	11
5	SwedishLeaf	12.40	24.60	13.40	**12.20**	500	625	128	15	1	0.9	6
6	50Words	21.11	33.11	28.22	**8.89**	450	455	270	50	1	0.5	1
7	Trace	1.00	**0.00**	9.00	**0.00**	100	100	275	4	0.01	0.3	2
8	TwoPatterns	**0.00**	**0.00**	**0.00**	**0.00**	1,000	4,000	128	4	1	0.1	1
9	FaceFour	8.33	25.00	12.50	**0.00**	24	88	350	4	1	0.1	1
10	Lightning-7	27.14	32.86	28.57	**7.14**	70	73	319	7	1	0.2	1
11	Adiac	38.97	40.51	39.49	**15.90**	390	391	176	37	1	0.5	8
12	Fish	13.71	26.29	17.14	**7.43**	175	175	463	7	0.1	0.5	4
13	Beef	66.67	53.33	66.67	**23.33**	30	30	470	5	0.1	0.4	2
14	OliveOil	16.67	13.33	16.67	**3.33**	30	30	570	4	0.01	0.3	2
15	ChlorineConc.	38.97	38.97	**38.76**	56.32	467	3,840	166	3	1	0.7	4
16	ECG_torso	12.50	32.50	25.00	**2.50**	40	1,380	1,639	4	1	0.4	3
17	Cricket_X	**18.46**	20.26	21.54	25.38	390	390	300	12	1	0.5	14
18	Cricket_Y	24.10	20.51	23.33	**19.23**	390	390	300	12	0.1,_1_	0.4	9
19	Cricket_Z	24.10	22.56	24.87	**23.08**	390	390	300	12	1	0.4	14
20	Diatom Red.	6.25	6.25	6.25	**0.00**	16	306	345	4	0.01,0.1,_1_	0.1	4
21	FacesUCR	2.50	10.00	5.50	**0.50**	200	2,050	131	14	1	0.8	9
22	Haptics	49.68	58.71	54.19	**29.68**	155	308	1,092	5	1	0.1	15
23	InlineSkate	50.00	59.00	49.00	**43.00**	100	550	1,882	7	1	0.5	3
24	MALLAT	5.45	5.45	5.45	**0.00**	55	2,345	1,024	8	1	0.1	1
25	MedicalImages	27.82	27.56	**26.51**	34.91	381	760	99	10	0.1	0.4	13
26	StarLightC.	10.70	9.60	13.80	**8.60**	1,000	8,236	1,024	3	0.1	0.5	14
27	Symbols	**0.00**	4.00	8.00	**0.00**	25	995	398	6	0.1	0.1	1
28	uWaveGest_X	25.78	29.35	26.67	**25.22**	896	3,582	315	8	0.1,_1_	0.5	11
29	uWaveGest_Y	**28.24**	37.05	33.93	34.60	896	3,582	315	8	1	0.7	13
30	uWaveGest_Z	29.24	33.59	31.25	**27.46**	896	3,582	315	8	1	0.2	12
31	WordsSynon.	22.10	36.33	28.46	**13.86**	267	638	270	25	1	0.8	13
32	ECGThorax1	18.17	20.11	17.83	**7.17**	1,800	1,965	750	42	1	0.5	15
33	ECGThorax2	10.83	14.17	11.72	**7.67**	1,800	1,965	750	42	1	0.5	14
34	Gun Point	**4.00**	18.00	8.00	8.00	50	150	150	2	0.01	0.3	2
35	Wafer	**0.10**	1.40	**0.10**	1.70	1,000	6,164	152	2	1	0.9	4
36	Lightning-2	16.67	13.33	13.33	**5.00**	60	61	637	2	0.01	0.1	15
37	ECG	14.00	23.00	18.00	**12.00**	100	100	96	2	1	0.8	2
38	Yoga	**12.00**	18.33	17.33	22.33	300	3,000	426	2	0.1	0.2	15
39	Coffee	25.00	**14.29**	25.00	21.43	28	28	286	2	0.01	0.3	1
40	ECGFiveDays	26.09	43.48	26.09	**0.00**	23	861	136	2	1	0.2	4
41	ItalyPowerDemand	**4.48**	**4.48**	5.97	5.97	67	1,029	24	2	0.1,_1_	0.9	2
42	MoteStrain	15.00	25.00	25.00	**0.00**	20	1,252	84	2	0.1	0.5	7
43	SonySurface1	10.00	20.00	15.00	**0.00**	20	601	70	2	1	0.1	1
44	SonySurface2	11.11	14.81	18.52	**3.70**	27	953	65	2	0.1	0.3	1
45	TwoLeadECG	4.35	8.70	4.35	**0.00**	23	1,139	82	2	0.01,0.1	0.1	4

used in time series and has been shown to provide excellent classification accuracy results [34], (2) ERP is a variant of DTW and Edit Distance fixing the non-metric property of DTW, and (3) MSM is also metric and has been shown to outperform ERP and DTW in terms of NN classification accuracy on several datasets. Their time complexity is quadratic with respect to the time series length.

The Cross Validation method works as follows: (1) for each dataset we computed the classification accuracy for DTW, ERP, and MSM on the training set using leave-one-out cross validation, and (2) the method outputs the classification accuracy on the test set of the measure with the best accuracy on the training set. If more than one measures provide the same highest classification accuracy on the training set, then the accuracy of Cross Validation is the accuracy on the test set of the measure that outperforms the other tied measure(s) on most datasets (on their training sets).

For each dataset, parameter c of MSM was selected from $\{0.01, 0.1, 1\}$ [31]. For each value the classification accuracy was found using leave-one-out cross-validation on the training set, and the value yielding the highest accuracy was selected. For the training of HMMs, for each of the 45 datasets, we varied M from 0.1 to $0.9 * |X|$ (step 0.1) and applied 15 refinement iterations (135 combinations). For each combination we measured the percentage of training time series for which the model producing the highest likelihood through the Forward algorithm was the correct one. Then, the combination leading to the highest accuracy was selected. If more than one combinations provided the same highest accuracy we chose the smallest M (for further ties smallest number of iterations).

Evaluation Measures. We first evaluated the performance of HMMs against DTW, ERP, and MSM on the training sets, and between MTSC and Cross Validation on the test sets in terms of *classification error rate*. This rate is defined as the percentage of time series misclassified using the NN classifier. Secondly, we evaluated the *efficiency* of MTSC and Cross Validation. Nonetheless, runtime measurements may depend on particular aspects of the hardware, implementation details, compiler optimizations, and programming language. To overcome these limitations, we also present per dataset the percentage of classes selected at the filter step $((K/z) * 100)$, and, more importantly, the percentage of training time series that are finally evaluated by MTSC, as the number of time series may (sometimes greatly) deviate among classes in the same dataset. MTSC was implemented in Matlab, while, for efficiency, DTW, MSM, ERP, and the Forward algorithm were implemented in Java. Experiments were performed on a PC running Linux, with Intel Xeon Processor at 2.8GHz.

4.2 Experimental Results

Next, we present our experimental findings for the methods and evaluation measures.

Classification Accuracy of HMMs. For each of the 45 datasets, we compared the classification error rates on the training set for MSM, DTW, ERP, and HMMs. The results are shown in Table 1. We observe that HMMs achieve better or equal error rate than that of the competitor distance measures in 34 datasets, out of which they outperform them in 30. The performance of HMMs is in many cases significantly better than all competitors. For example, for *ECGFiveDays* HMMs achieve an error rate of 0% as opposed to the next best which is 26.09% (achieved by both ERP and MSM), while for 18 datasets (e.g., *50Words*, *Lightning-7*, *Adiac*, *Beef*, *OliveOil*, and *ECG_torso*) the error rate of HMMs is at least two times lower than that of the competitors. These numbers show that modeling time series classes with HMMs is highly competitive and promising for NN classification.

Classification Accuracy of MTSC. We studied the classification error rates of MTSC against Cross Validation on the test sets of the 33 datasets with $z > 2$. The results are shown in Table 2. Note that if $z = 2$ with MTSC we can either select one or two models for the refine step. However, $K = 1$ would essentially exclude the refine step, since time series of only one class would be evaluated, which is meaningless. Hence, the classification error rate would depend solely on how well the HMMs represent and discriminate the classes of the dataset, which is not always the case due to their very compact representation of (sometimes large) classes. In addition, $K = 2$ would make

Table 2. NN classification error rates attained by MTSI and Cross Validation on the test set of 33 datasets from the UCR repository. The table also shows for each dataset: the classification error rate of MSM, DTW, and ERP on the test set, the number of HMM models used at the refine step of MTSI and the respective percentage of classes it corresponds to, the average number of training objects evaluated at the refine step per test object, and the percentage of training objects this average corresponds to. Numbers in bold indicate the smallest error rate.

ID	Dataset	MTSI	Cross Valid.	error rate (%) MSM	DTW	ERP	top K	% classes	avg train num.	% train obj.
1	Synthetic	**3.33**	3.70	2.67	0.70	3.70	2	33.33	100.00	33.33
2	CBF	3.67	**1.22**	1.22	0.30	0.30	2	66.67	20.45	68.16
3	FaceAll	18.99	**18.88**	18.88	19.20	20.20	5	35.71	200.00	35.71
4	OSU	22.73	**19.83**	19.83	40.90	39.70	5	83.33	169.00	84.50
5	SwedishLeaf	**9.60**	10.40	10.40	21.00	12.00	3	20.00	100.87	20.17
6	50Words	**19.56**	19.56	19.56	31.00	28.10	40	80.00	412.26	91.61
7	Trace	**0.00**	**0.00**	7.00	0.00	17.00	2	50.00	49.82	49.82
8	TwoPatterns	**0.05**	0.08	0.08	0.00	0.00	2	50.00	498.47	49.85
9	FaceFour	**4.55**	5.68	5.68	17.00	10.20	2	50.00	11.44	47.68
10	Lightning-7	**21.92**	23.29	23.29	27.40	30.10	6	85.71	64.18	91.68
11	Adiac	**33.76**	38.36	38.36	39.60	37.90	2	5.41	20.96	5.37
12	Fish	**7.43**	8.00	8.00	16.70	12.00	6	85.71	149.87	85.64
13	Beef	**46.67**	50.00	50.00	50.00	50.00	3	60.00	18.00	60.00
14	OliveOil	16.67	**13.33**	16.67	13.33	16.67	3	75.00	21.80	72.67
15	ChlorineConc.	40.42	**37.40**	37.27	35.20	37.40	2	66.67	362.14	77.55
16	ECG_torso	15.07	**10.29**	10.29	34.90	25.00	3	75.00	30.36	75.91
17	Cricket_X	25.90	27.18	27.18	22.30	29.23	5	41.67	162.54	41.68
18	Cricket_Y	**20.00**	20.80	16.67	20.80	21.28	5	41.67	162.38	41.64
19	Cricket_Z	**20.77**	**20.77**	21.54	20.77	24.36	10	83.33	330.00	84.62
20	Diatom Red.	**4.58**	**4.58**	4.58	3.30	5.23	3	75.00	14.67	91.67
21	FacesUCR	**3.20**	3.27	3.27	9.51	4.24	13	92.86	191.02	95.51
22	Haptics	**57.47**	59.42	59.42	62.30	57.47	3	60.00	98.00	63.22
23	InlineSkate	57.45	**56.91**	55.64	61.60	56.91	6	85.71	87.25	87.25
24	MALLAT	**6.74**	**6.74**	6.74	6.60	7.46	6	75.00	41.98	76.33
25	MedicalImages	**27.89**	**27.89**	24.74	26.30	27.89	7	70.00	334.98	87.92
26	StarLightC.	9.35	**9.30**	11.72	9.30	13.62	2	66.67	759.62	75.96
27	Symbols	3.12	**3.02**	3.02	5.00	5.83	5	83.33	21.00	83.98
28	uWaveGest_X	**22.28**	22.36	22.36	27.30	25.71	5	62.50	574.04	64.07
29	uWaveGest_Y	**30.35**	30.37	30.37	36.60	33.61	5	62.50	553.80	61.81
30	uWaveGest_Z	29.12	31.07	31.07	34.20	32.97	2	25.00	222.04	24.78
31	WordsSynon.	23.67	**23.51**	23.51	35.10	32.13	24	96.00	263.92	98.85
32	ECGThorax1	**18.37**	19.29	18.27	20.90	19.29	2	4.76	85.99	4.78
33	ECGThorax2	**10.89**	11.25	11.25	13.50	10.74	7	16.67	300.03	16.67

our approach perform brute-force search, which is undesirable. In columns "top K", "% classes", "avg train num.", "% train obj." we show the K value used at the filter step of MTSC (due to space limitations we present the smallest $K < z$ for which the accuracy could not be improved or provided a competitive error rate), the ratio $(K/z) *$ 100, the average number of training time series evaluated per query at the refine step, and the percentage of the "train size" to which this average corresponds to, respectively.

We observe that MTSC achieves at least as good or better error rates than Cross Validation in 23 datasets; it is better in 17 datasets and equal in 6. There are two reasons for such competitive performance of MTSC: a) the correct class of the test time series is among the K models selected at the filter step, and the distance measure applied at the refine step is able to better differentiate the correct class from the rest $K - 1$, since there are less training objects to throw away as being "bad" matches compared to brute-force search, and b) the HMMs of these 23 datasets are constructed exploiting a sufficient number of training time series comprising their classes, making the probability distribution of such classes effectively represent these (and similar) time series.

Carefully analyzing our experimental findings, we concluded that 16 training time series is a sufficient number to provide a good model representing a class of objects. This claim is supported by the following examples, where MTSC yields worse error rates than Cross Validation. Datasets with *ID* 14, 16, and 20 consist of 4 classes, but no class of the first two has more than 15 training time series, while the classes of the latter include only 1, 6, 5, and 4 time series. Moreover, none of the 6 classes of dataset with *ID* 27 has more than 8 time series, *WordsSynon*. (*ID* 31) with $z = 25$ has only 4 classes with more than 15 time series, as happens with 3 out of the 7 classes of *InlineSkate* (*ID* 23). The error rates for datasets *ChlorineConc*. and *StarLightC*. (*ID* 15 and 26) can be attributed to overfitting, since for the first no class comprises of less than 91 time series, while for the second all classes have more than 152 time series. In addition, building a histogram over the number of classes that include specific numbers of training time series, we observed that 167 out of the 399 classes (comprising the 33 datasets) have up to 15 training time series. As a result, we would like to emphasize the need for a sufficient number of training time series per class.

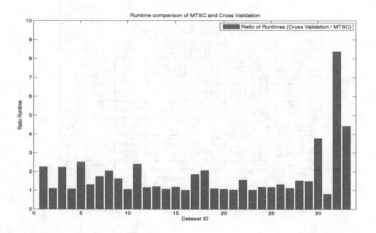

Fig. 1. Speedup of MTSC vs. Cross Validation for 33 datasets. Each bar represents the ratio of the Cross Validation avg. total runtime to that of MTSC, for NN classification of a test object.

Efficiency. In Figure 1 we present the average *speedup* per test time series when using MTSC instead of Cross Validation for 33 datasets. The speedup is the runtime ratio of Cross Validation over MTSC, and it intuitively depends on K. For example, we gain up to a factor of 9 in terms of runtime for dataset with *ID* 32, since MTSC selects only 2 out of 42 classes at the filter step, and its error rate is lower than that of Cross Validation. We observe that there are several datasets for which there is no significant speedup. This is mainly because, for these datasets, MTSC could not achieve a competitive error rate for large K, even for $z - 1$ in some cases. Thus, for such values its runtime converged to that of brute-force using the appropriate distance measure. Additionally, there may be cases where the length of the time series is not huge enough to provide a noticeable difference in the runtimes of the two competitors (*ID* 25), resulting in a slight speedup. The latter result may also happen when "train size" is small and/or the average number of training time series for the K selected models is much higher than that of the non-selected ones. The last claim holds, e.g., for datasets with *ID* 6, 15, 20, 25, 26, where the value of "% train obj." is significantly higher than that of "%

classes", showing that the training time series are not equally distributed to all classes. Additionally, the percentage of training time series that were evaluated ranges from just 4.78% (*ID* 32) to 98.85% (*ID* 31), which is the worst possible case since no smaller *K* could provide better accuracy. We have to point out, though, that out of the 23 datasets for which MTSC is better than or equal to Cross Validation there are 11 datasets for which less than 50% of their training set is evaluated.

Based on these results, we can argue that MTSC outperforms Cross Validation in classification error rate more often than not, while allowing for a speedup of up to a factor of 9. An acute reader may argue that the runtime comparison of the two methods is unfair since we could alternatively have used existing speedup methods for DTW[4], or even faster techniques such as cDTW with LB_Keogh [16]. Nonetheless, we argue that any speedup achieved by each method used by Cross Validation is also equally beneficial for MTSC. This is due to the fact that MTSC is using the exact same set of methods for the refine step, and thus any speedup obtained by Cross Validation is essentially exploited by MTSC as well (the filter step cost is negligible compared to that of the refine step).

5 Conclusions and Future Work

We presented an effective way of modeling classes of time series using HMMs and proposed MTSC, a filter-and-refine framework for NN classification of time series. Experimenting with 45 widely known time series datasets and three distance measures we observed that HMMs provide better or equal classification accuracies than the competitors on the training set in 34 datasets. MTSC has equal or better accuracy than Cross Validation in 23 out of 33 datasets, while achieving a speedup of up to a factor of 9. We plan to test MTSC on larger datasets with more classes, where we expect its performance to further improve.

References

1. Agrawal, R., Faloutsos, C., Swami, A.: Efficient similarity search in sequence databases. In: Lomet, D.B. (ed.) FODO 1993. LNCS, vol. 730, pp. 69–84. Springer, Heidelberg (1993)
2. Assent, I., Wichterich, M., Krieger, R., Kremer, H., Seidl, T.: Anticipatory dtw for efficient similarity search in time series databases. PVLDB 2(1), 826–837 (2009)
3. Athitsos, V., Hadjieleftheriou, M., Kollios, G., Sclaroff, S.: Query-sensitive embeddings. In: SIGMOD, pp. 706–717 (2005)
4. Athitsos, V., Papapetrou, P., Potamias, M., Kollios, G., Gunopulos, D.: Approximate embedding-based subsequence matching of time series. In: SIGMOD, pp. 365–378 (2008)
5. Baum, L.E., Petrie, T., Soules, G., Weiss, N.: A maximization technique occurring in the statistical analysis of probabilistic functions of markov chains. The Annals of Mathematical Statistics 41(1), 164–171 (1970)
6. Bellman, R.: The theory of dynamic programming. Bull. Amer. Math. Soc. 60(6), 503–515 (1954)
7. Chen, H., Tang, F., Tino, P., Yao, X.: Model-based kernel for efficient time series analysis. In: SIGKDD, pp. 392–400 (2013)

8. Chen, L., Ng, R.: On the marriage of l_p-norms and edit distance. In: VLDB, pp. 792–803 (2004)
9. Chen, L., Özsu, M.T.: Robust and fast similarity search for moving object trajectories. In: SIGMOD, pp. 491–502 (2005)
10. Chen, Y., Nascimento, M.A., Chin, B., Anthony, O., Tung, K.H.: Spade: On shape-based pattern detection in streaming time series. In: ICDE, pp. 786–795 (2007)
11. F. Ferraty and P. Vieu. Curves discrimination: a nonparametric functional approach. *Computational Statistics and Data Analysis*, 44(1-2):161–173, 2003.
12. Frentzos, E., Gratsias, K., Theodoridis, Y.: Index-based most similar trajectory search. In: ICDE, pp. 816–825 (2007)
13. Ghassempour, S., Girosi, F., Maeder, A.: Clustering multivariate time series using hidden markov models. International Journal of Environmental Research and Public Health 11(3), 2741–2763 (2014)
14. González, J., Muñoz, A.: Representing functional data using support vector machines. In: Ruiz-Shulcloper, J., Kropatsch, W.G. (eds.) CIARP 2008. LNCS, vol. 5197, pp. 332–339. Springer, Heidelberg (2008)
15. Hallak, A., Di-Castro, D., Mannor, S.: Model selection in markovian processes. In: ICML (2013)
16. Keogh, E.: Exact indexing of dynamic time warping. In: VLDB, pp. 406–417 (2002)
17. Keogh, E., Zhu, Q., Hu, B., Hao, Y., Xi, X., Wei, L., Ratanamahatana, C.: The UCR time series classification/clustering homepage,
   http://www.cs.ucr.edu/~eamonn/time_series_data/
18. Kotsifakos, A., Athitsos, V., Papapetrou, P., Hollmén, J., Gunopulos, D.: Model-based search in large time series databases. In: PETRA (2011)
19. Kotsifakos, A., Papapetrou, P., Hollmén, J., Gunopulos, D.: A subsequence matching with gaps-range-tolerances framework: A query-by-humming application. PVLDB 4(11), 761–771 (2011)
20. Kotsifakos, A., Papapetrou, P., Hollmén, J., Gunopulos, D., Athitsos, V.: A survey of query-by-humming similarity methods. PETRA, 5:1–5:4 (2012)
21. Kotsifakos, A., Papapetrou, P., Hollmén, J., Gunopulos, D., Athitsos, V., Kollios, G.: Huma-song: a subsequence matching with gaps-range-tolerances query-by-humming system. PVLDB 5(12), 1930–1933 (2012)
22. Kruskall, J.B., Liberman, M.: The symmetric time warping algorithm: From continuous to discrete. In: Time Warps. Addison-Wesley (1983)
23. Lafferty, J.D., McCallum, A., Pereira, F.C.N.: Conditional random fields: Probabilistic models for segmenting and labeling sequence data. In: ICML, pp. 282–289 (2001)
24. Lemire, D.: Faster retrieval with a two-pass dynamic-time-warping lower bound. Pattern recognition 42(9), 2169–2180 (2009)
25. Lin, J., Keogh, E., Lonardi, S., Chiu, B.: A symbolic representation of time series, with implications for streaming algorithms. In: SIGMOD Workshop DMKD, pp. 2–11 (2003)
26. Marteau, P.-F.: Time warp edit distance with stiffness adjustment for time series matching. Pattern Analysis and Machine Intelligence 31(2), 306–318 (2009)
27. Oates, T., Firoiu, L., Cohen, P.R.: Clustering time series with hidden markov models and dynamic time warping. In: In Proceedings of the IJCAI, pp. 17–21 (1999)
28. Pikrakis, A., Theodoridis, S., Kamarotos, D.: Classification of musical patterns using variable duration hidden Markov models. Transactions on Audio, Speech, and Language Processing 14(5), 1795–1807 (2006)
29. Rabiner, L.: A tutorial on hidden Markov models and selected applications in speech recognition. Proceedings of the IEEE 77(2), 257–286 (1989)
30. Sakoe, H., Chiba, S.: Dynamic programming algorithm optimization for spoken word recognition. Transactions on Acoustics, Speech and Signal Processing 26, 43–49 (1978)

Model-Based Time Series Classification

31. A. Stefan, V. Athitsos, and G. Das. The move-split-merge metric for time series. Transactions on Knowledge and Data Engineering (2012)
32. Vlachos, M., Kollios, G., Gunopulos, D.: Discovering similar multidimensional trajectories. In: ICDE, pp. 673–684 (2002)
33. Wang, S.B., Quattoni, A., Morency, L.-P., Demirdjian, D., Darrell, T.: Hidden conditional random fields for gesture recognition. In: CVPR, pp. 1521–1527 (2006)
34. Wang, X., Mueen, A., Ding, H., Trajcevski, G., Scheuermann, P., Keogh, E.J.: Experimental comparison of representation methods and distance measures for time series data. Data Mining and Knowledge Discovery 26(2), 275–309 (2013)
35. Ye, L., Keogh, E.: Time series shapelets: a novel technique that allows accurate, interpretable and fast classification. Data Mining and Knowledge Discovery 22(1-2), 149–182 (2011)

Fast Simultaneous Clustering and Feature Selection for Binary Data

Charlotte Laclau[1,2] and Mohamed Nadif[1]

[1] Université Paris Descartes, LIPADE, Paris, France
[2] University of Ottawa, Imagine Lab, Ottawa, Canada
{charlotte.laclau,mohamed.nadif}@parisdescartes.fr

Abstract. This paper addresses the problem of clustering binary data with feature selection within the context of maximum likelihood (ML) and classification maximum likelihood (CML) approaches. In order to efficiently perform the clustering with feature selection, we propose the use of an appropriate Bernoulli model. We derive two algorithms: Expectation-Maximization (EM) and Classification EM (CEM) with feature selection. Without requiring a knowledge of the number of clusters, both algorithms optimize two approximations of the *minimum message length* (MML) criterion. To exploit the advantages of EM for clustering and of CEM for fast convergence, we combine the two algorithms. With Monte Carlo simulations and by varying parameters of the model, we rigorously validate the approach. We also illustrate our contribution using real datasets commonly used in document clustering.

1 Introduction

Cluster Analysis is an important tool in a variety of scientific areas, including pattern recognition, information retrieval, microarrays and data mining. These methods organize the dataset into homogeneous classes or natural classes, in a way that ensures that objects within a class are similar to one another. Different approaches and algorithms are used. Most of them, however, can lead to clusters that are not relevant in a high-dimensional context. This difficulty is due to the noise introduced by some variables that, as we will show, are irrelevant for clustering. Clustering with feature selection therefore remains a challenge. With given variables, feature selection aims to find the features that best uncover classes from data. It has several advantages: it facilitates the visualization and understanding of data, reduces computational time, defies the curse of dimensionality, and improves clustering performance.

We tackle the problem of feature selection by using mixture models. Basing cluster analysis on mixture models has become a classical and powerful approach (see for instance Duda and Hart (1973) McLachlan and Peel (2000)). Mixture models assume that a sample is composed of subpopulations characterized by a probability distribution. The models are very flexible and can deal with a variety of situations, heterogeneous populations and outliers alike. Furthermore, their

H. Blockeel et al. (Eds.): IDA 2014, LNCS 8819, pp. 192–202, 2014.

associated estimators of posterior probabilities give rise to a fuzzy or hard clustering using the maximum a posteriori (MAP) principle. In general, the mixture model approach aims to maximize the likelihood over the mixture parameters, usually estimated by the EM algorithm (Dempster et al., 1977). In Pudil et al. (1995), the authors proposed a method of feature selection based on the approximation of class conditional densities by a mixture of parametrized densities inspired by Grim (1986) parametric models. In Law et al. (2004),the authors adopted this model for Gaussian mixtures and proposed *feature saliency* as a measure of the importance of each variable.

We chose to use the original model proposed by Grim (1986) for its flexibility and the derived EM algorithm proposed by Law et al. (2004) for its performance in terms of simultaneous clustering and feature selection for Gaussian mixture models. The EM algorithm was evaluated on real datasets with a very small number of features. Although EM can be slow and dependent on the initial position, these two difficulties can be overcome by using a classification version (CEM).

The object of this paper is threefold: first, we propose an extension of the simultaneous clustering and feature selection model for binary data. Second, we derive a scalable version of EM, Classification EM, in which the number of clusters is not fixed. In this version, the optimization criterion requires an approximation of the *minimum message length* (MML). Third, we show that combining EM and CEM algorithms allows the two difficulties of EM mentioned above to be overcome.

The rest of the paper is organized as follows: in Section 2, we provide background on the maximum likelihood (ML) and classification maximum likelihood (CML) approaches. Then we focus on binary data. In Section 3, we study the problem of simultaneous clustering and feature selection; we describe the model using a multivariate Bernoulli mixture model and present in detail our proposed CEM algorithm. Section 4 is devoted to numerical experiments on both synthetic and real datasets showing the appropriateness of our contribution to document clustering. The final section sums up the study and gives recommendations for further research.

2   Mixture Models

Let $\mathbf{x}$ denote an $n \times d$ matrix defined by $\mathbf{x} = \{(x_{ij}); i \in I$ and $j \in J\}$, where I is a set of n objects and J a set of d variables. The input data of $\mathbf{x}$ are binary and the partition obtained is noted $\mathbf{z}$. This partition can be represented by a matrix of elements in $\{0, 1\}^g$ satisfying $\sum_{k=1}^{g} z_{ik} = 1$. Notation $(z_1, \ldots, z_n)$ where $z_i \in \{1, \ldots, k, \ldots, g\}$ represents the cluster of the ith row will be also used.

2.1   Definition of the Bernoulli Mixture Model

Before tackling the problem of feature selection we first provide a brief description of the mixture model. In model-based clustering, it is assumed that the data are generated by a mixture of underlying probability distributions, where each component k of the mixture represents a cluster. The data matrix is then assumed to be an i.i.d sample $\mathbf{x} = (\mathbf{x}_1, \ldots, \mathbf{x}_i, \ldots, \mathbf{x}_n)$ where $\mathbf{x}_i = (x_{i1}, \ldots, x_{id}) \in \mathbb{R}^d$ from a probability distribution with density $p(\mathbf{x}_i; \theta) = \sum_{k=1}^{g} \pi_k p(\mathbf{x}_i; \alpha_k)$. The function $p(.; \alpha_k)$ is the density of an observation $\mathbf{x}_i$ from the kth component and the α_k's are the corresponding class parameters. These densities belong to the same parametric family. The parameter π_k is the probability that an object belongs to the kth component, and g, which is assumed to be known, is the number of components in the mixture. The parameter of this model is the vector $\theta = (\pi, \alpha)$ containing the mixing proportions $\pi = (\pi_1, \ldots, \pi_g)$ and the vector $\alpha = (\alpha_1, \ldots, \alpha_g)$ of parameters of each component. The likelihood of the observed data $\mathbf{x}$ can be expressed as $P(\mathbf{x}; \theta) = \prod_{i=1}^{n} \sum_{k=1}^{g} \pi_k p(\mathbf{x}_i; \alpha_k)$. For binary data with $\mathbf{x}_i \in \{0, 1\}^d$, using multivariate Bernoulli distributions for each component and considering the conditional independence model, we have $p(\mathbf{x}_i; \alpha_k) = \prod_{j=1}^{d} \alpha_{kj}^{x_{ij}} (1 - \alpha_{kj})^{1-x_{ij}}$ where $\alpha_k = (\alpha_{k1}, \ldots, \alpha_{kd})$ and $\alpha_{kj} \in (0, 1)$.

2.2   Model-Based Clustering

The problem of clustering can be studied with two different approaches: maximum likelihood (ML) and classification maximum likelihood (CML).

(i) The first approach estimates the parameters of the mixture; the partition on the objects is derived from these parameters using the maximum a posteriori principle (MAP). The maximum likelihood estimation of the parameters results in an optimization of the log-likelihood of the observed sample $L(\theta; \mathbf{x}) = \sum_i \log(\sum_k \pi_k p(\mathbf{x}_i; \alpha_k))$. This optimization can be achieved using the EM algorithm (Dempster et al., 1977). But first we have to define the complete data log-likelihood, also known as classification log-likelihood $L(\theta; \mathbf{x}, \mathbf{z}) = \sum_{i,k} z_{ik} \log(\pi_k p(\mathbf{x}_i; \alpha_k))$ where $\mathbf{z}$ represents an unobserved latent variable, which is the label of the objects. The EM algorithm maximizes the log-likelihood by maximizing iteratively the conditional expectation of the complete data log-likelihood $L(\theta; \mathbf{x}, \mathbf{z})$ given previous current estimates θ' and $\mathbf{x}$.

$$Q(\theta|\theta') = \mathbb{E}[L(\theta; \mathbf{x}, \mathbf{z})|\theta', \mathbf{x}] = \sum_{i,k} s_{ik}(\log \pi_k' + \log p(\mathbf{x}_i; \alpha_k')),$$

where $s_{ik} \propto \pi_k' p(\mathbf{x}_i; \alpha_k')$ denotes the conditional probability, given θ', that $\mathbf{x}_i$ arises from the mixture component with Bernoulli distribution $p(\mathbf{x}_i; \alpha_k')$. The steps of the EM algorithm are reported in Algorithm 1.

Under certain conditions, it has been established that EM always converges to a local likelihood maximum. This is simple to implement and it behaves well in clustering and estimation contexts. The problem is that it can be slow in some situations. But this drawback can be overcome by using a variant of EM derived from the CML approach described below.

(ii) The CML approach (Symons, 1981) is a fruitful one, displaying some of the statistical aspects of many classical clustering criteria. It estimates the parameters of the mixture and the partition simultaneously, by optimizing the

---

Algorithm 1. EM

---

input : $\mathbf{x}$ of size $n \times d$ and g the number of components (clusters).
initialization : θ
repeat
   E-step compute $s_{ik} \propto \pi'_k p(\mathbf{x}_i; \alpha'_k)$
   M-step compute θ maximizing $Q(\theta|\theta')$, for $k = 1, \ldots, g$ and $j = 1, \ldots, d$, we
   have $\pi_k = \frac{\sum_i s_{ik}}{n}$ and $\alpha_{kj} = \frac{\sum_i s_{ik} x_{ij}}{\sum_i s_{ik}}$.
until convergence
output: $\hat{\theta}$ and $\mathbf{z}$ defined by $z_{ik} = \arg\max_{k'=1\ldots,g} s_{ik}(\hat{\theta})$.

---

classification log-likelihood. This optimization can be performed using the Classification EM (CEM) algorithm (Celeux and Govaert, 1992), a variant of EM, which converts the s_{ik}'s to a discrete classification in a C-step before performing the M-step. This algorithm is faster than EM and is scalable. Next, we embed the clustering and feature selection in the mixture approach. We use the ML and CML approaches and derive EM and CEM algorithms with feature selection, assuming that the number of clusters is unknown. These algorithms are respectively known as EM-FS and CEM-FS.

3   Feature Selection

3.1   Definition of a New Bernoulli Mixture Model

We adopt an alternative formulation of the mixture model described in Section 2. First proposed by Grim (1986), the formulation assumes that each component density of the finite mixture consists of a common background to all classes, multiplied by a modifying parametric function. Thus, the model can be rewritten as

$$p(\mathbf{x}_i|\phi) = \sum_k \pi_k \prod_j p(x_{ij}, \alpha_{kj})^{\phi_j} q(x_{ij}; \lambda_j)^{1-\phi_j}$$

where $\phi_j = 1$ if the feature j is relevant and $\phi_j = 0$ otherwise. The function $q(x_{ij}; \lambda_j)$ is the common background Bernoulli distribution, $p(x_{ij}, \alpha_{kj})$ is the Bernoulli distribution of the jth feature in the kth component. Law et al. (2004) considered this model and introduced a key notion: the *feature saliency*. The *feature saliency* is the probability that a feature j is relevant i.e. $P(\phi_j = 1)$. In this context, ϕ_j can be considered as a hidden variable which determines which edges exist between the hidden class label and the individual features. We note $\Phi = (\phi_1, \ldots, \phi_d)$ the vector of binary parameters. Considering this new definition of the model, the mixture density can be written as $p(\mathbf{x}_i; \phi) = p(\mathbf{x}_i|\phi)p(\phi)$ and the marginal density is $p(\mathbf{x}_i) = \sum_\Phi P(\mathbf{x}_i; \phi)$. Assuming that the ϕ_j's are mutually independent and also independent from the hidden class label z_i, $p(\mathbf{x}_i)$

takes the following form:

$$p(\mathbf{x}_i) = \sum_k \pi_k \sum_\Phi \prod_j p(x_{ij}; \alpha_{kj})^{\phi_j} q(x_{ij}; \lambda_j)^{1-\phi_j} \prod_j \rho_j^{\phi_j}(1-\rho_j)^{1-\phi_j}$$

$$= \sum_k \pi_k \prod_j \sum_{\phi_j=0}^1 [\rho_j p(x_{ij}; \alpha_{kj})]^{\phi_j} [(1-\rho_j)q(x_{ij}; \lambda_j)]^{1-\phi_j},$$

and the likelihood of this model is given by:

$$P(\mathbf{x}; \boldsymbol{\theta}) = \prod_i \sum_k \pi_k \prod_j [\rho_j p(x_{ij}; \alpha_{kj}) + (1-\rho_j)q(x_{ij}; \lambda_j)]. \tag{1}$$

Next, we focus on the estimations of the parameters and clustering thanks to EM and CEM algorithms.

3.2   Estimation of the Parameters and Clustering

First, the expectation $Q(\boldsymbol{\theta}|\boldsymbol{\theta}')$ takes the following form:

$$Q(\boldsymbol{\theta}|\boldsymbol{\theta}') = \sum_{i,k} s_{ik} \log \pi_k' + \sum_{i,k,j} u_{ikj}[x_{ij} \log \frac{\alpha_{kj}'}{1-\alpha_{kj}'} + \log(1-\alpha_{kj}')]$$

$$+ \sum_{i,k,j} v_{ikj}[x_{ij} \log \frac{\lambda_j'}{1-\lambda_j'} + \log(1-\lambda_j')]$$

$$+ \sum_{i,k,j} u_{ikj} \log(\rho_j') + \sum_{i,k,j} v_{ikj} \log(1-\rho_j'),$$

where

$$s_{ik} = p[z_{ik}=1|\mathbf{x}_i, \boldsymbol{\theta}'] \propto \pi_k' \prod_{j=1}^d [\rho_j' p(x_{ij}; \alpha_{kj}') + (1-\rho_j')q(x_{ij}; \lambda_j')],$$

$$u_{ikj} = p[z_{ik}=1, \phi_j=1|\mathbf{x}_i, \boldsymbol{\theta}'] = p[z_{ik}=1|\mathbf{x}_i, \boldsymbol{\theta}'] \times p[\phi_j=1|z_{ik}=1, \mathbf{x}_i, \boldsymbol{\theta}']$$

$$= s_{ik} \times \beta_{ikj} \text{ where } \beta_{ikj} = \frac{\rho_j' p(x_{ij}; \alpha_{kj}')}{\rho_j' p(x_{ij}; \alpha_{kj}') + (1-\rho_j')q(x_{ij}; \lambda_j')},$$

$$v_{ikj} = p[z_{ik}=1, \phi_j=0|\mathbf{x}_i, \boldsymbol{\theta}']$$

$$= s_{ik} \times \gamma_{ikj} \text{ where } \gamma_{ikj} = \frac{(1-\rho_j')q(x_{ij}; \lambda_j')}{\rho_j' p(x_{ij}; \alpha_{kj}') + (1-\rho_j')q(x_{ij}; \lambda_j')}.$$

We then deduce the estimations of the parameters that are computed in the same way that for EM. In the M-step we have to estimate for each k and j the four parameters π_k, α_{kj}, λ_j and ρ_j. Fixing three of them, and setting the derivate of Q with respect to the fourth as zero, we have

$$\pi_k = \frac{\sum_i s_{ik}}{n}, \alpha_{kj} = \frac{\sum_i u_{ikj}x_{ij}}{\sum_i u_{ikj}}, \lambda_j = \frac{\sum_{i,k} v_{ikj}x_{ij}}{\sum_{i,k} v_{ikj}}, \rho_j = \frac{\sum_{i,k} u_{ikj}}{n} \tag{2}$$

At the convegence we deduce the clustering. Now, if we consider the CEM algorithm, in the M-step the computations are performed after the C-step; s_{ik} is replaced by $z_{ik} \in \{0,1\}$ in π_k. In addition, the posterior probabilities u_{ikj} and v_{ikj} depend on z_{ik} and we have $u_{ikj} = z_{ik}\beta_{ikj}$, $v_{ikj} = z_{ik}\gamma_{ikj}$.

Assessing the number of classes remains a challenging problem in clustering. In mixture modeling, different criteria need to be minimized, such as BIC (Schwarz, 1978) or ICL (Biernacki et al., 2000). The number of clusters is assessed at the convergence of standard EM. These criteria require the algorithms to be run with different numbers of classes. But in Law et al. (2004), the authors proposed an incremental EM algorithm optimizing the minimum message length (MML) criterion based on information/coding theory. They showed the usefulness of this criterion, but focused only on continuous data. Adapting this approach to our model, we now propose two algorithms and extend the model to binary data.

3.3 MML Criterion

Using the approximation of the Fisher information matrix of the log-likelihood, noted $\mathbf{I}$, by $\mathbf{I}_c$ the Fisher information matrix of the complete data log-likelihood, the MML criterion consists in minimizing, with respect to $\boldsymbol{\theta}$, the following cost function $-\log P(\boldsymbol{\theta}) - \log P(\mathbf{x}, \boldsymbol{\theta}) + \frac{1}{2}\log|\mathbf{I}_c(\boldsymbol{\theta})| + \frac{\mu}{2}(1 + \log\frac{1}{12})$, where μ is the number of parameters of th model. This leads to some calculations about the Fisher information matrix by differentiating the complete data log-likelihood and considering only the priors on the π_ks $P(\pi_1, \ldots, \pi_g) \propto \prod_k \pi_k^{-d/2}$ (see for details, (Figueiredo and Jain, 2002)). In addition, we approximate $P(\mathbf{x}, \boldsymbol{\theta})$ by $P(\mathbf{x}, \mathbf{z}, \boldsymbol{\theta})$. Note that this proposed approximation is also used in ICL (Biernacki et al., 2000) which corresponds to a classifying version of BIC (Schwarz, 1978). We propose a classifying version of MML. It is noteworthy that these two approximations of $\mathbf{I}$ and $P(\mathbf{x}, \boldsymbol{\theta})$ become exact in the limit of non-overlapping components. Finally, when we substitute $P(\boldsymbol{\theta})$, $|\mathbf{I}_c(\boldsymbol{\theta})|$ and $P(\mathbf{x}, \boldsymbol{\theta})$, and drop the order-one term, the cost function becomes

$$-\log P(\mathbf{x}, \mathbf{z}, \boldsymbol{\theta}) + \frac{1+d}{2}\log(n) + \frac{1}{2}\sum_{k,j}\log(n\pi_k\rho_j) + \frac{1}{2}\sum_j \log n(1-\rho_j).$$

Hence, we can propose two algorithms EM-FS and CEM-FS with detection of the number of clusters. In Algorithm 2, CEM FS is depicted. For EM-FS, criterion MML is based on $P(\mathbf{x}, \boldsymbol{\theta})$ instead of $P(\mathbf{x}, \mathbf{z}, \boldsymbol{\theta})$. In algorithmic terms, this removes the C-step and the M-step is performed with the same formulas but simply replacing z_{ik} by s_{ik}. We also propose a strategy which consists of running CEM-FS and, using the parameter estimation obtained to initialise EM-FS (CEM+EM-FS).

Algorithm 2. CEM-FS

---

   input : **x**, minimum number of components g_{min}.
   initialization : θ, Φ
   while $g > g_{min}$ **do**
      while not reach local minimum **do**
         E-step compute s_{ik}, u_{ikj}, v_{ikj}.
         C-step compute $z_{ik} = \arg\max_{k'=1,\dots,g} s_{ik}$.
         M-step compute θ using equations of (2) but using z_{ik} instead of s_{ik}.
         If π_k becomes 0, the k-th component is pruned.
      end while
      record : **z**, g, ρ_j, MML.
      remove : the component with the smallest weight.
   end while
   Return the parameters of the model with the MML and **z**.

---

4 Numerical Experiments

4.1 Synthetical Datasets

To study the performances of EM-FS, CEM-FS and CEM+EM-FS (EM-FS initialized by CEM-FS), we used simulated data of varying sizes. We selected binary data from three components, with 70% of the features considered as noise. For size, we took $n \times d = 1000 \times 100, 1000 \times 500, 1000 \times 1000$. We therefore selected nine types of data corresponding to three degrees of cluster overlap $(+, ++, +++)$, and three dimensions. Figure 1 shows data matrices with 500 features, 1000 objects, equal proportions and different degrees of separation of clusters. We can clearly see three cluster structures as well as the non salient features.

For each data structure, we generated 200 samples. For each sample, we ran EM-FS and CEM-FS 30 times in random situations. For each set of parameters in these trials, we summarized each error rate and NMI (Strehl and Ghosh, 2003) by its mean and its standard deviation.

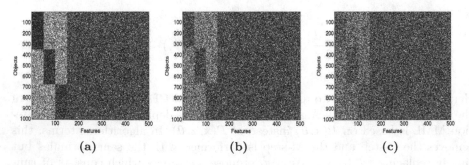

Fig. 1. Visualisation of reorgnized data matrix according the partition: dark point represents a 0 and white point a 1

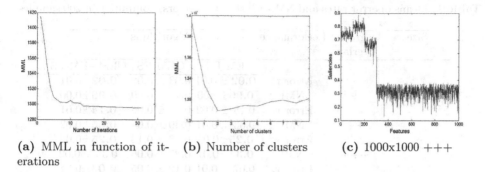

(a) MML in function of iterations   **(b)** Number of clusters   **(c)** 1000x1000 +++

Fig. 2. CEM-FS: behavior of MML (a), detection of the number of clusters (b) and estimation of the feature saliency (c) for dataset 1000×1000

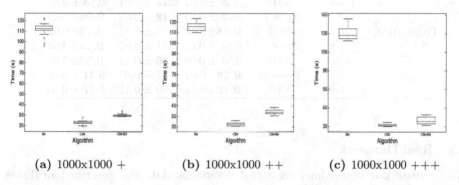

(a) 1000x1000 +   **(b)** 1000x1000 ++   **(c)** 1000x1000 +++

Fig. 3. Running time (Running time (including the selection of the number of clusters) for EM-FS versus CEM-FS and CEM+EM-FS for the three degree of separation + (a), ++(b) and +++ (c)

- The version of MML based on the complete data log-likelihood and minimized by CEM-FS appears particularly interesting in terms of convergence. Its behaviour is illustrated in Figure 2(a) for dataset 1000×1000.
- This model successfully detected the number of clusters in all situations. For instance, in Figure 2(b), the number of clusters is well detected (3 clusters). The number of relevant variables (30%) is also detected in Figure 2(c).
- EM-FS gives good results in terms of error rates and NMI (see Table 1). It outperforms CEM-FS but it suffers from slow convergence, as illustrated in Figure 3.
- The simple random initialization by EM-FS is often outperformed by CEM-FS and CEM+EM-FS. This gives encouraging results in terms of clustering (Table 1), with fast convergence compared to EM-FS. In Figure 3, it can be seen that CEM+EM-FS is about six times faster than EM-FS.

Table 1. Means of error rates and NMIs (± standard errors) computed on 200 samples

Size	Degree of overlap	Performance	Algorithms		
			EM-FS	CEM-FS	CEM+EM-FS
(1000, 100)	+	Error r.	**0.02** ±0.01	0.03 ± 0.02	**0.02**±0.01
		NMI	**0.96**±0.00	0.94 ± 0.00	**0.96**±0.00
	++	Error r.	0.15 ± 0.02	0.23 ± 0.06	**0.14**±0.01
		NMI	0.53 ± 0.01	0.30 ± 0.04	**0.55**±0.02
	+++	Error r.	**0.25**±0.05	0.38 ± 0.13	**0.25**±0.02
		NMI	**0.34**±0.01	0.27 ± 0.06	0.35 ± 0.02
(1000, 500)	+	Error r.	0.07 ± 0.01	0.12 ± 0.05	**0.03**±0.01
		NMI	0.75 ± 0.01	0.63 ± 0.05	**0.90**±0.01
	++	Error r.	**0.16**±0.02	0.23 ± 0.01	0.17 ± 0.07
		NMI	**0.52**±0.02	0.30 ± 0.05	0.51 ± 0.08
	+++	Error r.	**0.24**±0.05	0.32 ± 0.03	0.25 ± 0.07
		NMI	**0.20**±0.02	0.14 ± 0.01	0.18 ± 0.07
(1000, 1000)	+	Error r.	0.07 ± 0.01	0.12 ± 0.02	**0.06**±0.00
		NMI	**0.78**±0.12	0.76 ± 0.18	**0.78**±0.02
	++	Error r.	0.15 ± 0.03	0.21 ± 0.07	**0.14**±0.04
		NMI	0.52 ± 0.09	0.30 ± 0.11	**0.53**±0.06
	+++	Error r.	**0.28**±0.01	0.37 ± 0.05	0.31 ± 0.07
		NMI	**0.19**±0.12	0.09 ± 0.12	0.17 ± 0.05

4.2   Real Datasets

We evaluate our algorithms on 3 real document datasets described in Table 2. Each cell of these datasets denotes the number of occurrences of a word in a document. Note that these matrices are sparse and they are converted into binary data; each cell having a value higher to 1 is considered equal to 1 and 0 otherwise. In our approach the number of clusters is assumed to be unknown.

Table 2. Datasets Description

Datasets	# documents	#words	#class
CSTR	475	1000	4
Classic3	3891	4303	3
WebKB4	4199	8035	4

Tables 3 show the accuracy (ACC) which corresponds to (1-error rate), the normalized mutual information (NMI) and the adjusted rand index (ARI) (Hubert and Arabie, 1985). The results are averaged for 15 trials. It can be seen that both EM-FS, CEM-FS and CEM+EM-FS algorithms outperform the K-means and the Nonnegative Matrix Factorization (NMF) Berry et al. (2006).

Table 3. Accuracy, Normalized Mutual Information and Adjusted Rand Index

		\multicolumn{5}{c}{Algorithms}					Time
		Kmeans	NMF	CEM-FS	EM-FS	CEM+EM-FS	$\dfrac{time(\text{EM-FS})}{time(\text{CEM+EM-FS})}$
CSTR	ACC	0.74	0.67	0.75	0.88	**0.90**	
	NMI	0.59	0.58	0.58	**0.69**	**0.69**	7.4
	ARI	0.54	0.53	0.53	**0.67**	0.65	
Classic3	ACC	0.86	0.87	0.92	0.98	**0.99**	
	NMI	0.62	0.62	0.73	0.91	**0.93**	4.85
	ARI	0.64	0.66	0.77	0.94	**0.96**	
WebKB4	ACC	0.54	0.60	0.52	**0.65**	**0.65**	
	NMI	0.22	0.31	0.18	**0.33**	0.32	3.6
	ARI	0.14	0.30	0.08	**0.35**	**0.35**	

The accuracy, NMI and ARI of EM-FS and CEM+EM-FS are really close. For instance, in Classic3, CEM+EM-FS is slightly better than EM-FS, but the latter requires seven times more computation time. For the other datasets, CEM+EM-FS not only significantly outperforms K-means and NMF, it is also faster than EM-FS and the rate time $\dfrac{time(\text{EM-FS})}{time(\text{CEM+EM-FS})}$ is at least greater than 3.5.

5   Conclusion

We considered the clustering of binary data with feature selection using two approaches, ML and CML. We developed and studied two derived algorithms EM-FS and CEM-FS in detail. Since most classical clustering criteria can be viewed as complete data log-likelihood under constraints, the CEM-FS algorithm turns out to be a general clustering algorithm but with feature selection and with the assumption that the number of clusters is unknown. However, the clustering results are not better than with EM-FS. Then we proposed initializing EM-FS by the best result obtained by CEM-FS. The results in clustering and computing time are very encouraging. The use of CEM+EM-FS deserves to be extended to other models, such as multinomial mixtures, and compared to the model proposed by Li and Zhang (2008).

References

Berry, M.W., Browne, M., Langville, A.N., Pauca, V.P., Plemmons, R.J.: Algorithms and applications for approximate nonnegative matrix factorization. In: Computational Statistics and Data Analysis, pp. 155–173 (2006)

Biernacki, C., Celeux, G., Govaert, G.: Assessing a mixture model for clustering with the integrated completed likelihood. IEEE Transactions on Pattern Analysis and Machine Intelligence, 719–725 (2000)

Celeux, G., Govaert, G.: A classification em algorithm for clustering and two stochastic versions. Comput. Stat. Data Anal., 315–332 (1992)

Dempster, A.P., Laird, M.N., Rubin, D.B.: Maximum likelihood from incomplete data via the EM algorithm. Journal of the Royal Statistical Society: Series B (Statistical Methodology), 1–22 (1977)

Duda, R.O., Hart, P.E.: Pattern Classification and Scene Analysis. John Willey & Sons, New Yotk (1973)

Figueiredo, M.A.T., Jain, K.: Unsupervised learning of finite mixture models. IEEE Trans. Pattern Anal. Mach. Intell., 381–396 (2002)

Grim, J.: Multivariate statistical pattern recognition with nonreduced dimensionality. Kybernetika, 142–157 (1986)

Hubert, L., Arabie, P.: Comparing partitions. Journal of Classification, 193–218 (1985)

Law, M.H.C., Figueiredo, M.A.T., Jain, A.K.: Simultaneous feature selection and clustering using mixture models. IEEE Trans. Pattern Anal. Mach. Intell., 1154–1166 (2004)

Li, M., Zhang, L.: Multinomial mixture model with feature selection for text clustering. Know.-Based Syst., 704–708 (2008)

McLachlan, G.J., Peel, D.: Finite mixture models. New York (2000)

Pudil, P., Novovicová, J., Choakjarernwanit, N., Kittler, J.: Feature selection based on the approximation of class densities by finite mixtures of special type. Pattern Recognition, 1389–1398 (1995)

Schwarz, G.E.: Estimating the dimension of a model. Annal of Statistics, 461–464 (1978)

Strehl, A., Ghosh, J.: Cluster ensembles — a knowledge reuse framework for combining multiple partitions. J. Mach. Learn. Res., 583–617 (2003)

Symons, M.: Clustering criteria and multivariate normale mixture. Biometrics, 35–43 (1981)

Instant Exceptional Model Mining
Using Weighted Controlled Pattern Sampling

Sandy Moens[1,2] and Mario Boley[2,3]

[1] University of Antwerp, Belgium
firstname.lastname@uantwerpen.be,
[2] University of Bonn, Germany
firstname.lastname@uni-bonn.de,
[3] Fraunhofer IAIS, Germany
firstname.lastname@iais.fgh.de

Abstract. When plugged into instant interactive data analytics processes, pattern mining algorithms are required to produce small collections of high quality patterns in short amounts of time. In the case of Exceptional Model Mining (EMM), even heuristic approaches like beam search can fail to deliver this requirement, because in EMM each search step requires a relatively expensive model induction. In this work, we extend previous work on high performance controlled pattern sampling by introducing extra weighting functionality, to give more importance to certain data records in a dataset. We use the extended framework to quickly obtain patterns that are likely to show highly deviating models. Additionally, we combine this randomized approach with a heuristic pruning procedure that optimizes the pattern quality further. Experiments show that in contrast to traditional beam search, this combined method is able to find higher quality patterns using short time budgets.

Keywords: Controlled Pattern Sampling, Subgroup Discovery, Exceptional Model Mining.

1 Introduction

There is a growing body of research arguing for the integration of Local Pattern Mining techniques into instant, interactive discovery processes [3,5,9,10,14]. Their goal is to tightly integrate the user into the discovery process to facilitate finding patterns that are interesting with respect to her current subjective interest. In order to allow true interactivity, the key requirement for a mining algorithm in such processes, is that it is capable of producing high quality results within very short time budgets — only up to a few seconds.

A particularly hard task for this setting is Exceptional Model Mining (EMM) [8], i.e., the discovery of subgroups showing data models that highly deviate from the model fitted to the complete data. In the EMM setting, even fast heuristic methods that cut down the search space tremendously, e.g., beam search [8], can fail to deliver the fast response times necessary for the interactive setting. This

H. Blockeel et al. (Eds.): IDA 2014, LNCS 8819, pp. 203–214, 2014.

comes from the fact that every individual search step in the subgroup description space involves an expensive model induction step.

In this paper, we extend an alternative randomized technique to pattern discovery, Controlled Direct Pattern Sampling [4], and adapt it to EMM. As opposed to many other algorithmic approaches, direct pattern sampling does not traverse any part of the pattern search space. Instead, it defines an efficient sampling process that yields patterns according to a distribution, which overweights high-quality patterns. A previously published framework [6] allows to express distributions in terms of the pattern support. Here we extend it to specify distributions in terms of the weighted pattern support. We then develop a weighting scheme based on Principal Component Analysis, which leads to efficient sampling procedures particularily suitable for EMM tasks. As we show empirically, when combined with a lightweight local search procedure as post-processing step, the resulting EMM algorithms outperform both, pure local search as well as pure sampling strategies, and deliver high-quality results for short time budgets.

2   Exceptional Model Mining

Throughout this paper we assume that a **dataset** $D = \{d_1, \ldots, d_m\}$ consists of m **data records** $d \in D$, each of which is described by n **descriptive attributes** $A = \{a_1, \ldots, a_n\}$ and annotated by k **target attributes** $T = \{t_1, \ldots, t_k\}$. All attributes $f \in A \cup T$ assign to each data record a value from their **attribute domain** $\mathrm{Dom}(f)$, i.e., $f : D \to \mathrm{Dom}(f)$. In this paper we assume that all attributes $f \in A \cup T$ are either **numeric**, i.e., $\mathrm{Dom}(f) \subset \mathbb{R}$ and we use $\leq$ to compare attribute values, or **categoric**, i.e., $\mathrm{Dom}(f)$ is finite and its values are conceptually incomparable. We are interested in conjunctive **patterns** of simple binary propositions about individual data records. This is the standard setting in subgroup discovery and itemset mining. That is, a **pattern descriptor** p can be formalized as a set $p = \{c_1, \ldots, c_l\}$ where $c_j : \mathrm{Dom}(a_{i_j}) \to \{\mathrm{true}, \mathrm{false}\}$ is a **constraint** on the descriptive attribute a_{i_j} for $j = 1, \ldots, l$ (corresponding to item literals in frequent set mining). Correspondingly, the **support set** (or **extension**) of p is the subset of data records for which all constraints hold, i.e.,

$$\mathrm{Ext}(D, p) = \{d \in D : c_1(a_{i_1}(d)) \wedge \cdots \wedge c_l(a_{i_l}(d))\} \ ,$$

and the **frequency** of p is defined as the size of its extension relative to the total number of data records $\mathrm{frq}(D, p) = |\mathrm{Ext}(D, p)|/m$. We write $\mathrm{Ext}(p)$, resp. $\mathrm{frq}(p)$, when D is clear. For the constraints, one typically uses equality constraints if a_{i_j} is categorical, i.e., $c_j(v) \equiv v = w$ for $w \in \mathrm{Dom}(a_{i_j})$, and interval constraints if a_{i_j} is numeric, i.e., $c_j(v) \equiv v \in [l, u]$ for a few expressive choices of interval borders $l, u \in \mathrm{Dom}(a_{i_j})$ (e.g., corresponding to the quartiles of $\{a_{i_j}(d) : d \in D\}$).

Let C denote the **constraint universe** containing all the constraints that we want to use to express patterns. We are interested in searching the **pattern language** $L = \mathcal{P}(C)$ for descriptors $p \in L$ with a) a relatively high frequency and b) such that the target attributes behave differently in $\mathrm{Ext}(p)$ than in the complete data. This behavior is captured by how the target attributes are represented by a

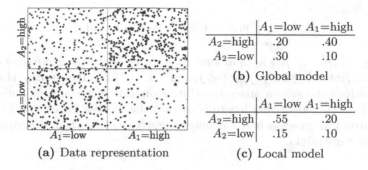

	A_1=low	A_1=high
A_2=high	.20	.40
A_2=low	.30	.10

(b) Global model

	A_1=low	A_1=high
A_2=high	.55	.20
A_2=low	.15	.10

(a) Data representation   **(c)** Local model

Fig. 1. Exceptional contingency table models for fictitious dataset with two numerical attributes: red+blue is the global data and red is a local pattern. (b) shows the global model and (c) the local model. The model deviation equals .35.

model of a certain **model class** M. That is, formally, a **model** $m(D') \in M$ can be induced for any subset of the data records $D' \subseteq D$, and there is a meaningful **distance measure** $\delta \colon M \times M \to \mathbb{R}_+$ between models. Then the **interestingness** of a pattern descriptor $p \in L$ is given as $\mathrm{int}(p) = \mathrm{frq}(p)\, \delta(m(D), m(\mathrm{Ext}(p)))$.

In this paper we focus on non-functional models that treat all target attributes symmetric. When all target attributes are numeric, the perhaps simplest example of a model class are the **mean models** $M_{\mathrm{mn}} = \mathbb{R}^k$ defined by

$$m(D') = (\bar{t}_1(D'), \ldots, \bar{t}_k(D'))$$

with $\bar{t}_i(D') = \sum_{d \in D'} t_i(d)/|D'|$. A useful distance measure between two mean models $m, m' \in M_{\mathrm{mn}}$ is for instance given by the **normalized Euclidean distance** $\delta_{\mathrm{nl2}}(m, m') = \sqrt{(m - m')^T S^{-1}(m - m')}$ where S denotes the diagonal matrix with entries $S_{i,i}$ equal to the standard deviation of target attribute t_i on the data. For categorical targets, a simple example are the **contingency table models** $M_{\mathrm{ct}} = \mathbb{R}^{V_{t_1} \times \cdots \times V_{t_k}}$ where each $m \in M_{\mathrm{ct}}$ represents the relative counts of all target value combinations, i.e.,

$$m(D')_v = |\{d \in D' \colon t_1(d) = v_1 \wedge \cdots \wedge t_k(d) = v_k\}|/|D'|$$

for all $v \in V_{t_1} \times \cdots \times V_{t_k}$. A meaningful distance measure between contingency tables $m, m' \in M_{\mathrm{ct}}$ is the **total variation distance** defined by $\delta_{\mathrm{tvd}}(m, m') = \sum_{v \in V_{t_1} \times \cdots \times V_{t_k}} |m_v - m'_v|/2$. See Figure 1 for an example.

Since EMM is a computationally hard problem and no efficient way is known to find a pattern descriptor $p \in L$ for a given dataset that maximizes the EMM interestingness, the standard algorithmic approach to EMM is heuristic **beam search**. This strategy is an extension of a greedy search, where on each search level (corresponding to a number of constraints in a pattern descriptor) instead of extending only one partial solution by a constraint that locally optimizes the interestingness, one considers $b \in \mathbb{N}$ best partial solutions. This parameter b is referred to as the **beam-width**. Formally, starting from **search level** $L'_0 = \{\emptyset\}$, level L'_{i+1} is defined as

$$L'_{i+1} = \bigcup_{i=1}^{b} \{p_i \wedge c \colon c \in C \setminus p_i, \mathrm{frq}(p_i \wedge c) \geq \tau\}$$

where $\{p_1, \ldots, p_b, \ldots, p_z\} = L'_i$ in some order consistent with decreasing interestingness, i.e., $\mathrm{int}(p_i) \geq \mathrm{int}(p_j)$ for $i < j$, and $\tau \in [0, 1]$ is a **frequency threshold** used to reduce the search space (possibly alongside other anti-monotone hard constraints). This algorithm has to construct the models for $\Theta(bl|C|)$ elements of the pattern language where l denotes the average length of descriptors that satisfy the constraints.

3   Sampling Exceptional Models

In this section we develop an alternative approach to EMM using Controlled Direct Pattern Sampling (CDPS). The key idea of this approach is that we create random patterns by a fast procedure following a controlled distribution that is useful for EMM, i.e., that favors patterns with a high frequency and a large model deviation. In contrast to beam search, sampling only requires to perform a model induction after a full descriptor is found. As we will argue later, it is most efficient to combine this sampling approach with a very lightweight local search procedure as post-processing step.

3.1   Weighted Controlled Direct Pattern Sampling

Boley et al. [6] gives a fast algorithm for CDPS that draws samples from a user-defined distribution over the pattern space using a simple two-step random experiment. Distributions that can be simulated with this approach are those that can be expressed as the product of frequency functions wrt to different parts of the data. Here we extend this idea by allowing to specify **utility weights** $w(d) \in \mathbb{R}^+$ for each data record $d \in D$.

With this we define the **weighted frequency** as the relative total weight of a pattern's extension, i.e., $\mathrm{wfrq}(D, p) = \sum_{d \in \mathrm{Ext}(p)} w(d) / \sum_{d \in D} w(d)$, and the **negative weighted frequency** equals $\overline{\mathrm{wfrq}}(D, p) = 1 - \mathrm{wfrq}(D, p)$. Let $D_i^+, D_j^- \subseteq D$ be subsets of the data for $i \in \{1, \ldots, a\}$ and $j \in \{1, \ldots, b\}$. Now we can define a random variable over the pattern space $\mathbf{p} \in L$ by

$$\mathbb{P}[\mathbf{p} = p] = \prod_{i=1}^{a} \mathrm{wfrq}(D_i^+, p) \prod_{j=1}^{b} \overline{\mathrm{wfrq}}(D_j^-, p)/Z, \tag{1}$$

with a normalization constant Z such that $\sum_{p \in L} \mathbb{P}[\mathbf{p} = p] = 1$. This distribution gives a high probability to patterns that have a high weighted frequency in D_i^+, further referred to as the **positive data portions**, and a low weighted frequency in D_j^-, further referred to as the **negative data portions**. As an example, when designing an algorithm for subgroup discovery in data with binary labels, data records with a positive label could be assigned to a positive data portion and data records with a negative label to a negative data portion. This results in a pattern distribution favoring patterns for which data records in their extension are assigned mainly a positive label. In subsequent sections we will use distributions from this family (Eq. 1) to construct effective EMM algorithms. However, we first show that realizations of $\mathbf{p}$ can be computed with a two-step framework similar to the one given in Boley et al. [6][1].

Let us denote by $\mathbb{D} = D_1^+ \times \cdots \times D_a^+ \times D_1^- \times \cdots \times D_b^-$ the Cartesian product of all data portions involved in the definition of $\mathbf{p}$ containing one representative record for each positive and each negative data portion. For a tuple of data records $r \in \mathbb{D}$ let

$$L_r = \{p \in L \colon r(i) \in \mathrm{Ext}(D_i^+, p), 1 \leq i \leq a \wedge r(j) \notin \mathrm{Ext}(D_{j-a}^-, p), a < j \leq a+b\}$$

denote the set of pattern descriptors having in their extensions all positive representatives $r(1), \ldots, r(a)$ and none of the negatives $r(a+1), \ldots, r(a+b)$. Then consider the random variable $\mathbf{r} \in \mathbb{D}$ defined by

$$\mathbb{P}[\mathbf{r} = r] = |L_r| \prod_{i=1}^{a+b} w(r(i))$$

In the following proposition we note that in order to simulate our desired distribution $\mathbf{p}$ it is sufficient to first draw a realization r of $\mathbf{r}$ and then to uniformly draw a pattern from L_r.

Proposition 1. *For a finite set X denote by $\mathbf{u}(X)$ a uniform sample from X. Then $\mathbf{p} = \mathbf{u}(L_\mathbf{r})$.*

Proof. Denote by $\mathbb{D}^p = \{r \in \mathbb{D} \colon p \in L_r\}$. Noting that $\mathbb{D}^p$ is equal to

$$\mathrm{Ext}(D_1^+, p) \times \cdots \times \mathrm{Ext}(D_a^+, p) \times \left(D_1^- \setminus \mathrm{Ext}(D_1^-, p)\right) \times \cdots \times \left(D_b^- \setminus \mathrm{Ext}(D_b^-, p)\right)$$

it follows that

$$\mathbb{P}[\mathbf{u}(L_\mathbf{r}) = p] = \sum_{r \in \mathbb{D}^p} \mathbb{P}[\mathbf{u}(L_r) = p | \mathbf{r} = r]\mathbb{P}[\mathbf{r} = r]$$

$$= \sum_{r \in \mathbb{D}^p} \frac{1}{|L_r|} \frac{|L_r| \prod_{i=1}^{a+b} w(r(i))}{Z} = \frac{1}{Z} \sum_{r \subset \mathbb{D}^p} \prod_{i=1}^{a+b} w(r(i))$$

$$= \frac{1}{Z} \prod_{i=1}^{a} \sum_{d \in \mathrm{Ext}(D_i^+, p)} w(d) \prod_{j=1}^{b} \sum_{d \in D_j^- \setminus \mathrm{Ext}(D_j^-, p)} w(d) \tag{2}$$

$$= \frac{1}{Z} \prod_{i=1}^{a} \mathrm{wfrq}(D_i^+, p) \prod_{j=1}^{b} \overline{\mathrm{wfrq}}(D_j^-, p) = \mathbb{P}[\mathbf{p} = p]$$

□

---

[1] Note that the algorithm given in Boley et al. [6] also allows to specify modular prior preferences for pattern descriptors as well as to avoid descriptors of length 1 and 0. We omit both additions here for the sake of simplicity and note that they could be included in exactly the same way as in the original algorithm.

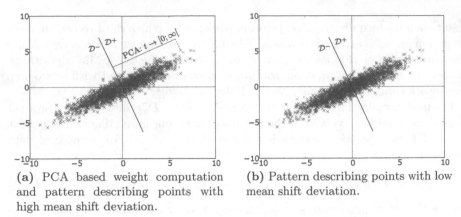

(a) PCA based weight computation and pattern describing points with high mean shift deviation.

(b) Pattern describing points with low mean shift deviation.

Fig. 2. Example weighting scheme for mean shift models for fictitious dataset

Efficient implementations of $\mathbf{r}$ and $\mathbf{u}(L_r)$ for $r \in \mathbb{D}$ can be performed by using coupling from the past and sequential constraint sampling, respectively, for which we refer to Boley et al. [6]. In the remainder of this paper, we focus on utilizing the resulting pattern sampler for EMM.

3.2 Application to Exceptional Model Mining

We start with the case of the contingency table models M_{ct}. Let $\mathbb{V} = V_{t_1} \times \cdots \times V_{t_k}$ be the set of all cells of a contingency table $m \in M_{ct}$. We give an instantiation of Eq. 1 using exactly one positive and one negative frequency factors, i.e., $a = b = 1$ with disjoint data portions D^+, D^- that partition D. The idea is that we try to oversample patterns with an extension lying mostly in contingency table cells with small counts. For that we can sample a random subset of the table cells $\mathbf{W} \subseteq \mathbb{V}$ with $|W| = |\mathbb{V}|/2$ such that $\mathbb{P}[v \in \mathbf{W}] = m(D)_v^{-1} Z^{-1}$ and assign each $d \in D$ to D^+ if $v(d) \in \mathbf{W}$ and to D^- otherwise (where $v(d)$ denotes the contingency table cell of d). Note that more focused versions of this distribution can be achieved by simply replicating the two frequency factors described here. Also, for this simple instantiation we did not use utility weights for the data records (i.e., they are chosen uniform).

Now turning to the case of high mean shift deviation M_{mn}, we will give another instantiation of Eq. 1 that also uses weights in addition to defining suitable positive and negative data portions. Since we only have one weight vector for the data records, we are interested in the direction in which the largest target deviation from the mean can be achieved. By applying a centralized Principal Component Analysis (PCA) we can find a linear transformation of the target data vectors that maximizes the variance among the data points. The first component then gives the direction of interest. Let us denote by $PCA_1(d) \in \mathbb{R}$ the first component of $(t_1(d), \ldots, t_k(d))$, i.e., the length of the target vector of a data record in the direction of highest variance. We define $d \in D^+$ if $PCA_1(d) \geq 0$ and $d \in D^-$ otherwise. This idea is shown in Figure 2a: the black line shows the first component, the points to the right are assigned to D^+ and the points to the left are assigned to D^-. Note that in practice we can randomly choose which side is D^+ or D^-. For the computation of weights, recall that our task is finding descriptors with a high mean shift (see again Figure 2a). As such, data points in the extension of a pattern that are in D^+, should be far away from the mean. While data point in the extension of a pattern that are in D^-, should be very close to the mean, such that the mean gets minimally shifted towards the center. Hence, it is sensible to use $w(d) = |\mathrm{PCA}_1(d)|$ as weights for data records. This means that in the positive part, data records far away from the mean will contribute a lot to $\mathrm{wfrq}(D^+, p)$ and in the negative part, points close to the center—having small absolute weights—will contribute a lot to $\overline{\mathrm{wfrq}}(D^-, p)$.

Finally, we propose to combine the EMM pattern sampler with a pruning routine, in order to further optimize the quality of sampled patterns. Our method then becomes a two-step framework: (1) optimizes the model deviation through direct sampling and (2) optimizes the interestingness via pruning. We employ heuristic optimization on patterns to optimize wrt the interestingness. First, we generate a random permutation of constraints. Then we remove each constraint one by one. If the quality increases, we replace the pattern. For a pattern p, the pruning step constructs models for $\Theta(k)$ patterns where $k = |p|$. The total cost of our sampling procedure is $\Theta(l + 1)$, where l is the average length of descriptors that satisfy the constraints. This is a theoretical advantage over beam search when model induction is expensive (e.g., when there are a lot of data points for which contingency tables have to be computed).

4 Experiments

Table 1. Overview of dataset characteristics

dataset	#attributes	#data records	time budget (ms)
Adult	15	30,163	300
Bank Marketing	17	45,211	300
Twitter	34	100,000	2,000
Cover Type	10	581,012	2,000

In the previous section we introduced a method for sampling exceptional models. We show now that our method is able to outperform beam search when given short time budgets. Throughout the experiments we used datasets available from the UCI Machine Learning Repository [2]. Their main characteristics together with the individual mining times are summarized in Table 1. The mining times do not taking into account loading the data, since in interactive systems the data is already loaded. For each dataset we removed lines with missing values. For Twitter we used only the first two measurements and for Cover Type we used the first 10 attributes. At last, both techniques run on Java 7.

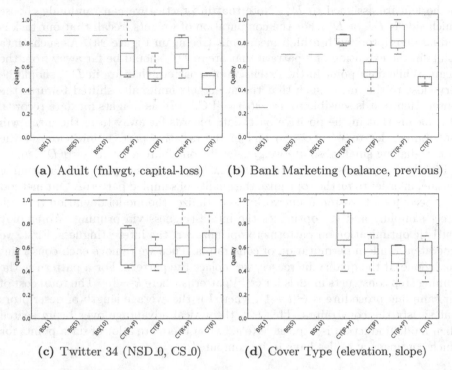

(a) Adult (fnlwgt, capital-loss)   (b) Bank Marketing (balance, previous)

(c) Twitter 34 (NSD_0, CS_0)   (d) Cover Type (elevation, slope)

Fig. 3. Max qualities for 2 target exceptional contingency table models

4.1 Contingency Table Quality

In this experiment we analyze contingency table models found by our sampler and compare them to models found by beam search. Throughout the experiments we used short time budgets found in Table 1. For the quality assessment we used the interestingness from Section 2.

We ran the algorithms to find exceptional contingency tables with 2 pre-defined targets. For beam search (BS) we only reported runs with beam widths 1, 5 and 10 since the others behave similar. For the sampling process Equation 1 with 2 positive and 1 negative factor and no weighting strategy. We fixed four settings: $CT(IP + P)$ – inverse probability for sampling D^+ and pruning, $CT(IP)$ – inverse probability without pruning, $CT(R + P)$ – uniform selection of D^+ in combination with pruning $CT(R)$ – uniform selection without pruning. Moreover, we ran each algorithm 10 times and extracted the highest quality patterns found for each run. We then normalized the results by the best pattern found over all algorithms. Aggregated results are shown in Figures 3.

Generally, sampling is able to find higher quality patterns using short time budgets. The problem with beam search is that it has to start from singleton patterns every time and evaluate them individually. The sampling process, in contrast, immediately samples larger seeds with high deviation. It then locally optimizes the interestingness by pruning. Therefore, it can quickly find high

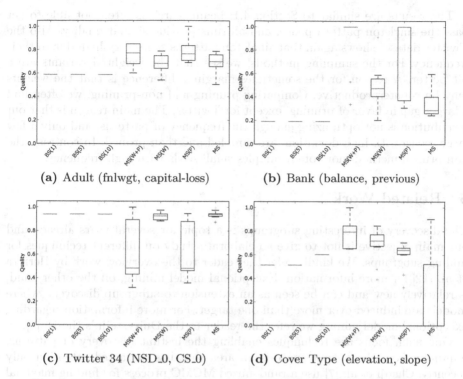

(a) Adult (fnlwgt, capital-loss)   (b) Bank (balance, previous)

(c) Twitter 34 (NSD_0, CS_0)   (d) Cover Type (elevation, slope)

Fig. 4. Max qualities for 2 target exceptional mean shift deviation models

deviation patterns with more than 1 descriptor, while beam search often is still enumerating patterns with 1 descriptor. Surprisingly, especially for Twitter 34, singleton patterns show already high interestingness, because of their frequency.

As expected, the unpruned versions performs slightly worse, because the frequency for sampled patterns is lacking. Comparing uniform (R) to pseudo-randomized (IP), we see that neither of the two is really able to outperform the other. One could argue that the pseudo-randomized version is a bit better at providing qualitative patterns more consistently.

4.2 Mean Model Quality

Here we used the same setup as before: i.e., we assume an interactive system with limited time budgets (see Table 1) and ran algorithms with 2 pre-defined targets. We ran beam search (BS) with beam sizes of 1, 5 and 10. For sampling we used 2 weighted positive and 1 weighted negative factor using the PCA method (implementation provided by WEKA [11]). We used 4 settings: $MS(W + P)$ – weighting and pruning, $MS(W)$ – weighting and no pruning, $MS(P)$ – no weights and pruning and, at last, MS – no weights and no pruning. Aggregated results of maximum quality patterns over 10 runs are summarized in Figure 4.

The results are similar to Section 4.1: beam search is often not able to get past the singleton pattern phase, and all runs provide similar quality. Also the Twitter dataset shows again that singleton patterns score very high due to their frequency. For the sampling methods, we see that the weighted variants are a bit better. A reason for the sometimes marginal difference is that the weights may be counterproductive. Comparing pruning and non-pruning, we often find a larger gap in favor of pruning, except for Twitter. The main reason is that our distribution is not optimizing enough the frequency of patterns, and only a few large seeds, with high deviation, are sampled and then pruned. In contrast, the non-pruned method more often samples small seeds with high frequency.

5   Related Work

The discovery of interesting subgroups is a topic for several years already and our main objective is not to give an elaborate study on different techniques for finding subgroups. We kindly refer the reader to the overview work by Herrera et al. [12] for more information. Exceptional model mining, on the other hand, is relatively new and can be seen as an extension to subgroup discovery, where models are induced over more than one target. For more information regarding exceptional model mining we refer the reader to the Duivesteijn's thesis [8].

Our main focus are techniques enabling the instant discovery of patterns. Sampling from the output space is an area that has attracted attention only recently. Chaoji et al. [7] use a randomized MCMC process for finding maximal subgraphs in graph databases. Their method is biased towards larger subgraphs, but they use heuristics to overcome this bias. Also on graphs, Al Hasan and Zaki [1] use Metropolis-Hastings to enable uniform sampling of maximal graphs. Moens and Goethals [13] proposed a method similar to the one by Chaoji et al. for sampling the border of maximal itemsets.

At last, we give a short overview of recent exploratory data mining tools, that have high demands wrt responsiveness. MIME [10] allows a user to interact with data directly by letting her create patterns and pattern collections that are evaluated on-the-fly. Moreover, different data mining algorithms can be applied and as their results become input to the user, she can adapt the results at will. Boley et al. [5] propose a framework combining multiple data mining algorithms in a black box environment, alleviating the user from the process of choosing pattern mining methods to apply. They employ a user preference model, based on user interactions, which influence running times for the black boxes. Dzuyba et al. [9] use beam search as their underlying method for finding interesting subgroups. Users then provide feedback on generated patterns, to give more/less importance to specific branches in the search tree.

6   Conclusion and Future Work

Existing methods for finding exceptional models, fail to produce instant results required for interactive discovery processes. In this work, we extended Controlled

Direct Pattern Sampling with weights for individual data records and used the framework to directly sample exceptional models using short time budgets.

We showed in our experiments that sampling is able to find better quality patterns in settings that where previously out of reach for beam search. We also showed that by optimizing sampled patterns locally, the quality of patterns can be improved even more. Moreover, we showed that our new weighting scheme can push sampled models into higher quality parts of the search space. However, the weighting can also have a negative effect, when instantiated improperly.

At last we point out future research directions for this research. An important step is extending the mean shift model to more than 3 attributes. The current framework uses the first component by PCA to obtain the highest variance direction, and next samples patterns that lie on the poles of this direction. However, when increasing the number of attributes, using only the first component is not enough and using more components is not optimizing the deviation enough in practice. Different strategies for partitioning the data in positive and negative parts with proper weight assignments is an important issue.

Acknowledgements. This work was supported by the German Science Foundation (DFG) under reference number 'GA 1615/2-1' and Research Foundation Flanders (FWO).

References

1. Hasan, M.A., Zaki, M.J.: Output space sampling for graph patterns. In: Proc. VLDB Endow, pp. 730–741 (2009)
2. Bache, K., Lichman, M.: UCI machine learning repository (2013)
3. Blumenstock, A., Hipp, J., Kempe, S., Lanquillon, C., Wirth, R.: Interactivity closes the gap. In: Proc. ACM SIGKDD 2006 Workshop on Data Mining for Business Applications (2006)
4. Boley, M., Lucchese, C., Paurat, D., Gärtner, T.: Direct local pattern sampling by efficient two–step random procedures. In: Proc. ACM SIGKDD 2011 (2011)
5. Boley, M., Mampaey, M., Kang, B., Tokmakov, P., Wrobel, S.: One click mining: Interactive local pattern discovery through implicit preference and performance learning. In: Proc. ACM SIGKDD 2013 Workshop IDEA, pp. 27–35. ACM (2013)
6. Boley, M., Moens, S., Gärtner, T.: Linear space direct pattern sampling using coupling from the past. In: Proc. ACM SIGKDD 2012, pp. 69–77. ACM (2012)
7. Chaoji, V., Hasan, M.A., Salem, S., Besson, J., Zaki, M.J.: Origami: A novel and effective approach for mining representative orthogonal graph patterns. In: Stat. Anal. Data Min., pp. 67–84 (2008)
8. Duivesteijn, W.: Exceptional model mining. PhD thesis, Leiden Institute of Advanced Computer Science (LIACS), Faculty of Science, Leiden University (2013)
9. Dzyuba, V., van Leeuwen, M.: Interactive discovery of interesting subgroup sets. In: Tucker, A., Höppner, F., Siebes, A., Swift, S. (eds.) IDA 2013. LNCS, vol. 8207, pp. 150–161. Springer, Heidelberg (2013)
10. Goethals, B., Moens, S., Vreeken, J.: Mime: a framework for interactive visual pattern mining. In: Proc. ACM SIGKDD 2011, pp. 757–760. ACM (2011)

214 S. Moens and M. Boley

11. Hall, M., Frank, E., Holmes, G., Pfahringer, B., Reutemann, P., Witten, I.H.: The weka data mining software: An update. SIGKDD Explor. Newsl. 11(1), 10–18 (2009)
12. Herrera, F., Carmona, C.J., González, P., del Jesus, M.J.: An overview on subgroup discovery: Foundations and applications. Knowl. Inf. Syst., 495–525 (2011)
13. Moens, S., Goethals, B.: Randomly sampling maximal itemsets. In: Proc. ACM SIGKDD 2013 Workshop IDEA, pp. 79–86 (2013)
14. Škrabal, R., Šimůnek, M., Vojíř, S., Hazucha, A., Marek, T., Chudán, D., Kliegr, T.: Association Rule Mining Following the Web Search Paradigm. In: Flach, P.A., De Bie, T., Cristianini, N. (eds.) ECML PKDD 2012, Part II. LNCS, vol. 7524, pp. 808–811. Springer, Heidelberg (2012)

Resampling Approaches to Improve News Importance Prediction

Nuno Moniz[1], Luís Torgo[1], and Fátima Rodrigues[2]

[1] LIAAD-INESC TEC / FCUP-DCC, University of Porto
[2] GECAD - ISEP/IPP, ISEP-DEI, Polytechnic of Porto
nmmoniz@liaad.up.pt, ltorgo@dcc.fc.up.pt, mfc@isep.ipp.pt

Abstract. The methods used to produce news rankings by recommender systems are not public and it is unclear if they reflect the real importance assigned by readers. We address the task of trying to forecast the number of times a news item will be tweeted, as a proxy for the importance assigned by its readers. We focus on methods for accurately forecasting which news will have a high number of tweets as these are the key for accurate recommendations. This type of news is rare and this creates difficulties to standard prediction methods. Recent research has shown that most models will fail on tasks where the goal is accuracy on a small sub-set of rare values of the target variable. In order to overcome this, resampling approaches with several methods for handling imbalanced regression tasks were tested in our domain. This paper describes and discusses the results of these experimental comparisons.

1 Introduction

The Internet is becoming one of the main sources for users to collect news concerning their topics of interest. News recommender systems provide help in managing the huge amount of information that is available. A typical example of these systems is Google News, a well-known and highly solicited robust news aggregator counting several thousands of official news sources. Although the actual process of ranking the news that is used by Google News is not known, official sources state that it is based on characteristics such as freshness, location, relevance and diversity. This process, based on the Page Rank algorithm explained in Page et al. [16] and generally described in Curtiss et al. [4], is of the most importance as it is responsible for presenting the best possible results for a user query on a set of given terms. However, some points have been questioned such as the type of documents that the algorithm gives preference to and its effects, and some authors conclude that it favours legacy media such as print or broadcast news "over pure players, aggregators or digital native organizations" [6]. Also, as it seems, this algorithm does not make use of the available information concerning the impact in or importance given by real-time users in an apparent strategy of deflecting attempts of using its capabilities in ones personal favour.

This paper describes some initial attempts on a task that is part of a larger project that tries to merge the news recommendations provided by two types

H. Blockeel et al. (Eds.): IDA 2014, LNCS 8819, pp. 215–226, 2014.

of sources: (i) official media that will be represented by Google News; and (ii) the recommendations of Internet users as they emerge from their social network activity. Concerning this latter source, the idea is to use Twitter as the source of information for checking which news are being shared the most by users. We will use the number of tweets of a given news as a kind of proxy for its impact on consumers of news, on a given topic[1]. The workflow solved by the system we plan to develop within this project is the following:

1. At a time t a user asks for the most interesting news for topic X
2. The system returns a ranking of news (a recommendation) that results from aggregating two other rankings:
   - The ranking provided by the official media (represented by Google News)
   - The ranking of the consumers of news (represented by the number of times the news are tweeted)

This workflow requires that at any point in time we are able to anticipate how important a news will be in Twitter. If some time has already past since the news publication this can be estimated by looking at the observed number of times this news piece was tweeted. However, when the news is very recent, the number of already observed tweets will potentially under-estimate the attributed importance of the news item. In this context, for very "fresh" news we need to be able to estimate their future relevance within Twitter, i.e. their future number of tweets. Moreover, given that we are interested in using this estimated number of tweets as a proxy for the news relevance, we are only interested in forecasting this number accurately for news with high impact, i.e. we are interested in being accurate at forecasting the news that will have a high number of tweets. This is the goal of the work presented in this paper. We describe an experimental comparison of different approaches to forecasting the future number of tweets of news, when the goal is predictive accuracy for highly popular news. This latter aspect is the key distinguishing aspect of our work when compared to existing related work, and it is motivated by the above-mentioned long-term project of news recommendation integrating different types of rankings. In this context, the main contribution of this work is a study and proposal of approaches to the problem of predicting highly popular news upon their publication.

2   Previous Work

In our research, although we did not find work that deals with the prediction of rare cases of highly tweeted news events, we did find important work focused on the general prediction of the number of tweets a news events will obtain in a given future.

---

[1] We are aware that this may be a debatable assumption, still this is something objective that can be easily operationalised, whilst other alternatives would typically introduce some subjectivity that would also be questionable.

Leskovec et al. [15] suggests that popular news take about four days until their popularity stagnates. Our own research using the data collected for this paper suggests that tweeting of a news item very rarely occours after two days.

Related to the subject in this paper, Asur and Huberman [1] use Twitter to forecast the box-office revenues for movies by building linear regression models, as well as demonstrate the utility that sentiment analysis has in the improvement of such objectives.

In Bandari et al. [2] classification and regression algorithms are examined in order to predict popularity, translated as the number of tweets, of articles in Twitter. The distinguishing factor of this work from others [19, 13, 20, 11, 14] that attempt to predict popularity of items, is that this work attempts to do this prior to the publication of the item. To this purpose the authors used four features: source of the article, category, subjectivity in the language and named entities mentioned. Furthermore, the authors conclude that the source of a given article is one of the most important predictors.

Regression, classification and hybrid approaches are used by Gupta et al. [7] also to predict event popularity. However, in this work the authors use data from Twitter, such as the number of followers/followees. The objective is the same in the work of Hsieh et al. [9], but the authors approach the problem by improving crowd wisdom with the proposal of two strategies: combining crowd wisdom with expert wisdom and reducing the noise by removing "overly talkative" users from the analysis.

Recently, a Bayesian approach was proposed by Zaman et al. [25] where a probabilistic model for the evolution of the retweets was developed. This work differs from the others in a significant manner as it is focused on the prediction of the popularity of a given tweet. The authors conclude that it is possible to predict the popularity of a given tweet after 10 minutes of its publication. They state that the number of tweets after two hours of its publication should improve by roughly 50% in relation to the first ten minutes. The test cases include both famous and non-famous twitter accounts. This work is preceded by others also using the retweet function as predictor having as the objective result an interval [8] or the probability of being retweeted.

3   Problem Description and Approach

This work addresses the issue of predicting the number of tweets of very recent news events with a focus on the predictive accuracy at news that are highly tweeted. This is a numeric prediction task, where we are trying to forecast this number based on some description of the news. However, this task has one particularity: we are only interested in prediction accuracy at a small sub-set of the news - the ones that are tweeted the most. These are the news that the public deems as highly relevant for a given topic and these are the ones we want to put at the top of our news recommendation. The fact that we are solely interested on being accurate at a low frequency range of the values of the target variable (the number of tweets) creates serious problems to standard prediction methods.

In this paper we describe and test several approaches that try to improve the predictive performance on this difficult task.

3.1 Formalization of the Data Mining Task

Our goal of forecasting the number of tweets of a given news is a numeric prediction task, usually known as a regression problem. This means that we assume that there is an unknown function that maps some characteristics of the news into the number of times this news is tweeted, i.e. $Y = f(X_1, X_2, \cdots, X_p)$, where Y is the number of tweets in our case, $X_1, X_2, \cdots, X_p$ are features describing the news and $f()$ is the unknown function we want to approximate. In order to obtain an approximation (a model) of this unknown function we use a data set with examples of the function mapping (known as a training set), i.e. $D = \{\langle \mathbf{x}_i, y_i \rangle\}_{i=1}^n$.

The standard regression tasks we have just formalized can be solved using many existing algorithms, and most of them try to find the model that optimizes a standard error criterion like the mean squared error. What sets our specific task apart is the fact that we are solely interested in models that are accurate at forecasting the rare and high values of the target variable Y, i.e. the news that are highly tweeted. Only this small sub-set of news is relevant for our overall task of providing a ranking of the most important news for a given topic. In effect, predictive accuracy at the more common news that have a small number of tweets is completely irrelevant because only the top positions of the ranking of recommended news are really relevant for the user, and these top positions are supposed to be filled by the news that have a very high number of tweets.

3.2 Handling the Imbalanced Distribution of the Number of Tweets

Previous work [17, 22, 23] has shown that standard regression tools fail dramatically on tasks where the goal is accuracy at the rare extreme values of the target variable. The main goal of the current paper is to compare some of the proposed solutions to this type of imbalanced regression tasks in the particular problem of forecasting the number of tweets of news.

Several methodologies were proposed for addressing this type of tasks. Resampling methods are among the simplest and most effective. Resampling strategies work by changing the distribution of the available training data in order to meet the preference bias of the users. Their main advantage is that they do not require any special algorithms to obtain the models - they work as a pre-processing method that creates a "new" training set upon which one can apply any learning algorithm. In this paper we will experiment with two of the most successful resampling strategies: (i) SMOTE [3] and (ii) under-sampling [12]. These methods were originally developed for classification tasks where the target variable is nominal. The basic idea of under-sampling is to decrease the number of observations with the most common target variable values with the goal of better balancing the ratio between these observations and the ones with the interesting target values that are less frequent. SMOTE works by combining under-sampling of the frequent classes with over-sampling of the minority class. Namely, new cases of the minority class are artificially generated by interpolating between existing cases. Recently, Torgo et al. [23, 24] extended these methods for regression tasks as it is the case of our problem. We have used the work of these authors to create two variants of each of our datasets. The first variant uses the SMOTEr algorithm [23] to create a new training set by over-sampling the cases with extremely large number of tweets, and under-sampling the most frequent cases, thus balancing the resulting distribution of the target variable. The second variant uses the under-sampling algorithm proposed by the same authors to decrease the number of cases with low number of tweets, hence the most common, once again resulting in a more balanced distribution. In our experiments we will apply and compare these methodologies in order to check which one provides better results in forecasting accurately the number of tweets of highly popular news items. As explained in detail in Section 4.3 these resampling strategies are framed in an utility-based regression framework (Torgo and Ribeiro [22] and Ribeiro [17]) which maps the values of the target variable into a $[0, 1]$ scale of relevance. By defining a relevance threshold, this scale is then used to define the sub-range of target variable values as rare and common. The SMOTEr and under-sampling strategies use this information to over- and under-sampling the values to balance the distribution.

4   Materials and Methods

4.1   The Used Data

The experiments that we will describe are based on news concerning four specific topics: economy, microsoft, obama and palestine. These topics were chosen due to two factors: its actual use and because they report to different types of entities (sector, company, person and country). For each of these four topics we have constructed a dataset with news mentioned in Google News during a period of 13 days, between 2013-Nov-15 and 2013-Nov-28. Figure 1 shows the number of news per topic during this period.

For each news obtained from Google News the following information was collected: title, headline and publication date. For each of the four topics a dataset was built for solving the predictive task formalized in Section 3.1. These datasets were built using the following procedure. For obtaining the target variable value we have used the Twitter API[2] to check the number of times the news was tweeted in the two days following its publication. These two days limit was decided based on the work of Leskovec et al. [15] that suggests that after a few days the news stop being tweeted. Despite Leskovec et al. [15] statement that this period is of four days, some initial tests on our specific data sets have shown that after a period of two days the number of tweets is residual, and therefore we chose this time interval. In terms of predictor variables used to describe each news we have selected the following. We have applied a standard bag of words

---

[2] Twitter API Documentation: https://dev.twitter.com/docs/api

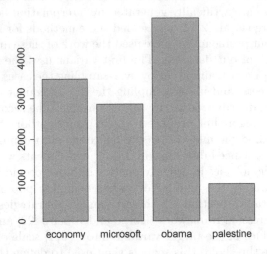

Fig. 1. Number of news per topic

approach to the news headline to obtain a set of terms describing it[3]. Some initial experiments we have carried out have shown that the headline provides better results than the title of the news item. We have not considered the use of the full news text as this would require following the available link to the original news site and have a specific crawler to grab this text. Given the wide diversity of news sites that are aggregated by Google News, this would be an unfeasible task. To this set of predictors we have added two sentiment scores: one for the title and the other for the headline. These two scores were obtained by applying the function **polarity()** of the R package **qdap** [18] that is based on the sentiment dictionary described by Hu and Liu [10]. Summarizing, our four datasets are built using the information described on Table 1 for each available news.

As expected, the distribution of the values of the target variable for the four obtained datasets is highly skewed. Moreover, as we have mentioned our goal is the accuracy at the low frequency cases where the number of tweets is very high. We will apply the different methods described in Section 3.2 to our collected data. This will lead to 12 different datasets, three for each of the selected topics: (i) the original imbalanced dataset; (ii) the dataset balanced using SMOTEr; and (iii) the dataset balanced using under-sampling. The hypothesis driving the current paper is that by using the re-sampled variants of the four original datasets we will gain predictive accuracy at the highly tweeted news, which are the most relevant for providing accurate news recommendations.

---

[3] We have used the infra-structure provided by the R package **tm** [5].

Table 1. The variables used in our predictive tasks

Variable	Description
NrTweets	The number of times the news was tweeted in the two days following its publication. This is the target variable.
$T_1, T_2, \cdots$	The term frequency of the terms selected through the bag of words approach when applied to all news headlines.
SentTitle	The sentiment score of the news title.
SentHeadline	The sentiment score of the news headline.

4.2 Regression Algorithms

In order to test our hypothesis that using resampling methods will improve the predictive accuracy of the models on the cases that matter to our application, we have selected a diverse set of regression tools. Our goal here is to try to make sure our conclusions are not biased by the choice of a particular regression tool.

Table 2 shows the regression methods and tools that were used in our experiments. To make sure our work can be easily replicable we have used the implementations of these tools available at the free and open source R environment. All tools were applied using their default parameter values.

Table 2. Regression algorithms and respective R packages

ID	Method	R package
RF	Random forests	randomForest
LM	Multiple linear regression	stats
SVM	Support vector machine	e1071
MARS	Multivariate adaptive regression splines	earth

4.3 Evaluation Metrics

It is a well-known fact that when the interest of the user is a small proportion of rare events, the use of standard predictive performance metrics will lead to biased conclusions. In effect, standard prediction metrics focus on the "average" behaviour of the prediction models and for these tasks the user goal is a small and rare proportion of the cases. Most of the previous studies on this type of problems was carried out for classification tasks, however, Torgo and Ribeiro [22] and Ribeiro [17] have shown that the same problems arise on regression tasks when using standard metrics like for instance the Mean Squared Error. Moreover, these authors have shown that discretizing the target numeric variable into a nominal variable followed by the application of classification algorithms is also prone to problems and leads to sub-optimal results.

In this context, we will base our evaluation on the utility-based regression framework proposed in the work by Torgo and Ribeiro [22] and Ribeiro [17]. The metrics proposed by these authors assume that the user is able to specify what is the sub-range of the target variable values that is most relevant. This is done by specifying a relevance function that maps the values of the target variable into a [0, 1] scale of relevance. Using this mapping and a user-provided relevance threshold the authors defined a series of metrics that focus the evaluation of models on the cases that matter for the user. In our experiments we have used as relevance threshold the value of 0.9, which leads to having on average 7% to 10% of the cases tagged as rare (i.e. important in terms of number of tweets) depending on the topic.

In our evaluation process we will mainly rely on two utility-based regression metrics: a variant of the mean squared error weighed by relevance, and the F-Score. The variant of the mean squared error (`mse_phi` in our tables of results) is calculated by multiplying each error by the relevance of the true number of tweets. This means that the errors on the most relevant news will be amplified. The main problem of this metric is that it does not consider situations were the models forecast a high number of tweets for a news that ends up having a low number of tweets, i.e false positives. On the contrary, the F-Score is able to take into account both problems. This is a composite measure that integrates the values of precision and recall according to their adaptation for regression described in the above mentioned evaluation framework.

5   Experimental Comparison

5.1   Experimental Methodology

Our data (news items) has a temporal order. In this context, one needs to be careful in terms of the process used to obtain reliable estimates of the selected evaluation metrics. This means that the experimental methodology should make sure that the original order of the news is kept so that models are trained on past data and tested on future data to avoid over-optimistic estimates of their scores. In this context, we have used Monte Carlo estimates as the experimental methodology to obtain reliable estimates of the selected evaluation metrics for each of the alternative methodologies. This methodology randomly selects a set of points in time within the available data, and then for each of these points selects a certain past window as training data and a subsequent window as test data, with the overall train+test process repeated for each point. All alternative approaches are compared using the same train and test sets to ensure fair pairwise comparisons of the obtained estimates. Our results are obtained through 10 repetitions of a Monte Carlo estimation process with 50% of the cases used as training set and 25% used as test set. This process is carried out in R using the infra-structure provided by the R package **performanceEstimation** [21].

5.2   Results

Our results contemplate four topics, as referred before: economy, microsoft, obama and palestine. Tables 3 and 4 present a summary of our results, with other results ommited due to space economy reasons though the general trends are similar. For each regression algorithm the best estimated scores are denoted in italics, whilst the best overall score is in bold. Table 3 presents all estimated metric scores for the palestine topic data set. The mean squared error estimates are provided for highlighting once again that this type of metrics may mislead users when the focus is accuracy on rare extreme values of the target variable. The last four metrics are the most interesting from the perspective of our application, particularly the last three (precision, recall and F-measure) because they also penalise false positives (i.e. predicting a very high number of tweets for a news that is not highly tweeted). These results clearly show that in most setups all algorithms are able to take advantage of resampling strategies to clearly boost their performance. The results obtained with both random forests and SVMs are particularly remarkable, moreover taking into account that all methods were applied with their default parameter settings. With precision scores around 60% we can assume that if we use the predictions of these models for ranking news items by their predicted number of tweets, the resulting rank will match reasonably well the reading preferences of the users.

Table 3. Prediction Models Results - Topic Palestine

	mse	mse_phi	prec	rec	F1
lm	*4622.81*	*1651.31*	0.28	0.14	0.18
lm+SMOTE	763343.67	115671.61	0.50	0.13	0.20
lm+UNDER	240787.13	36782.38	*0.57*	*0.16*	*0.24*
svm	**1681.23**	1543.91	0.00	0.00	0.00
svm+SMOTE	7845.29	662.88	*0.59*	**0.57**	*0.57*
svm+UNDER	7235.22	*645.46*	0.57	0.56	0.56
mars	*2334.96*	*1490.63*	0.41	0.10	0.16
mars+SMOTE	17238.67	1971.88	0.43	0.25	0.31
mars+UNDER	15291.91	1514.40	*0.49*	*0.36*	*0.41*
rf	*1770.16*	1438.93	0.23	0.04	0.06
rf+SMOTE	6309.61	636.77	0.49	0.50	0.49
rf+UNDER	8170.99	**618.81**	**0.61**	*0.56*	**0.58**

Table 4 shows the overall estimated precision, recall and F1 scores for all alternatives in the four topics. Once again we confirm that using resampling strategies provides very good results in the task of predicting highly tweeted news for these four diverse topics. These results are also in accordance with the findings by Torgo et al. [23] where it was reported that similar gains were obtained with both under-sampling and SMOTEr. These results show that in all experimental settings we have considered, the use of resampling was able to

Table 4. The F1 estimated scores for all topics

	economy			microsoft			obama			palestine		
	prec	rec	F1	prec	rec	F1	prec	rec	F1	prec	rec	F1
lm	0.15	0.12	0.13	0.30	0.11	0.16	0.15	0.07	0.10	0.28	0.14	0.18
lm+SMOTE	0.30	0.02	0.04	0.60	0.04	0.07	**0.55**	0.04	0.07	0.50	0.13	0.20
lm+UNDER	0.41	0.02	0.04	0.58	0.03	0.05	0.38	0.03	0.05	0.57	0.16	0.24
svm	0.00	0.00	0.00	0.37	0.03	0.05	0.00	0.00	0.00	0.00	0.00	0.00
svm+SMOTE	0.41	**0.51**	0.45	0.46	0.39	0.42	0.50	**0.56**	**0.53**	0.59	**0.57**	0.57
svm+UNDER	0.37	0.47	0.41	0.48	0.41	0.44	0.48	0.55	0.51	0.57	0.56	0.56
mars	0.16	0.10	0.12	0.35	0.10	0.15	0.09	0.04	0.06	0.41	0.10	0.16
mars+SMOTE	0.40	0.20	0.27	0.56	0.23	0.32	0.40	0.17	0.24	0.43	0.25	0.31
mars+UNDER	0.42	0.33	0.36	0.58	0.31	0.40	0.41	0.27	0.32	0.49	0.36	0.41
rf	0.11	0.05	0.07	0.29	0.06	0.10	0.02	0.01	0.01	0.23	0.04	0.06
rf+SMOTE	0.45	0.48	0.46	0.54	0.46	0.50	0.34	0.43	0.38	0.49	0.50	0.49
rf+UNDER	**0.54**	0.49	**0.51**	**0.63**	**0.52**	**0.56**	0.46	0.47	0.46	**0.61**	0.56	**0.58**

clearly improve the precision of the models at identifying the news that will be highly tweeted by users.

Overall, the main conclusion from our comparisons is that resampling methods are very effective in improving the predictive accuracy of different models for the specific task of forecasting the number of tweets of highly popular news. These methods are able to overcome the difficulty of these news being infrequent. This is particularly important within our application goal that requires us to be able to accurately identify the news that are more relevant for the users in order to be able to improve the performance of news recommender systems.

6   Conclusions

This paper describes an experimental analysis of different methods of predicting the number of times a news item will be tweeted. Being able to forecast accurately this number for news that will be highly tweeted is very important for effective news recommendation. These news are rare and this poses difficult challenges to existing prediction models. We evaluate recently proposed methods for addressing these problems in our particular task.

The results of our experimental comparisons clearly confirm the hypothesis that using resampling methods is an effective and simple way of addressing the task of predicting when a news item will be highly tweeted. Our results, under different experimental settings and using different prediction algorithms, clearly indicate that resampling is able to boost the accuracy of the models on cases that are relevant for this application. In particular, we have observed a marked increase of the precision of the models, which means that most of the times when they forecast that a news will be highly tweeted, that will happen. This is very important for this application as it means that rankings produced based on these predictions will be useful for users as they will suggest news items that are effectively interesting for them.

Future work will include the addition of more information such as the source of the news article and the use of the predictions of the different alternative ways of forecasting the number of tweets to produce actual news rankings. Moreover, we will use methods for comparing rankings in order to correctly measure the impact of the use of resampling methods on the quality of news recommendation.

Acknowledgments. This work is supported by SIBILA Project "NORTE-07-0124-FEDER-000059", financed by the North Portugal Regional Operational Programme (ON.2 - O Novo Norte), under the National Strategic Reference Framework (NSRF), through the European Regional Development Fund (ERDF), and by national funds, through the Portuguese funding agency, Fundação para a Ciência e a Tecnolo-gia (FCT). The work of N. Moniz is supported by a PhD scholarship of the Portuguese government (SFRH/BD/90180/2012).

References

1. Asur, S., Huberman, B.A.: Predicting the future with social media. In: Proc. of the 2010 IEEE/WIC/ACM International Conference on Web Intelligence and Intelligent Agent Technology, WI-IAT 2010, vol. 1, pp. 492–499. IEEE Computer Society (2010)
2. Bandari, R., Asur, S., Huberman, B.A.: The pulse of news in social media: Forecasting popularity. CoRR (2012)
3. Chawla, N.V., Bowyer, K.W., Hall, L.O., Kegelmeyer, W.P.: Smote: Synthetic minority over-sampling technique. JAIR 16, 321–357 (2002)
4. Curtiss, M., Bharat, K., Schmitt, M.: Systems and methods for improving the ranking of news articles. US Patent App. 10/662,931 (March 17, 2005)
5. Feinerer, I., Hornik, K., Meyer, D.: Text mining infrastructure in r. Journal of Statistical Software 5(25), 1–54 (2008)
6. Filloux, F., Gassee, J.: Google news: The secret sauce. Monday Note (2013), URL: http://www.mondaynote.com/2013/02/24/google-news-the-secret-sauce/
7. Gupta, M., Gao, J., Zhai, C., Han, J.: Predicting future popularity trend of events in microblogging platforms. In: ASIS&T 75th Annual Meeting (2012)
8. Hong, L., Dom, B., Gurumurthy, S., Tsioutsiouliklis, K.: A time-dependent topic model for multiple text streams. In: Proc. of the 17th ACM SIGKDD, KDD 2011. pp. 832–840. ACM (2011)
9. Hsieh, C., Moghbel, C., Fang, J., Cho, J.: Experts vs. the crowd: examining popular news prediction performance on twitter. In: Proc. of ACM KDD Conference (2013)
10. Hu, M., Liu, B.: Mining opinion features in customer reviews. In: Proc. of the 19th National Conference on Artificial Intelligence, AAAI 2004 (2004)
11. Kim, S., Kim, S., Cho, H.: Predicting the virtual temperature of web-blog articles as a measurement tool for online popularity. In: Proc. of the 2011 IEEE 11th International Conference on Computer and Information Technology, CIT 2011, pp. 449–454. IEEE Computer Society (2011)
12. Kubat, M., Matwin, S.: Addressing the curse of imbalanced training sets: One-sided selection. In: Proc. of the 14th Int. Conf. on Machine Learning, Nashville, TN, USA, pp. 179–186. Morgan Kaufmann (1997)

13. Lee, J.G., Moon, S., Salamatian, K.: An approach to model and predict the popularity of online contents with explanatory factors. In: Proc. of the 2010 IEEE/WIC/ACM International Conference on Web Intelligence and Intelligent Agent Technology, WI-IAT 2010, vol. 1, pp. 623–630. IEEE Computer Society (2010)

14. Lerman, K., Hogg, T.: Using a model of social dynamics to predict popularity of news. In: Proc. of the 19th International Conference on World Wide Web, WWW 2010, pp. 621–630. ACM (2010)

15. Leskovec, J., Backstrom, L., Kleinberg, J.: Meme-tracking and the dynamics of the news cycle. In: Proc. of the 15th ACM SIGKDD International Conference on Knowledge Discovery and Data Mining, KDD 2009, pp. 497–506. ACM (2009)

16. Page, L., Brin, S., Motwani, R., Winograd, T.: The pagerank citation ranking: Bringing order to the web. Technical report, Stanford University (1998)

17. Ribeiro, R.: Utility-based Regression. PhD thesis, Dep. Computer Science, Faculty of Sciences - University of Porto (2011)

18. Rinker, T.W.: qdap: Quantitative Discourse Analysis Package. University at Buffalo/SUNY (2013)

19. Szabo, G., Huberman, B.A.: Predicting the popularity of online content. Commun. ACM 53(8), 80–88 (2010)

20. Tatar, A., Leguay, J., Antoniadis, P., Limbourg, A., Amorim, M.D.d., Fdida, S.: Predicting the popularity of online articles based on user comments. In: Proc. of the International Conference on Web Intelligence, Mining and Semantics, WIMS 2011, pp. 67:1–67:8. ACM (2011)

21. Torgo, L.: An Infra-Structure for Performance Estimation and Experimental Comparison of Predictive Models (2013),
https://github.com/ltorgo/performanceEstimation

22. Torgo, L., Ribeiro, R.: Utility-based regression. In: Kok, J.N., Koronacki, J., Lopez de Mantaras, R., Matwin, S., Mladenič, D., Skowron, A. (eds.) PKDD 2007. LNCS (LNAI), vol. 4702, pp. 597–604. Springer, Heidelberg (2007)

23. Torgo, L., Ribeiro, R.P., Pfahringer, B., Branco, P.: SMOTE for regression. In: Correia, L., Reis, L.P., Cascalho, J. (eds.) EPIA 2013. LNCS, vol. 8154, pp. 378–389. Springer, Heidelberg (2013)

24. Torgo, L., Branco, P., Ribeiro, R., Pfahringer, B.: Re-sampling strategies for regression. Expert Systems (to appear, 2014)

25. Zaman, T., Fox, E.B., Bradlow, E.T.: A Bayesian Approach for Predicting the Popularity of Tweets. Technical Report arXiv:1304.6777 (April 2013)

An Incremental Probabilistic Model
to Predict Bus Bunching in Real-Time

Luis Moreira-Matias[1,2], João Gama[2,4],
João Mendes-Moreira[2,3], and Jorge Freire de Sousa[5,6]

[1] Instituto de Telecomunicações, 4200-465 Porto, Portugal
[2] LIAAD-INESC TEC, 4200-465 Porto, Portugal
[3] DEI-FEUP, U. Porto, 4200-465 Porto, Portugal
[4] Faculdade de Economia, U. Porto 4200-465 Porto, Portugal
[5] UGEI-INESC TEC, U. Porto, 4200-465 Porto, Portugal
[6] DEGI-FEUP, U. Porto, 4200-465 Porto, Portugal
{luis.m.matias,jgama,joao.mendes.moreira}@inescporto.pt,
jfsousa@fe.up.pt

Abstract. In this paper, we presented a probabilistic framework to predict Bus Bunching (BB) occurrences in real-time. It uses both historical and real-time data to approximate the headway distributions on the further stops of a given route by employing both offline and online supervised learning techniques. Such approximations are incrementally calculated by reusing the latest prediction residuals to update the further ones. These update rules extend the Perceptron's delta rule by assuming an adaptive beta value based on the current context. These distributions are then used to compute the likelihood of forming a bus platoon on a further stop - which may trigger an threshold-based BB alarm. This framework was evaluated using real-world data about the trips of 3 bus lines throughout an year running on the city of Porto, Portugal. The results are promising.

Keywords: supervised learning, probabilistic reasoning, online learning, perceptron, regression, bus bunching, travel time prediction, headway prediction.

1 Introduction

The bus has become a key player in highly populated urban areas. Inner-city transportation networks are becoming larger and consequently, harder to monitor. The large-scale introduction of GPS-based systems in the bus fleets opened new horizons to be explored by mass transit companies around the globe. This technology made it possible to create highly sophisticated control centers to monitor all the vehicles in real-time. However, this type of control often requires a large number of human resources, who make decisions on the best strategies for each case/trip. Such manpower requirements represent an important slice of the operational costs.

H. Blockeel et al. (Eds.): IDA 2014, LNCS 8819, pp. 227–238, 2014.

It is known that there is some schedule instability, especially in highly frequent routes (10 minutes or less) [1–5]. In this kind of routes it is more important the headway (time separation between vehicle arrivals or departures) regularity than the fulfilment of the arrival time at the bus stops. In fact, a small delay of a bus provokes the raising of the number of passengers in the next stop. This number increases the dwell time (time period where the bus is stopped at a bus stop) and obviously, it also increases the bus's delay. On the other hand, the next bus will have fewer passengers, shorter dwell times without delays. This will continue as a snow ball effect and, at a further point of that route, the two buses will meet at a bus stop, forming a platoon as it is illustrated in Fig. 1. This phenomenon is denominated as **Bus Bunching(BB)** [3,6].

The emergence of these events is completely stochastic as you *never know* when or where they may occur. However, there are some behavioural patterns that may anticipate its occurrence such as *consecutive* headway reductions and travel times longer than expected. Such patterns uncover some regularities on the causes that may be explored by Machine Learning algorithms to provide decision support. It can be done by mining not only the historical location-based data on the daily trips but also on their real-time tracking. Consequently, the problem complexity turns the off-the-shelf learning methods as inadequate to predict BB events in real-time.

By predicting these events, we can not only automatically forecast where a BB occurrence may emerge but also which is the problematic trip/vehicle and **how** can we prevent it from happening. In this work, we introduce a complex framework to predict BB occurrences in a short-term time horizon. This event detection is build over a stepwise methodology which starts by performing an 1) offline regression to predict the Link Travel Times (the travel time between two consecutive stops) which is **incrementally updated** by considering the

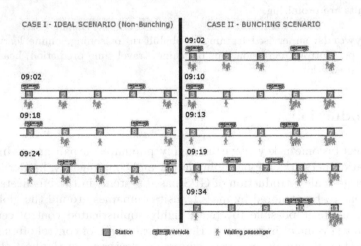

Fig. 1. Bus Bunching illustration

2) error measured from trip to trip and from 3) stop to stop as seeds for a Perceptron-based update rule. Then, a 4) **probabilistic** framework is devised to express the **likelihood** of a pair of buses to form a platoon on a given stop. Finally, 5) these probabilities are used to compute a *Bunching score* which, given a certain context-based threshold, triggers an alarm on a BB occurrence. Our main contributions are threefold:

1. we introduce a novel data driven approach to predict the emergence of BB events in a short-term horizon. More than maintaining the headway stable on the network in exchange of some schedule unreliability, it aims to anticipate last-resource contexts where a corrective action must be took;
2. by producing numerical scores rather than BUNCHING/NO_BUNCHING labels, we favour the framework's interpretability and, consequently, its ability to adapt to different scenarios;
3. we validated such framework using a large-scale dataset containing times-tamped trip records of three distinct bus routes running on the city of Porto, Portugal, during an one-year period.

2 Problem Overview

The Public Transportation (PT) companies operate on high competitive scenarios where there are many options to perform this short connections such as other bus companies, trains, light trams or even private transportation means. The service reliability is key to maintain their profitability. By guaranteeing on-time arrivals, the passengers' perception of the service quality will rise and, consequently, they will pick it often. On the other hand, an unreliable schedule may decrease the number of customers running on that company and therefore, lead to important profit losses [7, 8]. One of the most visible characteristics of an unreliable service is the existence of BB events. Two (or more) buses running together on the same route is an undeniable sign that something is going terribly wrong with the company's service.

To avoid such occurrences, the PT companies installed advanced Control centers where experienced operators are able to monitor the network operations in real-time. Their goal is to suggest **corrective actions** to the bus drivers able to prevent such occurrences. There are four typical methods employed as real-time control strategies [6, 9]:

1. **Bus Holding**: It consists of forcing the driver to increase/reduce the dwell time[1] on a given bus stop along the route;
2. **Speed Modification**: This strategy forces the driver to set a maximum cruise speed on its course (lower than usual on that specific route);
3. **Stop-Skipping**: Skip one or more route stops; also known as *short-cutting* when it requires a path change to reduce the original length of the route.

---

[1] The time spent by a bus stopped on a given stop.

4. **Short-Turning**: This complex strategy consists of causing a vehicle to skip the remaining route stops (usually at its terminus) to fill a large service gap in another route (usually, the same route but in the opposite direction). In a worst case scenario, the passengers may be subjected to a transfer.

By studying the BB phenomenon, we expect not only to anticipate **when** it may occurs but also which is the most adequate corrective action to employ in each situation. However, these actions must be took as a last resource as they also affect negatively the schedule reliability (even if they do it in a smaller scale). The idea is to be able to automatically perform the following decisions: 1) *when does it worth to take an action?* 2) *which is the action to employ?* 3) *which is the bus/pair of buses to be affected by such action?* Such framework will represent considerable savings to any PT company by reducing the manpower needs on the control department.

The most important variable regarding the BB events is the *distance* (in time) between two consecutive buses running on the same route. Such distance is denominated as **headway**. Let the trip k of a given bus route be defined by $T_k = \{T_{k,1}, T_{k,2}, ..., T_{k,s}\}$ where $T_{k,j}$ stands for the arrival time of the bus running the trip k to the bus stop j and s denotes the number of bus stops defined for such trip. Consequently, the headways between two buses running on consecutive trips $k, k+1$ be defined as follows

$$H = \{h_1, h_2, ..., h_s\} : h_i = T_{k+1,i} - T_{k,i} \tag{1}$$

Theoretically, the headway between two consecutive trips should be *constant*. However, due to the stochastic events (e.g. traffic jams, unexpected high demand on a given stop, etc.) arose during a bus trip, the headway suffers some variability. Such variability can provoke other events that may decrease the existing headway following a *snowball effect* (as illustrated in Fig. 1). The BB occurs not only when a bus platoon is formed but sooner, when the headway becomes **unstable**. The headway between two consecutive buses is defined as unstable whenever it is strictly necessary to apply a corrective action in order to recover the headway value to acceptable levels. Such threshold is usually defined in function of the frequency $f = h_1$ (the time between the departure of two consecutive buses) [6]. Let the BB occurrence be expressed as an boolean variable defined as follows

$$\text{BUNCHING} = \begin{cases} 1 \text{ if } \exists\, h_i \in H : h_i < f/4 \\ 0 \text{ otherwise} \end{cases} \tag{2}$$

Consequently, a relationship between the BB occurrences, the headway and the arrival time $T_{k,i}$ can be established. Let the arrival time be defined as $T_{k,i+1} = T_{k,i} + dw_{k,i} + CTT_{k,i,i+1}$ where $dw_{k,i}$ denotes the dwell time on the stop i and $CTT_{k,i,i+1}$ stands for the Cruise Travel Time between those two consecutive stops. Therefore, it is possible to anticipate the occurrence of BB events if we are able to predict the value of $dw_{k,i} + CTT_{k,i,i+1}$, which is often denominated by **Link Travel Time** [10]. In this work, we develop a probabilistic method to detect BB events that settles on Link Travel Time predictions based on the data described below.

2.1   Case Study

The source of this data was STCP, the Public Transport Operator of Porto, Portugal. It describes the trips from three distinct lines (A, B, C) during 2010. Each line has two routes – one for each way {A1, A2, B1, B2, C1, C2}. Line A is common urban line between *Viso* (an important neighbourhood in Porto) and *Sá da Bandeira*, a downtown bus hub. Line B is also an urban line but it is an arterial one. It traverses the main interest points in the city by connecting two important street markets: *Bolhão* - located in downtown - and *Mercado da Foz*, located on the most luxurious neighbourhood in the city. Line C connects the city downtown to the farthest large-scale neighborhood on the region (*Maia*).

This dataset has one entry for each stop made by a bus running in the route during that period. It has associated a timestamp, the weekday (MON to SUN) and a day type (1 for work days, 2-6 for other day types i.e.: holidays and weekends). Table 1 presents some statistics about the set of trips per route considered and the BB events identified. The BB Avg. Route Position represents the percentage of route accomplished when these events typically arise.

3   Travel Time Prediction

Let the Link Travel Time Prediction be defined as an offline **regression** problem where the target variable is the cruising time between two consecutive bus stops. Such predictions are computed in a daily basis (the forecasting horizon) using the θ most recent days (the learning period) to train our model. Consequently, we obtain a set of predictions for all the t trips of the day denoted as $\mathbb{P} = \bigcup_{i=1}^{t} P_i = \{P_{1,1}, P_{1,2}, ..., P_{t,s}\}$. These predictions are then incrementally refined in two steps: 1) trip-based and 2) stop-based. Both steps are based on the Perceptron's Delta Rule [11] by reusing each prediction's residuals to improve the further ones.

Let e denote the last trip completed before the current trip starts (i.e. c). The trip-based refinement consists into comparing the predictions to e $P_e =$

Table 1. Descriptive statistics for each route considered. The frequencies are in minutes.

	A1	A2	B1	B2	C1	C2
Number of Trips	20598	20750	20054	19361	26739	26007
Nr. of Stops	26	26	32	32	45	45
Min. Daily Trips	44	45	56	57	65	71
Max. Daily Trips	76	76	85	84	100	101
Min. Frequency	10	11	12	13	10	10
Max. Frequency	112	100	103	120	60	60
Nr. of Trips w/ BB	682	553	437	634	1917	1702
Nr. of HD events detected	63.22%	74.86%	58.31%	68.54%	49.71%	53.63%

$\{P_{e,1}, P_{e,2}, ..., P_{e,s}\}$ with the real times T_e to update P_c. Firstly, we compute
the **residuals** as $R_e = T_e - P_e$ and then its average value as $\nu_e = \sum_{i=1}^{s} \frac{R_{e,i}}{s}$.
Secondly, an user-defined parameter $0 < \alpha << 1$ is employed to set a threshold th
able to identify trips where the error is larger than expected. Consequently, $th = \alpha * f_e$ where f_e stands for the current frequency on this route (i.e. the difference
between the departure time of c and e). Three other variables are then defined:
$\vartheta_p = 0, \vartheta_n = 0$ and $\beta' = \beta$. The first two are counters that are incremented
whenever the prediction error is going to the same way (positive/negative) on
consecutive trips (e.g. if $\mu_e > th$ ϑ_p is incremented; otherwise, $\vartheta_p = 0$). The beta
value stands for the residual's percentage to be added to P_c (its initial value β is
user-defined). It is initialized with another user-defined parameter $0 < \beta << 1$
and updated according to a user-defined learning rate $0 < \kappa <= 1$. Consequently,
if ϑ_p or ϑ_n are incremented, the P_c and β' are updated as $P'_c = P_c \pm (\beta' \times P_c)$
and $\beta' = beta' + \vartheta * (1 + \kappa) * \beta$, respectively. If both ϑ stay the same, β' resumes
its original value as $\beta' = \beta$. These updates are performed **incrementally** (i.e.
whenever a real travel time for a given trip on one of its links arrives) to every
trips available in the dataset. Note that the residuals are always calculated over
the regression results P_c and not over the updated arrays P'_c. Thereby, its calculus
is iterative but not recursive.

Given the updated predictions of two consecutive trips (P'_c, P'_{c+1}), it is possible
to obtain the predicted headways $E_c = P'_{c+1} - P'_c$ while the real one is obtained
as $H_c = T_{c+1} - T_c$. The calculus of E_c works as an *offline* prediction as it does not
use information about the current headway experienced between the two trips.
The second refinement uses the headway residuals $HR_c = H_c - E_c$ to update
E_c stop-by-stop. Incrementally, we can obtain *online* headway predictions as
$E'_{c,i} = H_{c,i-1} + E_{c,i} - E_{c,i-1}, \forall i \in \{2, s\}$. The problem is to update the headway
online prediction for the next stop $E'_{c,i}$ given the value of $HR_{c,i-1}$. Let $\gamma' = \gamma$
be the residual's percentage to add to the prediction where its initial value for
each trip $(0 < \gamma << 1)$ is an user-defined parameter. $E'_{c,i}$ can be updated
as $E''_{c,i} = E'_{c,i} + (HR_{c,i-1} * \gamma')$. Finally, γ' is also updated by comparing the
residuals of E_c and E'_c (HR_c and HR'_c, respectively). If $|HR_c| > |HR'_c|$, then
$\gamma' = \gamma' * (1 - \gamma)$. Otherwise, $\gamma' = \gamma' * (1 + \gamma)$. The progression of γ' is bounded by
an user-defined domain $[\gamma_{min}, \gamma_{max}]$. The value of $E''_{c,i}$ is also used to update the
offline predictions for further stops as $E'_{c,j} = E''_{c,j-1} + E_{c,j} - E_{c,j-1} \wedge j = i + 1$
and $E'_{c,j} = E'_{c,j-1} + E_{c,j} - E_{c,j-1}, \forall j \in [i + 2, s]$. Again, whenever a newer
headway value H_c, i arrives, the entire headway array $E'_{c,q}, q \in \{i + 1, s\}$ is
updated accordingly. This scheme introduces a certain **flexibility** to handle the
real-time stochastic usually associated to this variable.

By performing these two steps, it is possible to seize distinct levels of information to approximate the real-time link travel times incrementally. The propagation of our updates for further stops on the trip is the key to anticipate
BB occurrences. The probabilistic framework devised to do so is detailed in the
following section.

4   Event Detection

Let M denote a $l \times (s - 1)$ matrix containing the l most recent residuals[2] for headway predictions from 1 to $s-1$ stops ahead of the current one (c) (where s is the number of stops) on a specific route, where l is an user-defined parameter to set the size of the sliding window to be employed. Consequently, $M[, i]$ represents a vector containing the most recent residuals on headway predictions i stops ahead. Departing from M, it is possible to build a rough approximation to the probability density function ($p.d.f.$) that describes the headway on a bus stop located i stops ahead. We do it so by assuming that all these distributions are **Gaussian**[3], being described by a function as $X_i = f_i(\mu, \sigma)$. μ is given by $E'_{c,i}$ or $E''_{c,1}$ while σ is given by the median value of $M[, i]$ (i.e. $\tilde{M}[, i]$). Considering the hypothesis of arising a BB event on this specific stop (i.e. H_i), we can express its likelihood as $Pr_i(X_i \leq f/4 \mid H_i)$. Such definition allows to quantify the statistical significance (i.e. p-value) of occurring a BB event on that specific stop. Using this framework, it is possible to quantify a Bunching likelihood for all the remaining stops in the route (and also to update them each time we obtain a newer value for the headway).

Using such estimations, it is possible to predict incrementally the BB occurrences in three simple steps: 1) calculate/update the Bunching likelihoods; 2) estimate a Bunching Score (BS) and 3) test if it is greater than the predefined threshold. These steps are performed each time a new headway value arrives (i.e. for each bus stop). Let j represent the latest bus stop for which the headway value is known. BS is calculated as follows: let m_j be an ordered vector (descendent) containing the likelihoods for the remaining bus stops and $n_j = 3 - ((j - 1) \times 3/s) : n_j \in \mathbb{N}$ be the number of likelihoods to be used to compute BS. Finally, we have that $BS_j = m_j[1 : n_j]$ as the mean likelihood of the n_j greater ones. The BS threshold is defined in function of the frequency as $th_{BS} = 0.3 + [(f \mod \rho) * 0.1] : 0 < th_{BS} \leq 1$ where ρ is an user-defined parameter to set how many threshold levels should be defined for the frequency. Therefore, a BB event is detected if $BS_i \geq th_{BS}$. The alarm is triggered on the nearest stop where $m_i \geq th_{BS}$.

This probabilistic framework allows an *incremental* detection of the BB events by refining the headway predictions reusing not only its latest true values but also the most recent residuals. Experiments were conducted to validate this methodology. They are extensively described in the following section.

5   Experiments

On the offline regression problem, a state-of-art algorithms was employed: Random Forest (RF). We did so by following previous work on this topic which used

---

[2]   E.g. given the newest headway value known, H_c, the residuals for the stops ranged between c and $c - l \geq 1$ are used

[3]   a D'Agostino K-Squared test [12] was conducted on the headways experienced on every stop using previous data.

data from the same source [13]. The experiments were conducted using the R Software [14]. A sensitivity analysis was conducted on the regression parameters based on a simplified version of Sequential Monte Carlo method (the reader can consult the survey in [15] to know more about this topic) on previous data. The goal was to identify the best parameter setting to optimize the regression task. The best parameter setting was `mtry=3` and `ntrees=750`. The learning period used was $\theta = 7$ days. The error threshold to trigger the inter-trip update rule was set to $\alpha = 0.05$ while the initial value for the residual's percentage to be employed is $\beta = 0.01$. The learning rate $kappa$ was set to 0.3. The initial residual's percentage employed on the stop-based update rule is $\gamma = 0.1$ while its domain is $\gamma \in [0.005, 0.3]$. Finally, the ρ was set to 360 seconds.

It is possible to divide the evaluation of our framework on two distinct contexts: (i) the mean absolute error and (ii) the BB detection accuracy. On the first one, we employed a prequential evaluation [16] by evaluating *just* the prediction made for the Link Travel Time performed for the next bus stop. We did so by using the Mean Absolute Error (MAE) on (1) the offline regression output and then on the (2) inter-trip and (3) intra-trip refinement. On the BB detection context, the Accuracy, the Precision and the Recall as evaluation metrics. An weighted accuracy was also employed by weighting the trips where a BB event emerge ten times more than the remaining ones. Such cost-based evaluation was done to address the different value on performing a false negative on detecting BB event - which is largely higher than raising a false positive. The Average Number of Stops Ahead is also displayed to show which is the forecasting horizon that this framework can reach. The results of these experiments are presented in the next section.

5.1   Results and Discussion

The results are presented on Table 2. More than identifying just a problematic link or stop, this framework also identifies which is the **vehicle pair** where a corrective action must be taken. In the current dataset, it was able to detect BB events thirteen stops ahead (in average), which gives more than enough room to perform any of the four possible corrective actions. Nevertheless its achievements, this framework also presents some limitations, namely, on the regression task and on the parameters employed. The regression task was tested using only one algorithm. Even considering that it presented good results in similar data [13], we do not know if there is another that could perform better using a similar computational effort. On the other hand, both the prediction refinements and the event detection framework rely on a large set of parameters. To get a fair parameter setting can be a hard task - especially if the user has no expertise on the case study approached. This issue can be specially relevant on the parameters defining the learning rates and the residual's percentages (β, κ and γ). A large-scale sensitivity analysis on these parameters must be carried out as future work.

On the first span of Table 2, it is possible to observe that the two update rules have a significant impact on reducing the MAE produced by the headway prediction. The accuracy is high. However, the Precision is low (i.e. 52.51%). It demonstrates that our model triggers more BB alarms than necessary. This behavior can be partially justified by the **preventive** characteristics of this framework. Nevertheless its existence, it is not possible to quantify the negative impact it may have without regarding the corrective actions. By quantifying the BB probability along the route, our framework also quantifies the necessary *range* of the corrective action, which is given by $0 \leq BS_i - th_{BS} < 1$. This value can also be useful to determine which may be the corrective action to be applied in each case. The selection of a low-impact action can mitigate the effects of this over-prediction. However, such conclusions have to be validated by further experiments regarding such corrective actions (which are out of this paper scope).

6 Related Work

One of the first works to address the BB phenomenon was presented by Powell and Sheffi [17]. They devised a probabilistic model which built a set of recursive relationships to calculate the *p.d.f.* to validate the hypothesis of forming a platoon of vehicles on each stop. Nevertheless it has many similarities with the work presented here, both the relationships and the distributions were calculated based on a set of assumptions - and not on the real-time data. After this paper, many others works followed the *stability* concept (i.e. if we guarantee a stable headway, BB events will never emerge) by constantly introducing corrective actions on the system to avoid headway instability. Some examples are the work in [2], where each bus is an agent that negotiates with others the bus holding time on each station or in [4], where the negotiation is centered on the cruising speed. A more sophisticated approach to the *p.d.f.* estimation is done in [5] by accounting complex models to determine dwell times or even arrivals during such dwell times.

Table 2. Experimental results. The times are in seconds. The ALL column contains the average for the first two spans and the sum for the last one.

	A1	A2	B1	B2	C1	C2	ALL
MAE offline regression	1356.96	643.99	1475.22	1871.01	473.61	2776.57	1432.88
MAE inter-trip update	148.85	92.91	124.99	148.85	40.65	123.77	113.34
MAE incremental update	13.21	26.35	22.67	13.21	31.79	27.47	22.45
Accuracy	97.99%	96.34%	97.08%	97.83%	96.63%	93.83%	96.62%
Weighted Accuracy	93.97%	93.57%	94.57%	95.52%	95.73%	91.51%	94.14%
Precision	65.88%	40.85%	41.53%	45.70%	69.44%	51.67%	52.51%
Recall	81.81%	83.18%	83.07%	83.24%	94.48%	87.95%	85.62%
Avg. Nr. of Stops Ahead	11.85	14.78	13.88	15.01	12.96	14.52	13.83
Correct BB Predictions	558	460	363	303	1811	1497	4992
Real BB Events	682	553	437	364	1917	1702	5655

The employment of historical data to address this problem is very recent. In [3], a model to determine the optimal holding time in each station based on real-time location is presented. Delgado *et al.* [18] also suggested preventing passengers from boarding by establishing maximum holding times to maintain the headway stable. The efficiency of this type frameworks is usually demonstrated by simulations assuming i) stochastic demand and/or traffic events or 2) using historical data. Despite their usefulness, all these works do not account the historical and the real-time data. Moreover, they have a low interpretability because their outputs do not provide any clew on which is the best corrective action to took (usually, these works just pick one corrective action). The predictive method presented along this paper is able to deal with the network stochasticity, independently on which corrective action we want to take. Finally, it is important to highlight that the majority of the works on the literature try to maintain the headway stable at cost of some schedule uncertainty (introduced by the constant corrective actions), independently on the existing risk on forming a bus platoon on a further stop. By the abovementioned reasons, the authors believe that the proposed framework meets no parallel in the existing literature on this topic.

7 Final Remarks

In this paper, a probabilistic framework to anticipate the occurrence of BB events in real-time was presented. This framework employs Supervised Machine Learning techniques that incrementally refine predictions on the Link Travel Time of each bus trip. The residuals of such predictions are then used to build Gaussian Distributions on the headway values which can be used to estimate the Bunching likelihood on each bus stop. Experiments conducted on a real world data set of six bus routes running on the city of Porto, Portugal throughout an year validated this a framework as a step forward on automatizing the BB prediction task.

The present work is a proof of concept on the usefulness of predicting BB events instead of trying to maintain the headway stable at all cost. This work can be extended on three distinct axis: 1) the dataset, by including a larger dataset containing a set of lines representative of the entire network; 2) the parameter setting, by conducting a large-scale sensitivity analysis on their values and 3) on the corrective actions, by proposing a method to choose **where** and **when** a action should be took to avoid BB, as well as one to choose **which** is the best one to took in each case. Such issues comprise open research questions to be explored on future work.

Acknowledgments. The authors would like to thank STCP for the data supplied. This work was supported by the VTL: "Virtual Traffic Lights" (PTDC/ EIA-CCO/118114/2010), by MAESTRA (ICT-2013-612944), by I-CITY - "ICT for Future Mobility" (NORTE-07-0124-FEDER-000064) and by ERDF - European Regional Development Fund through the COMPETE Programme (operational programme for competitiveness) and also by the Portuguese Funds through the FCT (Portuguese Foundation for Science and Technology) within project FCOMP-01-0124-FEDER-037281.

References

1. Powell, J. W., Huang, Y., Bastani, F., Ji, M.: Towards reducing taxicab cruising time using spatio-temporal profitability maps. In: Pfoser, D., Tao, Y., Mouratidis, K., Nascimento, M.A., Mokbel, M., Shekhar, S., Huang, Y. (eds.) SSTD 2011. LNCS, vol. 6849, pp. 242–260. Springer, Heidelberg (2011)
2. Gershenson, C., Pineda, L.: Why does public transport not arrive on time? the pervasiveness of equal headway instability. PloS One 4(10), 72–92 (2009)
3. Daganzo, C.: A headway-based approach to eliminate bus bunching. Transportation Research Part B 43(10), 913–921 (2009)
4. Daganzo, C., Pilachowski, J.: Reducing bunching with bus-to-bus cooperation. Transportation Research Part B: Methodological 45(1), 267–277 (2011)
5. Bellei, G., Gkoumas, K.: Transit vehicles' headway distribution and service irregularity. Public Transport 2(4), 269–289 (2010)
6. Moreira-Matias, L., Ferreira, C., Gama, J., Mendes-Moreira, J., de Sousa, J.F.: Bus bunching detection by mining sequences of headway deviations. In: Perner, P. (ed.) ICDM 2012. LNCS, vol. 7377, pp. 77–91. Springer, Heidelberg (2012)
7. Wang, F.: Toward intelligent transportation systems for the 2008 olympics. IEEE Intelligent Systems 18(6), 8–11 (2003)
8. Mishalani, R., McCord, M., Wirtz, J.: Passenger wait time perceptions at bus stops: empirical results and impact on evaluating real-time bus arrival information. Journal of Public Transportation 9(2), 89 (2006)
9. Strathman, J., Kimpel, T., Dueker, K.: Transportation Northwest: Bus transit operations control: review and an experiment involving tri-met's automated bus dispatching system. Technical report, Transportation Northwest, Department of Civil Engineering, University of Washington (2000)
10. Chen, G., Yang, X., An, J., Zhang, D.: Bus-arrival-time prediction models: Link-based and section-based. Journal of Transportation Engineering 138(1), 60–66 (2011)
11. Rosenblatt, F.: The perceptron: A probabilistic model for information storage and organization in the brain. Psychological Review 65(6), 386 (1958)
12. D'Agostino, R.B.: Transformation to normality of the null distribution of g1. Biometrika, 679–681 (1970)
13. Mendes-Moreira, J., Jorge, A., de Sousa, J., Soares, C.: Comparing state-of-the-art regression methods for long term travel time prediction. Intelligent Data Analysis 16(3), 427–449 (2012)
14. R Core Team: R: A Language and Environment for Statistical Computing. R Foundation for Statistical Computing, Vienna, Austria (2012)
15. Cappé, O., Godsill, S., Moulines, E.: An overview of existing methods and recent advances in sequential monte carlo. Proceedings of the IEEE 95(5), 899–924 (2007)
16. Dawid, A.: Present position and potential developments: Some personal views: Statistical theory: The prequential approach. Journal of the Royal Statistical Society. Series A (General) 147, 278–292 (1984)

17. Powell, W., Sheffi, Y.: A probabilistic model of bus route performance. Transportation Science 17(4), 376–404 (1983)
18. Delgado, F., Muñoz, J.C., Giesen, R., Cipriano, A.: Real-time control of buses in a transit corridor based on vehicle holding and boarding limits. Transportation Research Record: Journal of the Transportation Research Board 2090(1), 59–67 (2009)

Mining Representative Frequent Patterns in a Hierarchy of Contexts

Julien Rabatel[1], Sandra Bringay[1,2], and Pascal Poncelet[1]

[1] LIRMM (CNRS UMR 5506), Univ. Montpellier 2
161 rue Ada, 34095 Montpellier Cedex 5, France
[2] Dpt MIAp, Univ. Montpellier 3
Route de Mende, 34199 Montpellier Cedex 5, France

Abstract. More and more data come with contextual information describing the circumstances of their acquisition. While the frequent pattern mining literature offers a lot of approaches to handle and extract interesting patterns in data, little effort has been dedicated to relevantly handling such contextual information during the mining process. In this paper we propose a generic formulation of the contextual frequent pattern mining problem and provide the CFPM algorithm to mine frequent patterns that are representative of a context. This approach is generic w.r.t. the pattern language (e.g., itemsets, sequential patterns, subgraphs, etc.) and therefore is applicable in a wide variety of use cases. The CFPM method is experimented on real datasets with three different pattern languages to assess its performances and genericity.

1 Introduction

In the data mining field as well as in every data-related domain, more and more data come with contextual information detailing the circumstances under which data have been acquired. A concrete example lies in the explosion of mobile phone usage accompanied by information about the user location and user profiles. Another omnipresent example is related to the Web, where users often give some information about themselves (e.g., on forums or social media) that can be used to better understand their Web usage.

While mining patterns of very various forms and structures (itemsets, sequences, episodes, subgraphs, spatio-temporal patterns, etc.) has been studied extensively in the past two decades [6], there has been little interest in fully exploiting the surrounding data, i.e., the so-called contextual data. Many machine learning approaches can however benefit from this contextual information to finely analyze or exploit the data. In the current paper, we study and propose a solution for mining frequent patterns in the presence of contextual information. More precisely, the contributions of this paper are twofold:

- **Providing a generic theoretical framework.** We propose a formalism for defining contextual frequent patterns that does not depend on a particular pattern language. This framework has the ability to generalize the frequent pattern mining problem to consider available contextual information.

H. Blockeel et al. (Eds.): IDA 2014, LNCS 8819, pp. 239–250, 2014.
© Springer International Publishing Switzerland 2014

– **Contextual frequent pattern mining algorithm.** We also propose a
new algorithm, so-called CFPM, which is generic w.r.t. the pattern language
and the underlying mining algorithm. This approach is based on relevantly
post-processing the output of existing algorithms, meaning that it can be applied in conjunction with any algorithm that aims at solving a transactional
frequent pattern mining problem and offers a great applicability range.

While seminal work has already defined the basis of contextual frequent pattern mining in the case of sequential patterns [12,13], the existing work has the
following drawbacks: *(1)* its formulation is only dedicated to sequential patterns,
while we are interested in providing a generic formulation applicable to most
frequent patterns definitions; *(2)* the algorithm designed to mine contextual frequent sequential patterns uses specific techniques that make it unusable for other
pattern languages. We build upon this previous work and show that the principles of mining contextual frequent patterns are not inherently associated to one
pattern language, or even to one mining method, and can be used in conjunction
with a lot of existing previous work for a great flexibility and applicability. Mining contextual frequent patterns only relies on pattern frequency and does not
relate to how a pattern frequency contrasts with the rest of the database. Such
patterns, found in the literature as *discriminative patterns*, *contrast patterns* or
correlated patterns [4] do not fall within the scope of this study.

The remaining is organized as follows. Section 2 defines the contextual frequent pattern mining problem. Then, Section 3 describes the proposed CFPM
algorithm. Experiments are conducted in Section 4 and some conclusions and
prospects are given in Section 5.

2   Contextual Data and Frequent Patterns

This section aims at formalizing the frequent pattern mining problem as well
as its extension for handling contextual information. According to [10], a large
family of pattern mining problems can be specified with the following formulation: given a database D, a class of patterns $\mathcal{P}$ called a pattern language and
a selection predicate q, it consists in finding the set $\{p \in \mathcal{P} \mid q(D,p) \text{ is true}\}$.
This definition is refined as follows to describe the *transactional frequent pattern
mining* problem addressed in this study.

Definition 1 (Transactional frequent pattern mining). *A **transaction** is
a couple $T = (tid, o_T)$, where tid is a unique transaction identifier, and o_T is
the transaction object, i.e., an object provided in an arbitrary description space.
A **transactional database** is a set of transactions.*

*The pattern language is associated with a **support operator** $\prec$, such that
a transaction T supports a pattern p (or p is supported by T) if $p \prec T$. From
the support operator, one can define the **frequency** of p in D as the fraction of
transactions in D supporting p: $Freq_D(p) = \frac{|\{T \in D \mid p \prec T\}|}{|D|}$.*

Frequent patterns are those whose frequency is above a user-specified **minimum frequency threshold**. *In other terms, given a transactional database D, a pattern language $\mathcal{P}$, a pattern support operator $\prec$, and a minimum frequency threshold σ, the* **transactional frequent pattern mining** *problem refers to finding the set $\mathcal{F}(\mathcal{P}, D, \prec, \sigma) = \{p \in \mathcal{P} | Freq_D(p) \geq \sigma\}$.*

As an example, the well-known *frequent itemset mining problem* [1] within this transactional setting can be: given an alphabet of items $\mathcal{I} = \{a, b, c, d, e\}$, transaction objects are itemsets, i.e., subsets of $\mathcal{I}$. The pattern language $\mathcal{P}$ is defined as $2^{\mathcal{I}}$ and the *support operator* $\prec$ as the set inclusion operator between patterns and transaction itemsets. For instance, given a pattern $p = \{a, d\}$ and a transaction $T = (t, \{a, b, d, e\})$, then $p \prec T$ because $\{a, d\} \subseteq \{a, b, d, e\}$.

Figure 1(a) provides an example of a transactional database of itemsets D using the alphabet $\mathcal{I}$. The first column gives the identifier of each of the 14 transactions, while the second one provides the corresponding transaction itemset. The third column is not used in the transactional pattern mining setting. By considering a minimum frequency threshold σ of 0.5, we notice that the pattern $p = \{a, b\}$ has a frequency $Freq_D(p) = 8/14$ and is therefore frequent with $Freq_D(p) \geq \sigma$.

tid	Itemset	Context
t_1	$\{a, b, e\}$	YS
t_2	$\{a, b, d\}$	YS
t_3	$\{a, b, e\}$	YS
t_4	$\{a, b, c\}$	YS
t_5	$\{a, b\}$	YS
t_6	$\{a, c, d\}$	YW
t_7	$\{a, b, d\}$	YW
t_8	$\{a, b\}$	YW
t_9	$\{a, b, c, e\}$	OS
t_{10}	$\{b, c, d\}$	OS
t_{11}	$\{b, d\}$	OS
t_{12}	$\{b, d, e\}$	OW
t_{13}	$\{b, d\}$	OW
t_{14}	$\{a, c, d, e\}$	OW

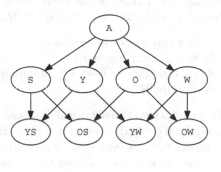

(a)   A   transactional database with associated contexts.

(b) A context hierarchy.

Fig. 1. A transactional contextual database composed of *(a)* a transactional database of itemsets with the associated minimal contexts, and *(b)* a context hierarchy

Such a theoretical framework is representative of a large fraction of frequent pattern mining approaches appeared in the literature in the past decades. These problems exploit a *transactional* view of the data, i.e., they are represented under the form of a collection of transactions and frequent patterns are those mapped to at least a given number of transactions. Among pattern mining problems that do not enter this family, an example is the relatively recent problem of mining patterns in one unique large graph or network, addressed for instance in [3].

We are interested in enriching this transactional setting for mining frequent patterns in the presence of contextual information, i.e., data describing some circumstances regarding each transaction. We therefore introduce the *context hierarchy* to manipulate this contextual information.

Definition 2 (Context hierarchy). *A context hierarchy $\mathcal{H}$ is a directed acyclic graph (DAG), denoted by $\mathcal{H} = (V_\mathcal{H}, E_\mathcal{H})$, such that*

- $V_\mathcal{H}$ *is a set of vertices also called* **contexts***,*
- $E_\mathcal{H} \subseteq V_\mathcal{H} \times V_\mathcal{H}$ *is a set of directed edges among contexts.*

$\mathcal{H}$ is naturally associated with a partial order $<_\mathcal{H}$ on its vertices, defined as follows: given $c_1, c_2 \in V_\mathcal{H}$, $c_1 <_\mathcal{H} c_2$ if there exists a directed path from c_2 to c_1 in $\mathcal{H}$. This partial order describes a specialization relationship: c_1 is said to be more *specific* than c_2 if $c_1 <_\mathcal{H} c_2$, and more *general* than c_2 if $c_2 <_\mathcal{H} c_1$.

A *minimal context* from $\mathcal{H}$ is a context such that no more specific context exists in $\mathcal{H}$, i.e., $c \in V_\mathcal{H}$ is minimal iff $\nexists c' \in V_\mathcal{H} \mid c' <_\mathcal{H} c$. The set of minimal contexts in $\mathcal{H}$ is denoted as $V_\mathcal{H}^-$.

A context hierarchy aims at offering more information about the elements of a transactional database D when mining frequent patterns. A transactional database and a context hierarchy are combined to produce a *contextual transactional database* $\mathcal{D}$, i.e., a triple $(D, \mathcal{H}, \delta)$ such that:

- D is a transactional database,
- $\mathcal{H}$ is a context hierarchy,
- δ is a function $\delta : D \mapsto V_\mathcal{H}^-$ mapping each transaction from D to a minimal context in $\mathcal{H}$.

In a contextual transactional database $\mathcal{D}$, a transaction T is explicitly mapped to a minimal context given by $c = \delta(T)$. By following the intuition that a transaction associated to a very specific context also is part of the more general contexts, we define the *database induced by c*, denoted by $D(c)$, as the subset of D which is associated with c. More formally, $D(c) = \{T \in D | (\delta(T) <_\mathcal{H} c) \vee (\delta(T) = c)\}$.

Example. To illustrate the contextual database notions, let first consider Figure 1(b) which provides a visual representation of a context hierarchy. Contexts are the labels of vertices, such as contextual information is given by the age (*young* or *old*, respectively shorten to Y and O) and the season (*summer* or *winter*, respectively shorten to S and W). Some examples of contexts provided by this context hierarchy shown in Figure 1(b) are (from the more specific to the more general) YS, S, or A, respectively corresponding to "*transactions associated to young people in summer*", "*transactions associated to summer (regardless of the age of people)*" and "*all (A) the transactions regardless of the age and season*". The third column of Figure 1(a) describes the δ function by mapping each transaction to a minimal context, such as the first transaction identified by t_1 associated with the context YS, i.e., $\delta(t_1) = YS$. Figure 1 hence provides a contextual transactional database $\mathcal{D}$. From this δ mapping, we notice for instance that the database induced by YS (i.e., $D(YS)$) is the set of transactions of identifiers $t_1, \ldots, t_5$, while $D(A)$ is the set of all transactions.

2.1   When Should a Pattern Be Associated with a Context?

The contextual frequent pattern mining problem aims at discovering patterns whose the property of being frequent is context-dependent. In order to study and highlight the interest of exploiting contextual information within the frequent pattern mining process, we below isolate two patterns from our running example.

Case 1. Itemset $\{a, b\}$ is frequent in D ($Freq_D(\{a, b\}) = 8/14$). However, considering the database and its fragmentation given by contextual information, one can notice that $\{a, b\}$ is frequent in the fraction of D corresponding to young people (Y) with a frequency of $7/8$ while it is infrequent in the rest of D, i.e., old people. In the following, we will state that this pattern, while *frequent in A*, is not *general in A* because it is not frequent in every context contained in A. On the other hand, this pattern is *general in Y* because, in addition to be frequent in Y, it is also frequent in YS ($Freq_{YS}(p) = 5/5$) and YW ($Freq_{YW}(p) = 2/3$), i.e., all contexts contained in Y.

Case 2. $p' = \{b, d\}$ is not frequent in D ($Freq_D(p') = 6/14$), while it however is frequent in O ($Freq_O(p) = 5/6$) as well as in the contexts contained in O: its frequency is $2/3$ both in OS and OW.

Case 1 shows that simply mining frequent patterns within a context does not necessarily provide representative patterns. In addition, Case 2 shows a pattern that is representative of a given context, but mining frequent patterns in the whole database could not make such patterns emerge as the context they represent is not large enough relatively to the whole database.

The next section exploits these intuitions to formally define what types of frequent patterns are mined in a contextual database and how they relate to the context hierarchy.

2.2   Contextual Frequent Patterns: A Formal Definition

The current section applies the contextual transactional setting defined above to first reformulate the notion of frequent pattern within a context and then introduce the notions related to the contextual frequent patterns (CFPs).

Definition 3 (c-frequent pattern). *A pattern p is frequent in c, or c-frequent, iff p is frequent in $D(c)$, i.e., if $Freq_{D(c)}(p) \geq \sigma$. For the sake of readability, we denote $Freq_{D(c)}(p)$ with $Freq_c(p)$.*

As discussed in the previous section, we are interested in patterns being representative of a context, i.e., such that their frequency property holds for all the descendants of this context. Such patterns are called *general patterns* and are used to define CFPs.

Definition 4 (c-general pattern). *A pattern p is general in c, or c-general, iff p is c-frequent and p is c'-frequent $\forall c' \mid c' <_{\mathcal{H}} c$.*

Definition 5 (Contextual frequent pattern). *A contextual frequent pattern is a couple $\alpha = (c, p)$, such that p is c-general. α is said to be **generated by** p. (c, p) is **context-maximal** if there does not exist another context c' more general than c and such that (c', p) is a CFP, i.e., $\nexists c' \in V_{\mathcal{H}} \mid (c <_{\mathcal{H}} c')$ and p is c'-general.*

The **CFP mining problem** consists in enumerating all the context-maximal CFPs given a contextual database $\mathcal{D}$ and a minimum frequency threshold σ.

The set of CFPs that are context-maximal constitutes an exact condensed representation of the set of CFPs, as no CFP cannot be derived from a *context-maximal* one. Indeed, following Definition 4, one may notice that if a pattern is c-general, then it is also general in all descendants of c. Therefore, mining all CFPs in $\mathcal{D}$ is equivalent to mining context-maximal CFPs only.

The CFP mining framework also has the advantage of associating to each context in $\mathcal{H}$ less patterns than what a typical transactional frequent pattern miner would provide (as being frequent in a context is only one of the requirements for a pattern to generate a CFP). To some extent,

3   Mining Contextual Frequent Patterns

This section describes how CFPs are mined given a contextual transactional database, by defining two approaches: *(1)* a baseline approach that makes direct use of the definitions given in Section 2, and *(2)* a more efficient approach that relies on theoretical properties emerging from the CFP mining framework.

3.1   A Baseline Approach

By relying on the requirements listed in Definition 4, Algorithm 1 provides a baseline approach to extract context-maximal CFPs. This approach relies on the following steps: *(1)* extracting frequent patterns from every possible context (lines 2-4), *(2)* for each context c and each pattern p frequent in c, checking the c-generality and context-maximality of (c, p) (lines 5-11).

Mining all the contexts in order to enumerate all their frequent patterns obviously is very time-consuming, as it requires to run an external pattern miner for each context separately. We therefore study in the following some theoretical properties in order to allow a more efficient extraction of CFPs.

Algorithm 1. A Baseline Approach

Require: A contextual database $\mathcal{D}$, a minimum frequency threshold σ
Ensure: Set of contextual frequent patterns in $\mathcal{D}$
1. $\mathcal{C} \leftarrow \emptyset$ `//` `initialize the set of discovered CFPs`
2. **for** $c \in \mathcal{D}$ **do**
3. $\mathcal{F}_c \leftarrow$ frequent patterns in c
4. **end for**
5. **for** $c \in \mathcal{D}$ **do**
6.   **for** $p \in \mathcal{F}_c$ **do**
7.   **if** p is c-general and context-maximal **then**
8. $\mathcal{C} \leftarrow \mathcal{C} \cup \{(p,c)\}$ `//` `generate and store the contextual pattern`
9.   **end if**
10.   **end for**
11. **end for**

3.2 CFPM: A More Efficient Post-processing Approach

The approach described in Algorithm 1 trivially exploits the definition of contextual data and CFPs and leads to costly calculation, in particular by first extracting frequent patterns from each possible context. We highlight some interesting properties to reduce redundant calculations, in particular by reducing the executions of the frequent pattern miner. As opposed to [12,13], we focus in this paper on providing a generic post-processing algorithm that, from the output of existing frequent pattern miners, generates the CFPs.

Additional properties of contextual general patterns. A context can be uniquely described by its minimal descendants in $\mathcal{H}$. To this end, we consider the *decomposition* of a context c in $\mathcal{H}$ as the set of minimal contexts in $\mathcal{H}$ being more specific than c, i.e., $decomp(c, \mathcal{H}) = \{c' \in V_{\mathcal{H}}^{-} | (c' <_{\mathcal{H}} c) \vee (c' = c)\}$. For instance, $decomp(Y) = \{YS, YW\}$ and $decomp(YS) = \{YS\}$.

Property 1. *p is c-general iff p is c'-frequent $\forall c' \in decomp(c)$.*

Property 1 (whose proof can be found in [12] and adapted to the current framework) is essential by allowing the reformulation of the c-generality property w.r.t. minimal contexts only. The checking of c-generality requirements for a context thus becomes much simpler: a pattern p is c-general if and only if the set of minimal contexts where p is frequent includes the decomposition of c. Extending this property to context-maximal CFPs is straightforward. CFPM, as presented in Algorithm 2, exploits this property. It can be decomposed into the following consecutive steps:

Mining. *(lines 2-4)* Frequent patterns are extracted from each minimal context. As opposed to Algorithm 1, CFPM does not mine non-minimal contexts.

Reading. *(lines 5-8)* Output files from previous step are read and patterns p are indexed by the set of minimal contexts where they are frequent, i.e., l_p.

Algorithm 2. CFPM: Contextual Frequent Pattern Mining

Require: A contextual database D, a minimum frequency threshold σ
Ensure: Set of contextual frequent patterns in D
1. $\mathcal{C} \leftarrow \emptyset$ // initialize the set of discovered CFPs
2. **for** $c \in V_{\mathcal{H}}^{-}$ **do**
3. $\mathcal{F}_c \leftarrow$ frequent patterns in c
4. **end for**
5. **for** $p \in \bigcup_{c \in V_{\mathcal{H}}^{-}} \mathcal{F}_c$ **do**
6. $l_p \leftarrow \{c \in V_{\mathcal{H}}^{-} \mid p \in \mathcal{F}_c\}$
7. $K[l_p] \leftarrow K[l_p] \cup \{p\}$
8. **end for**
9. **for** l a key in K **do**
10.   **for** $c \in maxContexts(l, \mathcal{H})$ **do**
11.   **for** $p \in K[l]$ **do**
12. $\mathcal{C} \leftarrow \mathcal{C} \cup (p, c)$ // generate and store a CFP
13.   **end for**
14.   **end for**
15. **end for**

Then, K is a hash table with keys being sets of minimal contexts and values being sets of patterns, such as $K[l]$ containing the patterns p such that $l_p = l$. The cost of this step mainly lies on intensive I/O processing.

Coverage Computation and Pattern Generation. *(lines 9-15)* During this step, each key of K is given to the *maxContexts* routine *(line 10)* which performs a bottom-up traversal of the vertices of $\mathcal{H}$ in order to return the set of maximal contexts among $\{c \in V_{\mathcal{H}} \mid decomp(c) \subseteq l\}$. This is the coverage computation step. Then, for each pattern p such that $l = l_p$ *(line 11)* and each context returned by *maxContexts* *(line 10)*, one context-maximal CFP is generated and stored *(line 12)*. Two patterns p and p' frequent in the same minimal contexts (i.e., $l_p = l_{p'}$) are general in the same contexts. They will generate the same result via the *maxContexts* routine. By using the hash table K to store the patterns that are frequent in the same minimal contexts, the number of calls to *maxContexts* is greatly reduced to the number of keys in K rather than the number of distinct patterns discovered during the *mining* step.

Discussion. Mining minimal contexts only is an essential advantage over the baseline approach. CFPM's post-processing oriented design also offers the possibility to use it with any transactional frequent pattern miner, whatever the structure of mined patterns (e.g., subgraphs, episodes, sequential patterns, itemsets, etc.). This genericity also is the main advantage over previous work [12,13].

4   Experimental Results

The implementation of the algorithm is divided into two parts. First, a Ruby wrapper is in charge of running external pattern miners, reading their output, and eventually generating the CFPs. Ruby's flexibility is particularly relevant for designing a generic approach, where final users should be able to add new components with very little effort to support new pattern languages. Second, a C++ module is in charge of the maxContexts routine of Algorithm 2, as it offers better performances without any drawbacks regarding genericity or usability.

The CFPM approach has been extensively experimented in order to assess its main features. Therefore, we have performed experiments implying real datasets and three common pattern languages, namely *itemsets* [1], *sequential patterns* [2] and *subgraphs* [9]. Each of these pattern languages involves different theoretical frameworks and algorithms. Experiments have been conducted on an Intel i7-3520M 2.90GHz CPU, with 16 GB memory.

Contextual Frequent Itemsets. First, in order to study the behavior of CFPM when considering the frequent itemset mining problem [1], we have used the APriori algorithm as implemented in [5]. The dataset used for this experiment initially comes from [8]. It consists of 100,000 product reviews published on the *amazon.com* website. Reviews have been lemmatized and grammatically filtered[1]. The remaining words compose the item alphabet and each review is represented as a transaction. Contextual information associated with the reviews is composed of the *type of product*, the *rating* given by the user and the *proportion of positive feedbacks* received. The resulting context hierarchy contains 210 contexts, whose 48 are minimal. The interested user may refer to [12] for details.

Contextual Frequent Sequential Patterns. The second pattern mining problem addressed with CFPM considers *frequent sequential patterns* as defined in [2]. To this end, CFPM uses the PrefixSpan [11] algorithm as implemented in the SPMF project [5]. The used dataset is the same as the one described above for frequent itemset mining, except that reviews have been converted to sequences of itemsets[2]. It is also the same as the one previously used in [12].

Contextual Frequent Subgraphs. In order to address the subgraph pattern language [9], we use a dataset that has been constructed to study the mutagenicity property of some molecules [7]. It contains 6,512 molecular graphs, such that each one is associated with a label that indicates whether it is mutagene or not. Contextual information is composed of the mutagenicity label, the molecular weight, and the source dataset from which the molecule has been extracted. The merging of these pieces of information produces a context hierarchy containing 39 contexts whose 10 are minimal.

Results. First of all, Figures 2(A1,B1,C1) show a large difference of runtimes when comparing the baseline approach with CFPM, for every used dataset and pattern language. This gap of runtime is mainly due to the fact that the baseline approach first requires to mine frequent patterns from each context, while CFPM

---

[1] Remaining terms are non-modal verbs, nouns, adjectives and adverbs.
[2] The conversion follows the principle that each sentence of the review is an itemset, and the order of itemsets in a sequence results from the order of sentences.

only mines minimal contexts, that are less numerous and smaller. Then, Figures 2 (A1,B1,C1) provide a view on how the runtime of CFPM is decomposed according to the algorithm steps. The largest fraction of time corresponds to the mining step, i.e., running pattern miners for each minimal context. This fraction systematically increases while the minimum frequency threshold decreases. Of course, this fraction of runtime also depends on the underlying implementations and algorithms. For instance, the time required to mine sequential patterns is relatively much larger (*cf.* Figure 2(B1)).

Figures 2(A2,B2,C2) show the total amount of patterns regarding their type. First, let consider the number of distinct patterns (i.e., the distinct frequent patterns discovered during the mining step) compared with the number of context-maximal CFPs. As expected, the latter is always greater than or equal to the

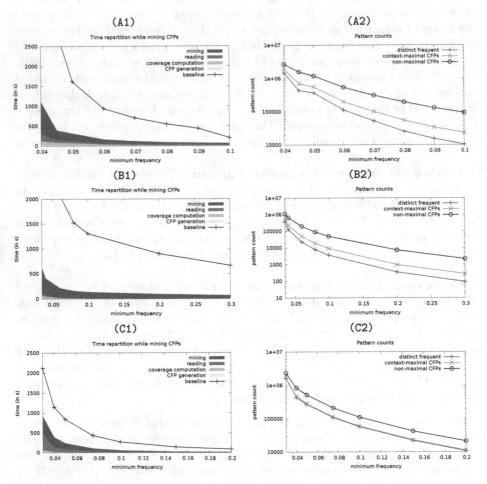

Fig. 2. Time consumption for mining CFPs for itemsets (A1), sequential patterns (B1) and subgraphs (C1), and number of distinct patterns and contextual patterns for itemsets (A2), sequential patterns (B2) and subgraphs (C2).

number of distinct patterns since every distinct pattern generates at least one CFP. On the other hand, the total number of CFPs (i.e., not necessarily context-maximal ones) is much higher, therefore demonstrating the interest of proposing a condensed representation of contextual frequent patterns.

5   Conclusion and Prospects

In this paper, we adapt the transactional setting for mining frequent patterns by considering the contextual information associated with transactions. We therefore define a theoretical framework for CFP mining and propose a relevant algorithm for mining such patterns in a totally generic way regarding the pattern language. By generalizing the typical transactional setting and by post-processing the output of existing frequent pattern miners, the proposed approach provides the benefit of being able to be used in conjunction with any such frequent pattern miner developed during the last decades. Such an approach provides opportunities to be exploited in numerous application domains where data are often accompanied with contextual information, e.g., the mining of mobile data where spatial and temporal information may be used as contextual data or user profiling on the Web, where user activities may be mined under the scope of contextual information about the user such as location, age, etc.

The contextual pattern mining approach offers numerous prospects, such as for instance adapting it to the case of relying on discriminative patterns rather than frequent patterns in the context hierarchy, or considering the case where contextual information is imprecise (i.e., some transactions are mapped to a non-minimal context).

References

1. Agrawal, R., Imieliński, T., Swami, A.: Mining association rules between sets of items in large databases. In: ACM SIGMOD Record, vol. 22, pp. 207–216. ACM (1993)
2. Agrawal, R., Srikant, R.: Mining sequential patterns. In: International Conference on Data Engineering, pp. 3–14. IEEE (1995)
3. Bringmann, B., Nijssen, S.: What is frequent in a single graph? In: Washio, T., Suzuki, E., Ting, K.M., Inokuchi, A. (eds.) PAKDD 2008. LNCS (LNAI), vol. 5012, pp. 858–863. Springer, Heidelberg (2008)
4. Bringmann, B., Nijssen, S., Zimmermann, A.: Pattern-based classification: A unifying perspective. In: LeGo Workshop, p. 36 (2009)
5. Fournier-Viger, P., Gomariz, A., Soltani, A., Lam, H., Gueniche, T.: Spmf: Open-source data mining platform (2014), http://www.philippe-fournier-viger.com/spmf
6. Han, J., Cheng, H., Xin, D., Yan, X.: Frequent pattern mining: current status and future directions. Data Mining and Knowledge Discovery 15(1), 55–86 (2007)
7. Hansen, K., Mika, S., Schroeter, T., Sutter, A., ter Laak, A., Steger-Hartmann, T., Heinrich, N., Müller, K.-R.: Benchmark data set for in silico prediction of ames mutagenicity. Journal of Chemical Information and Modeling (2009)

8. Jindal, N., Liu, B.: Opinion spam and analysis. In: International Conference on Web Search and Data Mining, pp. 219–230. ACM (2008)
9. Kuramochi, M., Karypis, G.: Frequent subgraph discovery. In: International Conference on Data Mining, pp. 313–320. IEEE (2001)
10. Mannila, H., Toivonen, H.: Levelwise search and borders of theories in knowledge discovery. Data Mining and Knowledge Discovery 1(3), 241–258 (1997)
11. Pei, J., Pinto, H., Chen, Q., Han, J., Mortazavi-Asl, B., Dayal, U., Hsu, M.-C.: Prefixspan: Mining sequential patterns efficiently by prefix-projected pattern growth. In: 2013 IEEE 29th International Conference on Data Engineering (ICDE), p. 0215. IEEE Computer Society (2001)
12. Rabatel, J., Bringay, S., Poncelet, P.: Contextual sequential pattern mining. In: International Conference on Data Mining Workshops, pp. 981–988. IEEE (2010)
13. Rabatel, J., Bringay, S., Poncelet, P.: Mining sequential patterns: A context-aware approach. In: Guillet, F., Pinaud, B., Venturini, G., Zighed, D.A. (eds.) Advances in Knowledge Discovery and Management. SCI, vol. 471, pp. 23–41. Springer, Heidelberg (2013)

A Deep Interpretation of Classifier Chains

Jesse Read and Jaakko Hollmén

Aalto University, Department of Information and Computer Science
PO Box 15400, FI-00076 Aalto, Espoo, Finland
Helsinki Institute for Information Technology (HIIT), Finland
{jesse.read,jaakko.hollmen}@aalto.fi

Abstract. In the "classifier chains" (CC) approach for multi-label classification, the predictions of binary classifiers are cascaded along a chain as additional features. This method has attained high predictive performance, and is receiving increasing analysis and attention in the recent multi-label literature, although a deep understanding of its performance is still taking shape. In this paper, we show that CC gets predictive power from leveraging labels as additional stochastic features, contrasting with many other methods, such as stacking and error correcting output codes, which use label dependence only as kind of regularization. CC methods can learn a concept which these cannot, even supposing the same base classifier and hypothesis space. This leads us to connections with deep learning (indeed, we show that CC is competitive precisely because it is a deep learner), and we employ deep learning methods – showing that they can supplement or even replace a classifier chain. Results are convincing, and throw new insight into promising future directions.

1 Introduction

Multi-label classification (MLC) is the supervised learning problem where an instance is associated with multiple binary class variables (i.e., *labels*), rather than with a single class, as in traditional classification problems ([12]). The typical argument (which this paper reanalyzes) is that, since these labels are often strongly correlated, modeling the dependencies between them allows MLC methods to improve their performance.

As in general classification scenarios, an n-th feature vector (instance) can be represented as $\mathbf{x}^{(n)} = [x_1^{(n)}, \ldots, x_D^{(n)}]$, where each $x_d \in \mathbb{R} | d = 1, \ldots, D$. In the traditional *binary* classification task, we are interested in having a model h to provide a prediction for test instances $\tilde{\mathbf{x}}$, i.e., $\hat{y} = h(\tilde{\mathbf{x}})$; where h, probabilistically speaking, seeks the expectation $\mathbb{E}[y|\mathbf{x}]$ of unknown $p(y|\mathbf{x})$. In MLC, there are L binary output class variables (*labels*), and we are interested in predictions

$$\hat{\mathbf{y}} = [\hat{y}_1, \ldots, \hat{y}_L] = h(\tilde{\mathbf{x}}) = \underset{\mathbf{y} \in \{0,1\}^L}{\operatorname{argmax}} \hat{p}(\mathbf{y}|\tilde{\mathbf{x}})$$

where $y_j = 1$ indicates the relevance of the j-th label; $j = 1, \ldots, L$.

H. Blockeel et al. (Eds.): IDA 2014, LNCS 8819, pp. 251–262, 2014.

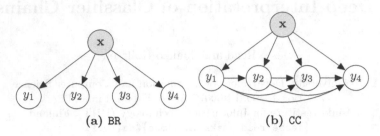

Fig. 1. BR (1a) and CC (1b) as graphical models, $L = 4$. Unlike many typical Bayesian networks, we have lumped $\mathbf{x} = [x_1, \ldots, x_D]$ into a single variable.

From N labelled examples (training data) $\mathcal{D} = \{(\mathbf{x}^{(n)}, \mathbf{y}^{(n)})\}_{n=1}^{N}$, we infer h. A most basic solution is to train L binary models. This method is called *binary relevance* (BR); illustrated graphically in Fig. 1a. BR classifies an $\tilde{\mathbf{x}}$ L times as $h_{\text{BR}}(\tilde{\mathbf{x}}) := [h_1(\tilde{\mathbf{x}}), \ldots, h_L(\tilde{\mathbf{x}})]$.

Practically the entirety of the multi-label literature points out that the independence assumption among the labels leads to suboptimal performance (e.g., [16,7,3,15,20] and references therein), and that for this reason BR cannot achieve optimal performance. A plethora of methods have been motivated by a perceived need to modelling this dependence and thus improve over BR. For example, Meta-BR (MBR, also known in the literature as 'stacked-BR' and '2BR') [7,3] stacks the output of one BR as input into a second (meta) BR[1], so as to learn to correct errors. For some $\tilde{\mathbf{x}}$,

$$h_{\text{MBR}}(\tilde{\mathbf{x}}) := h'_{\text{BR}}(h_{\text{BR}}(\tilde{\mathbf{x}}))$$

A related approach uses subset mapping (SM, e.g., as in [18]) to force infrequent label vector predictions to a more frequent ones,

$$h_{\text{SM}}(\tilde{\mathbf{x}}) := \operatorname*{argmin}_{\mathbf{y} \in \mathcal{Y}_{\text{train}}} \ell(\mathbf{y}, h_{\text{BR}}(\tilde{\mathbf{x}}))$$

where $\mathcal{Y}_{\text{train}}$ are all distinct $\mathbf{y}^{(n)}$ from the training data $\mathcal{D}$ and $\ell(\mathbf{y}, \hat{\mathbf{y}})$ is some penalty function typically rewarding small Hamming distance and high frequency in $\mathcal{D}$. SM is very closely related to error-correcting output code methods [6] and, like MBR, can be seen as a regularizer. The penalty goes to ∞ if $\mathbf{y} \notin \mathcal{Y}_{\text{train}}$, meaning that any predictions of label combinations not seen in the training set will be 'corrected'.

As an example, movie genres adult and family may be mutually exclusive in the training set, and having a regularization/correction component to avoid this classification at test time may lead to improved performance (over BR).

The classifier chains method (CC, [16]), illustrated in Fig. 1b, models label dependence by using binary label predictions as extra input attributes for the following classifier, in a chain, and therefore models labels and inputs together,

---

[1] There is no consensus in the literature as to whether it is best to also include the $\mathbf{x}$-space input again, or simply the label outputs $h(\mathbf{x})$, as input to the meta BR.

rather than correcting labels as a separate step. In the original formulation with greedy inference,

$$h_{\mathrm{CC}}(\tilde{\mathbf{x}}) := \Big[h_1(\tilde{\mathbf{x}}), h_2(\tilde{\mathbf{x}}, h_1(\tilde{\mathbf{x}})), \ldots, h_L(\tilde{\mathbf{x}}, h_1(\tilde{\mathbf{x}}), \ldots, h_{L-1}(\tilde{\mathbf{x}}))\Big].$$

CC variants have consistently performed strongly in the literature and there have been numerous extensions, variations and analyses, e.g., [2,20,15,11,5]. However, the reasons for its high performance are only recently being unravelled. In this paper, we throw new light on the subject.

Two focus points for improvement of CC have been the inference, and the order of the labels in the chain. Originally, [16] suggested an ensemble of randomly-ordered chains (ECC) with voting, whereas two recent high-performing CC methods [11] and [15], use beam and Monte Carlo search, respectively to obtain one well-ordered chain. We use the latter, which we denote MCC, in empirical comparisons, as well as ECC with 10 random chains.

2 Label Dependence in Multi-label Learning

The idea of leveraging label dependence to improve performance vs BR intuitively makes sense. However, the understanding behind this is only recently taking shape, with the authors of [2,5,4] opening an important discussion from a probabilistic perspective, noting the difference between

- marginal dependence, where $p(y_j|y_k) \neq p(y_j)$; and
- conditional dependence, where $p(y_j|y_k) \neq p(y_j|y_k, \mathbf{x})$.

Thus MBR and SM model marginal dependence, whereas CC models conditional dependence, by learning labels and input together.

An interesting point of debate is the following: given infinite data, can two *separate* binary models on labels y_j and y_k achieve as good performance as one that models them together (e.g., MBR, CC) – assuming the same base classifier (say, logistic regression)? Among others, [16,5] ponder if BR has been underrated and could equal CC's performance with enough training data. Indeed, [4] make the case that it should be possible make risk-minimizing predictions without any particular effort to detect or model label dependence. This seems to throw into doubt the bulk of the contributions to the multi-label literature.

It is also worth recalling here, that labels cannot only be learned together or separately, but also evaluated together or separately. A typical measure for the latter case is the HAMMING SCORE, which is widely used in MLC empirical evaluations. However, many MLC papers quietly overlook the fact that achieving statistically significant improvement over BR in this measure is difficult to obtain. This seems to add to [4]'s claim.

Proposition 1. *If we can predict $\mathbb{E}(Y_2|Y_1, \mathbf{x})$ to a certain degree of accuracy under some evaluation measure, then it is also possible to predict $\mathbb{E}(Y_2|\mathbf{x})$ with at least the same accuracy under the same measure.*

We elaborate on this in the following sections.

3 Analysis on Synthetic Datasets

Using the following synthetic datasets with the methods discussed above (Section 1), with logistic regression as a base classifier, we run some experiments to expand on the discussion from Section 2. Results are given in Tab. 1.

- Localization: a scenario (see Fig. 2) where labels correspond to pixels which represent floor tiles in a room, and are active/relevant ($y_j = 1$) if an object is on them. Light sensors signal detection (with 90% accuracy) $x_d = 1$ if an active tile lies between the sensor location and the light source. For each of 1000 instances (two thirds of which are used for training), a line of three tiles is activated ($y_j = 1$ for three j) in a random location in the grid, and another tile is activated in the furthest corner from that line; which is always a blind spot (undetectable by sensors). Based on the real-world deployment of [14].
- Logic: Two binary attributes, X_1, X_2, deterministically mapped to three labels Y_1, Y_2, Y_3, corresponding to AND(X_1, X_2), OR(X_1, X_2), XOR(X_1, X_2) (binary logical operations). We generate 20 X_1, X_2 randomly, and use 12 for training.

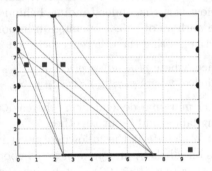

Fig. 2. Top-down view of a grid of L tiles in a room. There are D sensors, which signal detection ($x_d = 1$) with 90% probability if any of the active tiles ($y_j = 1$) lie between sensors (black semi circles) and the light source (black bar at the bottom). Here, three sensors have detected an object. Note the blind spot.

We use two standard, opposing evaluation methods,

$$\text{HAMMING SCORE} := \frac{1}{NL} \sum_{n=1}^{N} \sum_{j=1}^{L} \left[y_j^{(n)} = \hat{y}_j^{(n)} \right], \text{ and}$$

$$\text{EXACT MATCH} := \frac{1}{N} \sum_{n=1}^{N} \left[\mathbf{y}^{(n)} = \hat{\mathbf{y}}^{(n)} \right]$$

which are used in almost all MLC evaluations. HAMMING SCORE (which we mentioned in the previous section) rewards methods for predicting individual

Table 1. Predictive Performance on Toy Datasets with per-dataset (rank)
HAMMING SCORE

Dataset	BR	CC	SM	MBR	ECC	MCC
Localization	0.992 (1)	0.992 (1)	0.991 (4)	0.991 (4)	0.991 (4)	0.992 (1)
Logic	0.833 (5)	1.000 (1)	0.750 (6)	0.875 (3)	0.875 (3)	1.000 (1)

EXACT MATCH

Dataset	BR	CC	SM	MBR	ECC	MCC
Localization	0.412 (5)	0.491 (2)	0.455 (3)	0.408 (6)	0.447 (4)	0.497 (1)
Logic	0.500 (6)	1.000 (1)	0.625 (3)	0.625 (3)	0.625 (3)	1.000 (1)

labels well, whereas EXACT MATCH rewards a higher proportion of instances with *all* label relevances correct.

Given much of the discussion in the literature, one could understand that a method which models label dependence would excel on Localization. This could be claimed for EXACT MATCH; but the number of correct label relevances (HAMMING SCORE) is essentially identical across all methods. Despite the obvious label dependence here (wrt the position and shape of the 'relevant' pixels), conditioning on $\mathbf{x}$ (i.e., training BR) suffices for high label-wise precision.

The results on Logic indicate the opposite case: BR is clearly unable to learn the concept, even though we can be sure that $\mathbb{E}[y_3|y_2, y_1, \mathbf{x}] = \mathbb{E}[y_3|\mathbf{x}]$. Furthermore, although SM and MBR model label dependence, they cannot learn the concept either: their regularization cannot make up for the fact that their underlying BR models fail. CC and MCC score perfectly. ECC under-performs, so apparently CC only performs well under certain label orders (and got 'lucky'); confirmed by MCC, which finds one good order before final training. MBR does actually have a structure suitable for learning XOR, but apparently training cannot leverage it properly.

The authors of [5] uncovered a similar case with their *probabilistic classifier chains* (originally presented in [2], using Bayes-optimal inference instead of greedy inference for CC), putting it down to this method's "expanded hypothesis space" (it trials all 2^L combinations for $\hat{\mathbf{y}}$ at inference time). This, however, cannot be the complete answer, since the original CC makes L separate binary decisions just like BR; thus the same hypothesis space.

With real-world datasets it is difficult to postulate, but on the Logic dataset it is clear that BR's accuracy will *never* reach 1 even under infinite data, since its h_3 model will *never* learn XOR. The performance gap in Tab. 1 is a convincing 50 percentage points for EXACT MATCH. CC's h_3 is a perfect model, even with the same base classifier.

In the next section we explain how CC works as a deep structure, of up to L levels (let us simply state that any structure of more than one level is deep) and for this reason can outperform BR as well as the 'regularization'-type methods like MBR and SM.

4   Why Classifier Chains Works

Fig. 3 shows CC on Logic. It is clear how it learns the XOR label (Y_3): by leveraging off labels Y_1 and Y_2, which are acting like hidden units of a neural network. In terms of neural networks it is actually more than required; Fig. 3c shows the smallest neural network that can learn XOR as demonstrated in [17][2].

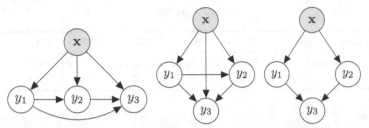

(a) CC standard depiction (b) redrawn wrt y_3 (c) Minimum network able to learn XOR function

Fig. 3. CC with three labels, as in the Logic dataset. Note we show (for now) $\mathbf{x} = [x_1, x_2]$ as a single variable.

In Fig. 4, as a probabilistic graphical models interpretation, are the junction trees (see [1]) of two of the models, showing that Fig. 3c can be tractable. Standard CC (Fig. 3a—3b) is fully connected and thus many forms of inference are intractable; a known issue [2,15].

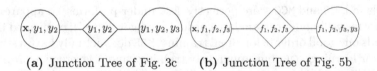

(a) Junction Tree of Fig. 3c   (b) Junction Tree of Fig. 5b

Fig. 4.   Junction Trees for different formulations

It is a straightforward interpretation of CC to think of labels being used as features to predict other labels. Let us not forget though that all estimated labels are derived from the input. Where labels are manually assigned to instances, then $y_j = f_j(\mathbf{x})$ are feature functions of the human mind, which (for a logistic regression base learner h_j) is being approximated by a sigmoid function on a linear combination of the input. From here we could get to conditional random fields (as in [2]), but we will continue through another route. From the point of view of y_3, there are three inputs ($\mathbf{x}$ still treated as a single variable),

$$y_3 = h_3(f_1(\mathbf{x}), f_2(\mathbf{x}, f_1(\mathbf{x})), \mathbf{x})$$

---

[2] Earlier, pessimistic results about solving the XOR problem with neural networks [13] resulted in the decline of neural networks research.

See Fig. 5a. It is clear that f_1 and f_2 are simply transformations of the input. In the case of CC with logistic regression as a base classifier,

$$f_j(\mathbf{x}) := \sigma(\mathbf{w}_j^\top [x_1, \ldots, x_D, f_{j-1}(\mathbf{x}), \ldots, f_1(\mathbf{x})])$$

where σ is the logistic/sigmoid function, but we can easily imagine arbitrary (possibly non-linear) transformations of the input, and an arbitrary number of such functions: $f_1^*(\mathbf{x}), \ldots, f_K^*(\mathbf{x})$; see Fig. 5b. Note independence among $f_k^*(\mathbf{x})$.

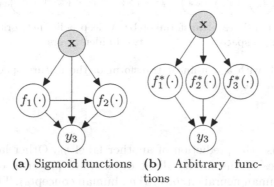

(a) Sigmoid functions   (b) Arbitrary functions

Fig. 5. As in Fig. 3b, but labels are shown as transformations of the input. Fig. 5a is easily equivalent to Fig. 5b in the case where $f_3^*(\mathbf{x}) = \mathbf{x}$, etc.

There is no reason to assume that the number of labels (L) equals the number of desired features (K). If we include the rest of the labels, expand $\mathbf{x}$ into D nodes, $X_1, \ldots, X_D$ and invert the graph such that the Y label variables are now at the top, the result is Fig. 6a. These last two changes are purely presentational, but important for what comes next.

Since f_k^* is an arbitrary function, and two hidden layers are enough for universal approximation ability [9] of any arbitrary function, Fig. 6a is therefore equivalent to the *deep* network of Fig. 6b with, e.g., a sigmoid function for all layers; i.e.,

$$z_k^{[1]} = \sigma\left(\sum_{d=1}^{D} x_d w_{dk}\right)$$

for the first layer, where w_{jk} is a weight on the link between x_d and hidden node z_k.

The two hidden layers can be learned by restricted Boltzmann machines (RBMs) [8]. This means that training BR on the top layer ($\mathbf{z}^{[1]}$ vectors) can theoretically be as competitive as CC trained on the input ($\mathbf{x}$ vectors). In fact they can be equivalent, except that the RBMs discover feature functions, instead of trying to approximate the human feature functions (labels) available. Indeed, we do obtain top performance (as with CC, MCC) on Logic (not shown).

In other words, since all labels are related to the input, with an adequate (possibly non-linear) binary model, we can predict a label y_j just as well as

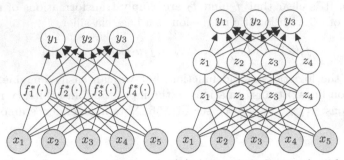

(a) A non-linear transform of the feature space

(b) A deep belief network with two hidden layers

Fig. 6. A network with a non-linear transform of the feature space (left) and two layers to approximate it (right)

we could given also the prediction of another label y_k. Other labels are simply additional features of the input, albeit often quite powerful ones, since they often represent human neural circuitry (i.e., human concepts). The true function behind the concept is of course typically not known, but given the true outputs in the training data, they can be approximated (standard supervised learning).

Whereas a typical basis function is deterministic, the $f_k^*(\mathbf{x})$ are not (necessarily), as reflected in the RBMs. Guided by this, in the next section we employ some deep learning methods and show them to be effective. But Tab. 2 already hints that random features can help in a classifier-chains approach (particularly when the chain is carefully ordered). Models with random activations have been considered in e.g., [19], or in 'extreme learning machines' [10] – but as a single hidden layer and not directly into the label space as we consider here.

Table 2. Per-label accuracy on the Music dataset (see, e.g., [12]), from 5×CV, with (+) and without 10 random labels (i.e., feature functions), of the form $y_k = \sigma(\sum_{d=1}^{D} w_{dk}x_d)$ for random $\mathbf{w}$

label	CC	CC+	MCC	MCC+
amazed	0.759	**0.793**	0.772	**0.776**
happy	0.688	**0.692**	0.722	**0.734**
relaxing	**0.764**	0.755	0.764	**0.781**
quiet	**0.895**	0.890	0.882	**0.895**
sad	**0.835**	0.819	0.793	**0.827**
aggressive	0.759	**0.814**	0.793	**0.819**

5 Deep Multi-label Learning

Since labels can be seen as high-level features of the input, other higher-level features should also positively affect predictive performance. For example, from an image, a feature for the presence of a grainy surface such as sand or pebbles, or for being adjacent to a (significant) body of water should help us predict **beach** just as much (or better) than label **urban**. We can use RBMs to learn layers of such hidden features, in an unsupervised fashion. These hidden layers can capture complex dependencies and structure from the input space.

If the features are powerful, the label variables become independent. This is intuitively attractive, because humans do not recognise beaches depending on the probability that what they see is urban or not. Unfortunately, learning high level features in an unsupervised fashion is not as easy as trying to approximate labels from training data. Powerful algorithms and computational resources are needed – a currently active field of research.

Tab. 3 show results, comparing baseline BR, MBR, and MCC, with deep learning approaches[3]: namely two RBMs plus a multi-label learner, either BR, MCC, or with back-propagation (BP) as in [8] but for MLC; all denoted with D. Also we included [21]'s BP multi-label learner (BPMLL); a multi-layer neural network *not* initialized using RBMs. All experiments in this paper are carried out with the WEKA-based MEKA framework[4] with a setup like [15] (the datasets are described there). We used a single parameter combination for RBMs for all datasets (namely 30 hidden units per layer, learning rate 0.1, momentum 0.2, 5000 iterations) chosen ad-hoc – to avoid intensive parameter tuning on many datasets. Implementations are available within MEKA. All base classifiers h_j are logistic regression (WEKA's implementation). We evaluated using HAMMING SCORE and EXACT MATCH described earlier, and additionally the micro averaged F-measure,

$$\text{Micro Averaged F1} := F_1([y_1^{(1)}, \ldots, y_L^{(N)}], [\hat{y}_1^{(1)}, \ldots, \hat{y}_L^{(N)}])$$

where $F_1(\mathbf{a}, \mathbf{b})$ returns the F_1 score of binary vectors $\mathbf{a}$ and $\mathbf{b}$.

Overall D·MCC performs best under EXACT MATCH, but not as well as D·BR or DBP (which are closely related) under HAMMING SCORE – a result which corresponds with our discussion; D·MCC provides extra depth with a CC, but with the RBMs underneath BR already becomes very competitive – especially compared directly to baseline BR. A well-ordered CC (MCC) is still very powerful for EXACT MATCH, but even better performance can be obtained with additional learned features. We could speculate that advances in deep learning should eventually reduce the effectiveness of CC, as higher-level features make labels more independent. Although, on the other hand, many kinds of CC models are more interpretable than RBMs (and usually faster to train), and may therefore still be interesting for many applications.

---

[3] Space does not permit a review of RBMs and deep learning, see e.g., [8] for details.
[4] http://meka.sourceforge.net

Table 3. Predictive performance on real datasets, with dataset-wise (rank)

EXACT MATCH

Dataset	BR	MBR	BPMLL	MCC	D-MCC	D-BR	DBP
music	0.193 (6)	0.193 (6)	0.252 (3)	0.208 (5)	0.218 (4)	0.267 (2)	0.287 (1)
scene	0.286 (6)	0.292 (5)	0.554 (2)	0.353 (4)	0.476 (3)	0.582 (1)	0.183 (7)
yeast	0.150 (5)	0.137 (7)	0.161 (4)	0.198 (2)	0.204 (1)	0.149 (6)	0.179 (3)
genbase	0.960 (3)	0.955 (4)	0.271 (7)	0.965 (1)	0.965 (1)	0.950 (5)	0.950 (5)
medical	0.439 (4)	0.457 (3)	0.194 (7)	0.474 (2)	0.361 (5)	0.200 (6)	0.521 (1)
enron	0.022 (5)	0.022 (5)	0.010 (7)	0.028 (4)	0.161 (1)	0.054 (2)	0.043 (3)
avg. rank	4.83	5.00	5.00	3.00	2.50	3.67	3.33

HAMMING SCORE

Dataset	BR	MBR	BPMLL	MCC	D-MCC	D-BR	DBP
music	0.761 (5)	0.762 (4)	0.776 (2)	0.742 (6)	0.726 (7)	0.772 (3)	0.791 (1)
scene	0.807 (4)	0.802 (6)	0.895 (1)	0.807 (4)	0.847 (3)	0.895 (1)	0.731 (7)
yeast	0.786 (3)	0.780 (5)	0.790 (2)	0.771 (7)	0.780 (5)	0.784 (4)	0.791 (1)
genbase	0.998 (3)	0.998 (3)	0.932 (7)	0.999 (1)	0.999 (1)	0.998 (3)	0.998 (3)
medical	0.980 (4)	0.981 (2)	0.969 (6)	0.981 (2)	0.971 (5)	0.967 (7)	0.984 (1)
enron	0.892 (6)	0.904 (5)	0.939 (3)	0.884 (7)	0.940 (2)	0.947 (1)	0.937 (4)
avg. rank	4.17	4.17	3.50	4.50	3.83	3.17	2.83

MICRO-AVERAGED F1

Dataset	BR	MBR	BPMLL	MCC	D-MCC	D-BR	DBP
music	0.570 (7)	0.571 (6)	0.603 (2)	0.574 (5)	0.580 (3)	0.577 (4)	0.629 (1)
scene	0.463 (5)	0.406 (6)	0.668 (1)	0.480 (4)	0.621 (2)	0.576 (3)	0.199 (7)
yeast	0.599 (6)	0.618 (3)	0.633 (2)	0.601 (5)	0.607 (4)	0.588 (7)	0.639 (1)
genbase	0.987 (1)	0.985 (2)	0.276 (5)	0.985 (2)	0.341 (4)	0.200 (6)	0.174 (7)
medical	0.665 (3)	0.655 (4)	0.315 (7)	0.681 (2)	0.557 (5)	0.487 (6)	0.771 (1)
enron	0.372 (5)	0.248 (7)	0.483 (2)	0.353 (6)	0.475 (3)	0.493 (1)	0.466 (4)
avg. rank	4.50	4.67	3.17	4.00	3.50	4.50	3.50

6   Conclusions

The high performance of the classifier chains (CC) approach can be seen as stemming from its leverage of labels as high-level features in a deep cascading structure across binary classifiers. This contrasts with many other approaches based on binary classifiers, that leverage label dependence in a regularization step, but provide limited additional learning power. We demonstrated several scenarios where CC can learn a concept where other methods fail.

We argued that if labels can be considered high-level features stemming from the input, then it is possible to learn such features independently of the training data. We employed deep-learning approaches (using restricted Boltzmann machines) to learn such higher-level features, and obtained provide strong performance, particularly when supplemented with a top-layer chain. Results indicate that further advances in multi-label classification will come from better models of features, and borrow from thus-related fields, rather than obsessive modelling of high-level label 'correlations'.

Many deep-learning methods have other important advantages, particularly in online and semi-supervised settings. We intend to investigate this, as well as produce further empirical study.

References

1. Barber, D.: Bayesian Reasoning and Machine Learning. Cambridge University Press (2012)
2. Dembczyński, K., Cheng, W., Hüllermeier, E.: Bayes optimal multilabel classification via probabilistic classifier chains. In: ICML 2010: 27th International Conference on Machine Learning, pp. 279–286. Omni Press, Haifa (2010)
3. Cheng, W., Hüllermeier, E.: Combining instance-based learning and logistic regression for multilabel classification. Machine Learning 76(2-3), 211–225 (2009)
4. Dembczyński, K., Waegeman, W., Cheng, W., Hüllermeier, E.: On label dependence and loss minimization in multi-label classification. Mach. Learn. 88(1-2), 5–45 (2012)
5. Dembczyński, K., Waegeman, W., Hüllermeier, E.: An analysis of chaining in multi-label classification. In: ECAI: European Conference of Artificial Intelligence. Frontiers in Artificial Intelligence and Applications, vol. 242, pp. 294–299. IOS Press (2012)
6. Ghani, R.: Using error-correcting codes for text classification. In: ICML 2000: 17th International Conference on Machine Learning, pp. 303–310. Morgan Kaufmann Publishers, Stanford (2000)
7. Godbole, S., Sarawagi, S.: Discriminative methods for multi-labeled classification. In: Dai, H., Srikant, R., Zhang, C. (eds.) PAKDD 2004. LNCS (LNAI), vol. 3056, pp. 22–30. Springer, Heidelberg (2004)
8. Hinton, G., Salakhutdinov, R.: Reducing the dimensionality of data with neural networks. Science 313(5786), 504–507 (2006)
9. Hornik, K., Stinchcombe, M., White, H.: Multilayer feedforward networks are universal approximators. Neural Networks 2(5), 359–366 (1989)
10. Huang, G.-B., Wang, D., Lan, Y.: Extreme learning machines: A survey. International Journal of Machine Learning and Cybernetics 2(2), 107–122 (2011)
11. Kumar, A., Vembu, S., Menon, A.K., Elkan, C.: Learning and inference in probabilistic classifier chains with beam search. In: Flach, P.A., De Bie, T., Cristianini, N. (eds.) ECML PKDD 2012, Part I. LNCS, vol. 7523, pp. 665–680. Springer, Heidelberg (2012)
12. Madjarov, G., Kocev, D., Gjorgjevikj, D., Džeroski, S.: An extensive experimental comparison of methods for multi-label learning. Pattern Recognition 45(9), 3084–3104 (2012)
13. Minsky, M., Papert, S.: Perceptrons — An Introduction to Computational Geometry. The MIT Press (1969)
14. Read, J., Achutegui, K., Miguez, J.: A distributed particle filter for nonlinear tracking in wireless sensor networks. Signal Processing 98, 121–134 (2014)
15. Read, J., Martino, L., Luengo, D.: Efficient monte carlo methods for multidimensional learning with classifier chains. Pattern Recognition 47(3) (2014)

16. Read, J., Pfahringer, B., Holmes, G., Frank, E.: Classifier chains for multi-label classification. Machine Learning 85(3), 333–359 (2011)
17. Rumelhart, D.E., McClelland, J.L., Research Group, P.D.P. (eds.): Parallel Distributed Processing: Explorations in the Microstructure of Cognition, Vol. 1: Foundations. MIT Press, Cambridge (1986)
18. Schapire, R.E., Singer, Y.: Improved boosting algorithms using confidence-rated predictions. Machine Learning 37(3), 297–336 (1999)
19. Thomas Miller III, W., Glanz, F.H., Gordon Kraft III, L.: CMAC: An associative neural network alternative to backpropagation. Proceedings of the IEEE 78(10), 1561–1567 (1990)
20. Zaragoza, J.H., Sucar, L.E., Morales, E.F., Bielza, C., Larrañaga, P.: Bayesian chain classifiers for multidimensional classification. In: 24th International Conference on Artificial Intelligence (IJCAI 2011), pp. 2192–2197 (2011)
21. Zhang, M.-L., Zhou, Z.-H.: Multilabel neural networks with applications to functional genomics and text categorization. IEEE Transactions on Knowledge and Data Engineering 18(10), 1338–1351 (2006)

A Nonparametric Mixture Model
for Personalizing Web Search

El Mehdi Rochd[1,2] and Mohamed Quafafou[1]

[1] Aix-Marseille University, LSIS UMR 7296, France
[2] Marketshot, Paris, France
{el-mehdi.rochd,mohamed.quafafou}@univ-amu.fr

Abstract. Probabilistic topic models were successfully used to achieve the personalization task using query logs. Thus, both users and previously clicked results are considered when estimating probability distributions in order to answer users'queries. However, the proposed models are generally parametric and require to define in advance the number of topics. Moreover, they can not deal with new users. To overcome these limitations, we propose a model called the Hierarchical personalized Dirichlet Processes (HpDP) that personalizes search and allows to automatically learn the number of latent topics. It also addresses the challenging problem of predicting results for new users. We compare our model, with recent topic models and use them to rank online products by their likelihood given a particular user/query pair. Experiments performed on data from a real online products comparator show the effectiveness of our approach.

1 Introduction

Building user profiles is an important component of personalization systems. In fact, in commercial applications, personalization relies on user profiles to help adapting the content of websites in order to propose information that best fits the user's interests. To achieve these ends, we should first gather information about users and build their profiles from the analysis of this information. Many reported approaches enable to get necessary knowledge about the users in order to build their profiles. An approach consists of considering information from the current search session to build short term profiles [10]. In [8], an approach attempted to build long-term user profiles. In [1], the authors have shown how these short and long-term profiles can be combined. Once prior interaction data are selected, the following step is to convert it into a user profile in order to perform a representation of the user's interests. Different techniques enable to generate these profiles. The authors of [6] adapted an approach using vectors of the original terms. Another approach, described in [7], aims to map the user's interests onto a set of topics, which can be defined by the users themselves. Then, an additional approach enables to extract these topics from large online ontologies of websites, such as the Open Directory Project [3].

H. Blockeel et al. (Eds.): IDA 2014, LNCS 8819, pp. 263–275, 2014.

A new technique that starts to arouse interest, consists in using latent topic models [9] to determine these topics instead of employing a human-generated ontology. Topic models are considered as a tool for exploratory and predictive analysis of text. The most used topic model is the latent Dirichlet allocation (LDA) [2]. It posits that a small number of distributions over words, called topics, can be used to explain the observed data.

It is in this perspective that a new model that extends the LDA for the analysis of the personalized search problem was proposed in [5]. A user/topic distribution was added in the graphical model of LDA, involving the user in the generative process. The experimental results were not satisfactory and have not allowed to conclude that personalization increases the performance. The authors hypothesized that this negative effect on the ranking lists, may be related to the integration of the user in the generative process, because it makes the user very influential in the model and can be overwhelming information derived from data, while this information can be more useful. Thereafter, in [4], a model was presented for personalized search from query logs using sets of latent topics derived directly from the log files themselves, where the user is not included in the generative process, but subtly introduced as part of the ranking formula, which is used to rank products for a given query. The authors concluded that there is an improvement in performance compared to non-personalized models. We will compare this system, called the PTM, with our proposed model. Two main shortcomings of the PTM are (1) it assumes a fixed prespecified number of topics regardless of the data and (2) it is unable to deal with new users.

We thus propose a new model, which enables to overcome this limit. Indeed, our HpDP model, is an extension of the HDP [13]. It allows to automatically learn the number of topics from the data. Once the topics have converged, we will be able to identify their number, and then introduce a user/topic distribution to determine the topical interests of users, and predict products for new users. We demonstrate the effectiveness of our approach through experiments conducted on web user sessions collected by a real online products comparator.

2   The HpDP Model

2.1   Background

Mixture models explicitly model the existence of K sub-populations in the data. Each sub-population is represented by a probability distribution:

$$p(\boldsymbol{w}|\theta,\phi_{1:K}) = \sum_{k=1}^{K} \theta_k f(\boldsymbol{w}|\phi_k)$$

where w is a data point, θ_k is the mixture proportion and $f(.|\phi_k)$ is the density function of the sub-population k. Under a Bayesian setting, prior distributions are specified for θ and ϕ_k. Since they are multinomial distributions, we use the Dirichlet distributions as their conjugate priors. Given the specification of the prior distributions, Bayesian mixture models specify likelihood of data point w as follows:

$$p(\boldsymbol{w}|\Delta) = \int_\theta \int_{\phi_{1:K}} \sum_{k=1}^K \theta_k f(\boldsymbol{w}|\phi_k) d\theta d\phi_{1:K}$$

where Δ is the set of hyperparameters used to specify the prior distributions for θ and ϕ_k. To derive the posterior distributions for θ and ϕ_k, we turn to approximate inference methods since exact inference is intractable. Thus, we use Gibbs sampling [9] by introducing a latent variable z_n for each data point w_n to specify which sub-populations or mixture component the data w_n belongs to. The distribution of w_n conditioned by the latent variable z_n can be expressed as:

$$p(\boldsymbol{w}|z_n = k, \Delta) = \int_{\phi_k} f(\boldsymbol{w}|\phi_k) d\phi_k$$

The limit of the mixture models introduced above is that it is necessary to specify in advance the number of sub-populations K. To overcome this limitation, we assume that K is infinite:

$$p(\boldsymbol{w}|\pi, \phi_k) = \sum_{k=1}^\infty \pi_k f(\boldsymbol{w}|\phi_k)$$

where θ is a draw from π. The next step is to evaluate the specifications of ϕ_k and the prior distribution over the mixture proportion π, which is infinite-dimensional. The theoretical basis of this approach is the hierarchical Dirichlet processes (Figure 1 (Right)). In fact, $DP(\gamma, H)$ is a distribution over a probability measure G_0. It is defined by 2 parameters: $\gamma > 0$ which is a concentration parameter and H which is a base measure used to generate the parameters ϕ_k of the sub-populations K. We note $G_0 \sim DP(\gamma, H)$.

2.2 Model Description

In this section, we present the generative process of our proposed model, the hierarchical personalized Dirichlet processes (HpDP) given in Figure 1 (Left).

Let w_{di} be the ith word token in the user's query which led to a click on product d, and z_{di} its chosen topic. The generative process of the HpDP follows the following steps:

1. $\pi|\gamma \sim \text{Beta}(1, \gamma)$
2. $z_{di}|\theta_d \sim \text{Multinomial}(\theta_d)$
3. $w_{di}|z_{di}, \phi_{z_{di}} \sim \text{Multinomial}(\phi_{z_{di}})$
   We place priors on the parameters θ_d and $\phi_{z_{di}}$:
4. $\phi_{z_{di}} \sim H$
5. $\theta_d|\alpha \sim \text{Dirichlet}(\alpha\pi)$
   After topics convergence, and for a fixed number of topics:
6. $u_{di}|z_{di}, \psi_{z_{di}} \sim \text{Multinomial}(\psi_{z_{di}})$
7. $\psi_{z_{di}}|\epsilon \sim \text{Dirichlet}(\epsilon)$

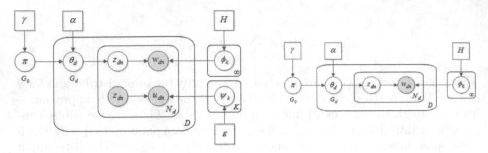

Fig. 1. (Left) Graphical Model of the HpDP, (Right) Graphical Model of the HDP

where π is the distribution over topics and H the distribution over the vocabulary (query items), α and γ are concentration parameters.

Since prior knowledge of the number of topics is difficult, we propose this model that can determine it automatically. In the HpDP, we have an infinite number of topics (θ_d and π are infinite-dimensional vectors), and we use a stick-breaking representation [12] for π: $\pi_k = \tilde{\pi}_k \prod_{l=1}^{k-1}(1 - \tilde{\pi}_l)$ for $k = 1, 2, \ldots$ where $\tilde{\pi}_l | \gamma \sim \text{Beta}(1, \gamma)$.

Using the notation of the Dirichlet process, we have: $G_d \sim DP(\alpha, G_0)$ and $G_0 \sim DP(\gamma, H)$ where: $G_d = \sum_{k=1}^{\infty} \theta_{dk}\delta_{\phi_k}$ and $G_0 = \sum_{k=1}^{\infty} \pi_k \delta_{\phi_k}$ are sums of point masses, and H is the base distribution.

2.3 Approximate Inference

We consider a product d with a probability distribution over words $z_{d1}, z_{d2}, \ldots, z_{dn_d}$ that make up the query that led to a click on product d. Since $G_d \sim DP(\alpha, G_0)$, we can characterize this distribution by describing how to generate $z_{d1}, z_{d2}, \ldots, z_{dn_d}$ using the Chinese Restaurant Process (CRP) [13]. In fact, the CRP considers n_d customers in a Chinese restaurant, with an unlimited number of tables. The first customer sits at the first table. The next customer sits at an occupied table with a probability proportional to the number of customers already present, or sits at an unoccupied table, with a probability proportional to α. Suppose customer i sits at table t_{di}, the conditional distributions are:

$$t_{di} | t_{d1}, \ldots, t_{di-1}, \alpha \sim \sum_t \frac{n_{dt}}{\sum_{t'} n_{dt'} + \alpha} \delta_t + \frac{\alpha}{\sum_{t'} n_{dt'} + \alpha} \delta_t^{new} \tag{1}$$

where n_{dt} is the number of customers currently at table t. When all customers have sat, we associate to table t a draw ζ_{dt} from G_0 and we set: $z_{di} = \zeta_{dt_{di}}$. We perform this process independently for each product d, we obtain all the $G_d(s)$ together with an assignment of each z_{di} to a sample $\zeta_{dt_{di}}$ from G_0, with the partition structure given by CRP(s). We note that all $\zeta_{dt}(s)$ are i.i.d draws from $G_0 \sim DP(\gamma, H)$. We apply the same CRP partitioning process to the $\zeta_{dt}(s)$. Suppose that the customer associated with ζ_{dt} sits at table k_{dt}, the conditional distributions are:

$$k_{dt}|k_{11}, ..., k_{1n_1}, k_{21}, ..., k_{dt-1}, \gamma \sim \sum_k \frac{m_k}{\sum_{k'} m_{k'} + \gamma} \delta_k + \frac{\gamma}{\sum_{k'} m_{k'} + \gamma} \delta_k^{new} \quad (2)$$

Now, we associate with table k a draw ϕ_k from H and we set: $\zeta_{dt} = \phi_{k_{dt}}$. Thus, the generative process for the z_{di}(s) is completed, and we marginalize out G_0 and all the G_d(s). This generative process is called the Chinese Restaurant Franchise (CRF). The CRF is defined by three variables: $t = (t_{di})$, $k = (k_{dt})$ and $\phi = \phi_k$. We describe an inference procedure based on Gibbs sampling t, k and ϕ given data points w. Let $f(.|\phi)$ and h be the density funtions for $F(\phi)$ and H respectively. The conditional probability of t_{di} given the other variables is proportional to the product of a prior and likelihood term. The prior term is given by (1) and the likelihood is given by $f(w_{di}|\phi_{k_{dt}})$ where for $t = t^{new}$, we can sample $k_{dt^{new}}$ using (2), and $\phi_{k^{new}} \sim H$. Thus, the distribution is:

$$p(t_{di} = t|t \backslash t_{di}, k, \phi, w) \propto \begin{cases} \alpha f(w_{di}|\phi_{k_{dt}}) & \text{if } t = t^{new} \\ n_{dt}^{-i} f(w_{di}|\phi_{k_{dt}}) & \text{if } t \text{ currently used} \end{cases}$$

where n_{dt}^{-i} is the number of $t_{di'}$ equal to t except t_{di}. In the same manner, the conditional distribution of k_{dt} is:

$$p(t_{dt} = k|t, k \backslash k_{dt}, \phi, w) \propto \begin{cases} \gamma \prod_{i:t_{di}=t} f(w_{di}|\phi_k) & \text{if } k = k^{new} \\ m_k^{-t} \prod_{i:t_{di}=t} f(w_{di}|\phi_k) & \text{if } k \text{ currently used} \end{cases}$$

where m_k^{-t} is the number of $k_{dt'}$ equal to k except k_{dt}.
Finally, the conditional distribution for ϕ_k is:

$$p(\phi_k|t, k, \phi \backslash \phi_k, w) \propto h(\phi_k) \prod_{di:k_{dt_{di}}=k} f(w_{di}|\phi_k)$$

For further details on the calculations, see [13].
Once topics have converged, we calculate the user/topic distribution ψ in order to consider the user profiles when ranking online products.

2.4 Calculation of the User/Topic Distribution

Since we are in the case where the variables are observed (topics which have converged in addition to users), we use the maximum likelihood method to estimate ψ (the user/topic distribution). Indeed, in the case of Bayesian estimation, the objective is to find the most likely parameters ψ given the observed data using a priori parameters. Bayes rule gives us:
$L = p(\psi|u, z) \propto p(u, z|\psi)p(\psi) \propto p(u|z, \psi)p(\psi)$
Since ψ is a multinomial distribution, its conjugate prior distribution is a Dirichlet distribution whose coefficient is ϵ. Thus, for K topics and U users, L becomes:

$$L = \prod_{k=1}^{K} \prod_{u=1}^{U} \psi_{uk}^{N_{uk}} \prod_{k=1}^{K} \prod_{u=1}^{U} \frac{\Gamma(U\epsilon)}{\Gamma(\epsilon)^U} \psi_{uk}^{\epsilon_k-1} = \frac{\Gamma(U\epsilon)}{\Gamma(\epsilon)^U} \prod_{k=1}^{K} \prod_{u=1}^{U} \psi_{uk}^{N_{uk}+\epsilon_k-1}$$

where Γ is the gamma function. Taking the logarithm of that term, we obtain:

$$\log L = \log \frac{\Gamma(U\epsilon)}{\Gamma(\epsilon)^U} + \sum_{k=1}^{K} \sum_{u=1}^{U} (N_{uk} + \epsilon_k - 1) \log \psi_{uk} \quad (3)$$

To simplify the calculations, we assume that the Dirichlet coefficients are equal:

$\epsilon_1 = \epsilon_2 = ... = \epsilon_K = \epsilon$. We know that: $\sum_{u=1}^{U} \psi_{uk} = 1$, thereby: $\psi_{Uk} = 1 - \sum_{u=1}^{U-1} \psi_{uk}$. By injecting the last two equations in equation (3), we obtain:

$$\log L = \log \frac{\Gamma(U\epsilon)}{\Gamma(\epsilon)^U} + \sum_{k=1}^{K} \left(\sum_{u=1}^{U-1} (N_{uk} + \epsilon - 1) \log \psi_{uk} + (N_{Uk} + \epsilon - 1) \log(1 - \sum_{u=1}^{U-1} \psi_{uk}) \right)$$

By taking the derivative of this term with respect to ψ_{uk}, we get:

$$\frac{\partial \log L}{\partial \psi_{uk}} = \frac{N_{uk} + \epsilon - 1}{\psi_{uk}} - \frac{N_{Uk} + \epsilon - 1}{1 - \sum_{u=1}^{U-1} \psi_{uk}} = \frac{N_{uk} + \epsilon - 1}{\psi_{uk}} - \frac{N_{Uk} + \epsilon - 1}{\psi_{Uk}}$$

By seting this term to zero, we get the maximum of ψ_{uk} that we denote $\widehat{\psi}_{uk}$:

$$\frac{N_{1k} + \epsilon - 1}{\widehat{\psi}_{1k}} = \frac{N_{2k} + \epsilon - 1}{\widehat{\psi}_{2k}} = ... = \frac{N_{Uk} + \epsilon - 1}{\widehat{\psi}_{Uk}} = \frac{\sum_{u=1}^{U} (N_{uk} + \epsilon - 1)}{\sum_{u=1}^{U} \widehat{\psi}_{uk}} = \sum_{u=1}^{U} (N_{uk} + \epsilon - 1)$$

Thus: $\frac{N_{uk} + \epsilon - 1}{\widehat{\psi}_{uk}} = \sum_{u=1}^{U} (N_{uk} + \epsilon - 1)$

Finally, we get the expression of the user/topic distribution:

$$\widehat{\psi}_{uk} = \frac{N_{uk} + \epsilon - 1}{\sum_{u=1}^{U} (N_{uk} + \epsilon - 1)} \tag{4}$$

This equation will be used to rank products according to the user's query.

2.5   Predicting Products for New Users

The limit of personalization systems is their inability to handle queries of new users. We propose the following approach to overcome this limitation:

1. For each new user, generate his/her distribution over the query items (vocabulary containing words composing all users queries) using LDA.
2. Calculate the probability distribution of old users over the query items (the same vocabulary size).
3. Calculate the KL divergence between a new user distribution over query items and each of old users distributions.
4. Select the old user u^{old} for which the KL divergence is the lowest.
5. Predict products for the new user using his/her query and the user/topic distribution of the selected u^{old}.

2.6   Ranking Online Products

In this section, we describe formulas for ranking products using the parameters that were estimated based on the HpDP. We aim to return to the user a ranked set of products ($d \in \mathcal{D}$) according to their likelihood given his/her query $q = \{w_1, w_2, ..., w_n\}$. The formula in the case of a non-personalized model (LDA) is:

$$p(d|q) \propto p(d)p(q|d) = p(d) \prod_{w \in q} p(w|d) = p(d) \prod_{w \in q} \sum_{z} p(w|z)p(z|d) \tag{5}$$

where: $p(d) = \frac{N_d}{N}$, N_d is the number of words composing the user's query, which led to a click on product d and N is the total number of words composing all users'queries.

The ranking formula consists of multiplying a prior on the probability of the product (which we denote $p(d)$) with the probability of the query given the product (which we denote $p(q|d)$). This latter quantity can be estimated by introducing latent topics. Indeed, topic models allow to estimate the probability of words given topics $p(w|z)$ and the probability of topics given products $p(z|d)$. By introducing the user in the graphical model, we have information about the queries issued by a user. Thus, the user's preferences can be included into the ranking formula. This means that we rank products according to their likelihood given both the query and the user as follows:

$$p(d|q,u) \propto p(d) \prod_{w \in q} p(w,u|d) = p(d) \prod_{w \in q} \sum_z p(w|z)p(u|z)p(z|d)$$

This model can be extended by introducing an additional parameter λ in the range zero to one, in order to weight the probability of a user given a particular topic $p(u|z)$ as follows:

$$\tilde{p}(d|q,u) = p(d) \prod_{w \in q} \sum_z p(w|z)p(u|z)^\lambda p(z|d) \qquad (6)$$

The introduction of this new parameter enables us to control the amount of influence that the user's topical interests may have on the ranking.

3 Experiments

3.1 Dataset

The dataset is from the query logs of a real products comparator[1] that connects potential buyers with major brands and distribution networks in the market of mobile telephony. We used two datasets, each one is based on a 1-month web log file. We have chosen to use data covering different periods to ensure that the model works regardless of the circumstances (promotion, flash sales, seasonal products, ...). The training data is generated automatically from log file without any human intervention.

For data cleaning, we have kept the queries which had resulted in a product selection. Then, we have selected only products for which more than 6 users had clicked on at least once. Finally, we selected only users with more than 6 remaining queries. This preprocessing step is carried out to ensure that users have made a significant number of queries and that products were also viewed reasonably. Table 1 gives a description of final corpus. Our log file is composed mainly of 7 attributes: the ID of the transaction, the ID of the user session, the mobile provider, the package, the package features, the user's query and the date when the query has been made. Table 2 shows an example of two transactions from this query log. In our experiments, we consider that a product is represented by the triplet: (Package, Mobile Provider, Package features).

---

[1] http://www.choisirsonforfait.com/

Table 1. Datasets features

Dataset 1		Dataset 2	
Training subset size	1,053	Training subset size	1,049
Testing subset size	60	Testing subset size	65
# Users	130	# Users	132
# Products	103	# Products	106
# Query items	100	# Query items	101

Table 2. Log file format

Id	Session	Package	Mobile Provider	Features	Date	User's query
3	73f08e	Mobile plan 1	Mobile Provider A	2-years contract	2013-01-15 11:57:22	sms & cell phone
2	ce77d6	Mobile plan 2	Mobile Provider B	2 hours plan	2013-01-15 11:57:15	1 hour of calls

3.2 Methodology

The cleaned data is separated in two subsets: training subset ($\sim 95\%$ of data) and testing subset ($\sim 5\%$ of data). We have selected the last queries of each user for testing, to respect the order in which the queries were made. Thus, the training and testing subsets follow the same chronological order. We ranked products according to scores values defined above. Concerning the parameter setting, we set the Dirichlet prior α to be $0.1/K$, where K is the number of topics used for experiments. We evaluate the rankings by calculating two standard measures in the field of information retrieval: the Mean Reciprocal Rank (MRR) and the Mean Average Precision (MAP). We report these measures up to rank 6, since in information retrieval, it is valuable that pertinent products appear early in the ranked list. We consider that a ranked product is relevant if it is the same product the user had actually viewed. In order to determine if the hierarchical process is improving the ranking performance, we report another metric that we call the hierarchical personalization gain (HP-Gain). This metric compares the number of times the HpDP improves the ranking (which we denote #*better*) to the number of times it worsens it (which we denote #*worse*). A simple expression of this equation is given by:

$$\text{HP-Gain} = \frac{\#better - \#worse}{\#better + \#worse}$$

When the value of this metric is 0, then there is no change between the HpDP and the other models, when it is positive, this means that our model improves the ranking and when it is negative, the ranking is deteriorated.

3.3 Results

Top K Products-Based Evaluation. Table 3 shows the results of the ranking experiments for the HpDP, the PTM and the LDA. An advantage key of our approach is that we do not have to vary the number of topics in order to obtain the optimum number of topics, since the HpDP enables to automatically determine them. In fact, we found 7 topics for the first dataset and 9 topics for the second one. However, to be fair with the two other models, we performed them by varying the number of topics. We notice an improvement compared to the PTM and the LDA. Moreover, we recall that the reciprocal rank of a query is the multiplicative inverse of the rank of the first correct answer and that the mean reciprocal rank is the average of the reciprocal ranks of results for a set of queries. This means that if the first proposed product to the user is relevant, then the reciprocal rank is equal to 100%, and if the first relevant product is second-ranked, then the reciprocal rank is equal to 50%. The mean reciprocal rank obtained by the HpDP is 69.47%, which means that the product it proposes to the user is broadly either ranked first or second. Therefore, we compute another metric which is $Precision@n$ to determine how much products should be proposed to the user so that the ranking will be the best.

Table 3. Ranking performance of the models on the test set over all queries ($\lambda = 0.10$): (Top) Results for the first dataset, (Bottom) Results for the second dataset

Measures	Models	Number of Topics										
		5	7	10	15	20	25	30	35	40	45	50
MRR (%)	HpDP	-	**69.47**	-	-	-	-	-	-	-	-	-
	PTM	65.17	63.13	61.79	61.42	55.70	57.22	56.11	62.17	59.70	60.69	61.63
	LDA	61.08	59.79	55.71	53.35	61.63	59.21	57.53	59.71	58.93	57.56	59.09
MAP (%)	HpDP	-	**64.15**	-	-	-	-	-	-	-	-	-
	PTM	57.96	58.18	58.65	55.49	53.85	52.11	55.55	54.00	54.46	55.54	53.01
	LDA	56.97	55.12	53.61	51.03	60.37	55.47	54.24	54.51	52.23	51.24	52.37
Measures	Models	Number of Topics										
		5	9	10	15	20	25	30	35	40	45	50
MRR (%)	HpDP	-	**68.04**	-	-	-	-	-	-	-	-	-
	PTM	60.28	60.73	61.67	58.53	62.78	57.07	60.58	58.39	57.64	61.45	60.05
	LDA	55.15	56.12	56.39	57.81	60.13	58.00	54.59	57.25	55.96	58.86	57.73
MAP (%)	HpDP	-	**64.10**	-	-	-	-	-	-	-	-	-
	PTM	53.56	55.53	59.84	54.30	58.89	54.59	56.49	56.53	54.80	57.43	54.00
	LDA	53.72	53.91	52.52	53.07	57.15	54.05	56.06	51.46	50.67	54.53	52.79

Table 4. Precision evolution according to the number of proposed products

Measures	HpDP		PTM		LDA	
	Dataset 1	Dataset 2	Dataset 1	Dataset 2	Dataset 1	Dataset 2
$p@1$	68.33 %	69.23 %	68.33 %	66.67 %	63.33 %	64.62 %
$p@2$	66.66 %	63.07 %	63.33 %	63.33 %	60.00 %	61.54 %
$p@3$	61.66 %	58.46 %	53.33 %	51.67 %	41.66 %	50.77 %

Table 4 shows the obtained result, which confirms our intuition about the relevance of the first and second ranked products, that are proposed to the user given his query. For the next experimentations, we will use 10 topics for PTM and 20 topics for LDA since their precisions are the best using these numbers of topics.

Influence of λ on the Gain. In this section, we highlight the influence of the parameter λ in terms of HP-Gain and hence on performance improvement.

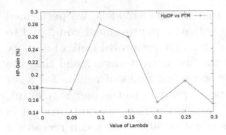 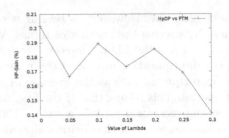

Fig. 2. The effect of varying the λ parameter in the ranking algorithm: (Left) Results for the first dataset, (Right) Results for the second dataset

In fact, the parameter λ plays an important role in the ranking formula for the HpDP, since it enables control over the amount of influence the user profile has on the products'scores. We tested the effect of this parameter within the range of $\{0, 0.05, 0.1, ..., 0.3\}$. When $\lambda = 0$, the estimates of HpDP are the same as those given by the HDP. Figure 2 shows an improvement performance, over all queries. The HP-Gain varies between 14% and 28%. We chose to perform our experiments using $\lambda = 0.10$ since for this value, inter alia, the gain is maximum.

Influence of the Click Entropy on the Gain. When a given user/query pair had been observed before, we can use this information about prior clicks by assuming that the user will again click on the same products as before. However, in almost cases, the user/query pair will be novel and we will not have such prior information to exploit. We will use a measure called the *click entropy* to identify such unambiguous queries. The click entropy of an observed query q is defined as follows:

$$H_q = \sum_{d \in D(q)} -p(d|q) \log_2 p(d|q)$$

where $D(q)$ is the set of clicked products given the query q and $p(d|q)$ is the probability of selecting product d given the query q. Since entropy values vary in the range zero to the logarithm of the number of distinct products clicked on for a query, then, the range of values depends on the query. This makes the comparison of click entropy values accross queries complicated. To deal with this issue, we will use, in our experiments, normalized entropy values instead, where the range of values is limited to $[0, 1]$, this new measure is defined as follows:

$$\widehat{H}_q = \frac{H_q}{\log_2 |D(q)|}$$

We calculated this measure for all queries. We separated these queries into two groups: queries for which this measure is lower than 0.5 and queries for which this measure is greater than 0.5. Then we calculated the HP-Gain for each of the two groups containing test queries. Figure 3 shows how the performance of the HpDP changes as the normalized click entropy of the queries evolves.

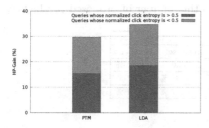 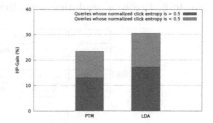

Fig. 3. The effect of query ambiguity in the ranking algorithm: (Left) Results for the first dataset, (Right) Results for the second dataset

We notice that the HP-Gain increases as the click entropy increases. In fact, it reaches 18% for queries, the normalized click entropy of which is greater than 0.5 and it drops to 14% for queries, the normalized click entropy of which is lower than 0.5.

Predictions for New Users. Unlike the first experiment where the personalization task required a particular separation of data (users in the test set must have appeared in the training set), in this section, we divide the data randomly ($\sim$ 95% for training, $\sim$ 5% for testing). Then, we apply the procedure described in section 2.5 and we compare HpDP to LDA (PTM can not perform this task). HpDP found 7 topics for the first dataset and 11 topics for the second dataset. Again, we compute the MRR and MAP for LDA by varying the number of topics. Table 5 shows the obtained results. We notice that the MAP obtained using HpDP is always greater that the LDA's. Otherwise, the MRR obtained using HpDP outperforms LDA's except when considering 20 topics for LDA. Thus, we will use 20 topics for LDA when evaluating the *Precision@n*, given in Table 6. We notice again a performance improvement using our approach.

Table 5. Ranking performance of the models on the test set over all queries ($\lambda = 0.10$): (Top) Results for the first dataset, (Bottom) Results for the second dataset

Measures	Models	Number of Topics										
		5	7	10	15	20	25	30	35	40	45	50
MRR (%)	HpDP	-	**64.99**	-	-	-	-	-	-	-	-	-
	LDA	61.23	61.84	63.35	61.58	**65.91**	62.29	62.21	57.50	61.37	62.38	60.25
MAP (%)	HpDP	-	**61.00**	-	-	-	-	-	-	-	-	-
	LDA	54.59	54.91	55.28	55.72	59.12	57.26	60.35	56.64	54.81	60.43	55.07

Measures	Models	Number of Topics										
		5	10	11	15	20	25	30	35	40	45	50
MRR (%)	HpDP	-	-	**63.90**	-	-	-	-	-	-	-	-
	LDA	50.07	54.41	55.78	58.14	**64.12**	53.90	59.98	60.09	59.74	58.06	62.93
MAP (%)	HpDP	-	-	**59.11**	-	-	-	-	-	-	-	-
	LDA	53.71	47.92	50.61	55.65	58.63	52.96	54.82	57.10	56.98	54.23	58.78

Table 6. Precision evolution according to the number of proposed products

	HpDP		LDA	
Measures	Dataset 1	Dataset 2	Dataset 1	Dataset 2
$p@1$	65.57 %	67.14 %	63.90 %	65.71 %
$p@2$	62.30 %	62.86 %	60.66 %	58.57 %
$p@3$	59.02 %	60.00 %	57.37 %	57.14 %

4   Conclusion

In this paper, we have proposed a nonparametric Bayesian model that builds user profiles for personalized search. The comparison with other approaches indicated that performance can be improved through personalization. We also addressed the prediction task for new users. The obtained results showed that our model can further improve ranked lists

In our future work, we plan to analyze other families of function, that allow to control the influence of the user's topical interests on the ranking, in order to improve the gain. In addition, we intend to introduce dynamics in our model, either under Markovian or non-Markovian fashion.

Acknowledgments. This work was supported by Marketshot.

References

1. Bennett, P.N., White, R.W., Chu, W., Dumais, S.T., Bailey, P., Borisyuk, F., Cui, X.: Modeling the impact of short- and long-term behavior on search personalization. In: Proceedings of the 35th International Conference on Research and Development in Information Retrieval, SIGIR (2012)
2. Blei, D., Ng, A., Jordan, M.I., Lafferty, J.: Latent Dirichlet allocation. Journal of Machine Learning Research 3, 993–1022 (2003)
3. Chirita, P.A., Nejdl, W., Paiu, R., Kohlschutter, C.: Using odp metadata to personalize search. In: Proceedings of the 28th Annual International Conference on Research and Development in Information Retrieval, SIGIR (2005)
4. Harvey, M., Crestani, F., Carman, M.: Building user profiles from topic models for personalised search. In: Proceedings of the 22nd ACM International Conference on Information and Knowledge Management, CIKM (2013)
5. Harvey, M., Ruthven, I., Carman, M.: Improving social bookmark search using personalised latent variable language models. In: Proceedings of the Fourth ACM International Conference on Web Search and Data Mining, WSDM (2011)
6. Matthijs, N., Radlinski, F.: Personalizing web search using long term browsing history. In: Proceedings of the Fourth International Conference on Web Search and Data Mining, WSDM (2011)
7. Pretschner, A., Gauch, S.: Ontology based personalized search. In: Proceeding of the International Conference on Tools with Artificial Intelligence, ICTAI (1999)
8. Qiu, F., Cho, J.: Automatic identification of user interest for personalized search. In: Proceedings of the 15th International Conference on World Wide Web, WWW (2006)

9. Steyvers, M., Griffiths, T.: Probabilistic Topic Models. In: Landauer, T., Mcnamara, D., Dennis, S., Kintsch, W. (eds.) Latent Semantic Analysis: A Road to Meaning. Laurence Erlbaum (2007)
10. White, R.W., Bailey, P., Chen, L.: Predicting user interests from contextual information. In: Proceedings of the 32nd International Conference on Research and Development in Information Retrieval, SIGIR (2009)
11. Nguyen, T., Phung, D., Gupta, S., Venkatesh, S.: Extraction of Latent Patterns and Contexts from Social Honest Signals Using Hierarchical Dirichlet Processes. In: The IEEE International Conference on Pervasive Computing and Communications, PerCom (2013)
12. Sethuraman, J.: A Constructive Definition of Dirichlet Priors. Statistica Sinica, 4 (1994)
13. Teh, Y., Jordan, M., Beal, M., Blei, D.: Sharing Clusters Among Related Groups: Hierarchical Dirichlet Processes. In: Neural Information Processing Systems 17, NIPS (2005)

Widened KRIMP:
Better Performance through Diverse Parallelism

Oliver Sampson and Michael R. Berthold

Chair for Bioinformatics and Information Mining
Department of Computer and Information Science
University of Konstanz, Germany

Abstract. We demonstrate that the previously introduced Widening framework is applicable to state-of-the-art Machine Learning algorithms. Using KRIMP, an itemset mining algorithm, we show that parallelizing the search finds better solutions in nearly the same time as the original, sequential/greedy algorithm. We also introduce Reverse Standard Candidate Order (RSCO) as a candidate ordering heuristic for KRIMP.

1 Introduction

Research into parallelism in Machine Learning has primarily focused on reducing the execution time of existing algorithms, e.g., parallelized K-MEANS [23,17,14,26] and DBSCAN [11,4,7]. There have been some exceptions, such as *metalearning* and *ensemble methods* [9], which have employed heterogeneous algorithms in parallel, and [3], which describes the application to simple examples. Recent work [2,15] describes *Widening*, a framework for employing parallel resources to increase accuracy. With Widening, measures of diversity are used to guarantee the parallel search paths' exploration of disparate regions within a solution space, thereby stepping around the common greedy algorithmic tendency to find local optima. Thus far, work has concentrated on a proof-of-concept and demonstrative application to algorithms for solving the SET COVER PROBLEM and the creation of Decision Trees. This document describes the same approach, but with a state-of-the-art algorithm, KRIMP [24].

KRIMP finds "interesting" itemsets from a transactional database via the Minimum Description Length (MDL) principle [21]. The authors summarize the method as "the best set of patterns [being] the set of patterns that describes the data best," where the best set of itemsets is the set that provides the highest compression using MDL. The algorithm not only provides a solution to the problem of pattern explosion, thereby greatly reducing the set of itemsets used to generate association rules, but provides exceptional performance in other applications such as classification [24].

This paper demonstrates that it is possible to apply Widening to find even more interesting sets of itemsets than those found by the standard KRIMP algorithm.

H. Blockeel et al. (Eds.): IDA 2014, LNCS 8819, pp. 276–285, 2014.

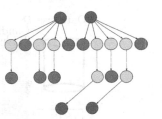

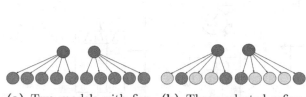

(a) Two models with five refinements

(b) Three selected refinements from each group of five

(c) Two selections (yellow) from the group of six, ready for the next iteration (green)

Fig. 1. Refine and Select with $l = 3$ and $k = 2$

2 Widening

Given the set of all models, $\mathcal{M}$, that describe the solution space for a typical greedy machine learning algorithm, $m(\cdot) \in \mathcal{M}$ is a model which describes a portion of the solution space. It is iteratively refined by a *refinement operator*, $r(\cdot)$, based on a subset, x, from a training dataset $\mathcal{T}$, i.e., $m'(\cdot) = r(m(\cdot), x), x \subseteq \mathcal{T}$. The derivation of a Decision Tree is one example of this process [15].

In contrast to that above, in the Widening framework a set of models, $M \subseteq \mathcal{M}$, is the result of a refinement operator based on data from $\mathcal{T}$ and a *diversity metric*, Δ, which describes some minimum difference between the resulting models, $\{m'_1, \cdots, m'_l\}$, i.e., $M = r_\Delta(m) = \{m'_1, \cdots, m'_l\}$. For clarity, the data elements from the training data are eliminated from the notation.

A *selection operator*, $s_{top-k}(\cdot)$ is employed to select the best k models at each step [15]. $M_{i+1} = s_{top-k}(r_\Delta(M_i))$. The results of the selection operation are further refined until some stop condition is met. This iterative refine-and-select process, as depicted in Figure 1, is conceptually similar to a *beam search* [19].

3 The KRIMP Algorithm

In the area of Itemset Mining, KRIMP finds the set of "most interesting" itemsets in a transaction database based on MDL, i.e., KRIMP defines the best model, m, as the model that maximizes the compression of a transaction database, $\mathcal{D}$, encoded with that model and the compression of the model itself [21,24].

Given a database $\mathcal{D}$ composed of transactions $t \in \mathcal{D}$, KRIMP finds the subset of itemsets, X, from the set of all itemsets, $\mathcal{X}$, that maximally compresses $\mathcal{D}$. KRIMP calculates the size of the encoded database using the codelengths of *prefix-free codes*, which are related to the frequency of the appearance of an itemset, $x \in X$, in the database and to the Shannon entropy: $L(x) = -\log_2 P(x)$, where $L(x)$ is the codelength measured in bits of an item or itemset in the database, and $P(x)$ is the relative frequency of the item's or itemset's appearance in the database [24].

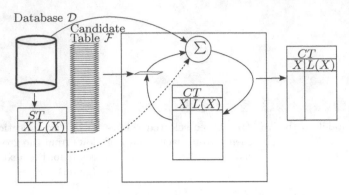

Fig. 2. The Krimp algorithm

Krimp begins with the generation of the Standard Code Table, ST, which is merely a code table comprised of only the individual items from the set of all single items, $\mathcal{I}$. The codelength of a given transaction, $L(t)$ is the sum of its compositional codelengths.

Krimp then iterates through a list of candidate itemsets, $\mathcal{F}$, generated by an algorithm external to Krimp, such as Afopt [18] or Apriori [1]. Each of the candidate itemsets from $\mathcal{F}$ is temporarily inserted into the code table CT, where all relative frequencies are determined and the compression evaluated. If it provides better compression, it is kept as part of CT and if not, it is discarded [24]. A general flow diagram is depicted in Figure 2.

The size of the encoded database, $L(\mathcal{D}|CT)$, is the sum of the encoded lengths of all transactions. The size of the encoded Code Table, $L(CT|\mathcal{D})$, is the size of each code plus the lengths of the encoded itemsets, for which the single items from ST are used. The compressed MDL size of the database is the size of the encoded database plus the size of the encoded code table. $L(\mathcal{D}, CT) = L(\mathcal{D}|CT) + L(CT|\mathcal{D})$ [24]

Both $\mathcal{F}$ and CT are ordered heuristically to maximize compression. $\mathcal{F}$ is ordered according to the *Standard Candidate Order*, which orders primarily by the itemsets' *support* in descending order, secondarily by cardinality in descending order, and tertiarily by lexicographical order, as a tie-breaker. The rationale is that itemsets with larger support are likelier to cover more transactions and are evaluated first. Itemsets with the same support are sorted secondarily by cardinality, because larger itemsets cover more items in each transaction, reducing the number of itemsets or items required to cover a transaction [24].

CT is ordered using the *Standard Cover Order*, which orders primarily by descending cardinality, secondarily by descending support, and tertiarily lexicographically, again as a tie-breaker. The rationale is that larger itemsets are preferred for their ability to cover more of each transaction. Of those, the ones with a larger support are more likely to cover more transactions in the database, thereby providing shorter codes [24].

Krimp also includes a post-processing step for each iteration called Pruning. If the relative frequency of any itemset in CT decreases as a result of adding a new itemset, CT is re-evaluated with each itemset in CT singly removed. If any of the itemsets' temporary absence from CT enables a better compression, it is discarded. Results in [24] indicate that PRUNING improves overall compression performance marginally, but can dramatically reduce the number of itemsets providing that level of compression.

4   Widenend Krimp

A given path through a KRIMP solution space is based on two things: 1) the order with which the candidate itemsets from $\mathcal{F}$ are evaluated, because the acceptance of a particular itemset into CT influences which itemsets are accepted in later iterations, and 2) the order in which the itemsets in CT are used to cover the database. Varying either of these two heuristics' orderings varies the solution path through the solution space and introduces diversity from the other paths taken.

For use as Δ in the refining function, $r_\Delta(\cdot)$, two *explicit* measures and one implicit measure of diversity are investigated here. Explicit measures *p-dispersion-min-sum* and *p-dispersion-sum* select maximally diverse subsets of candidates from the candidate table. Implicit method, *Directed Placement*, is investigated with respect to the ordering of the itemsets evaluated for covering the transactions in CT.

p-dispersion-min-sum maximizes the sum of minimum distances between pairs of members of the selected subset [20].

Definition 1 *p-**dispersion-min-sum**.*[1] *Given a set* $\mathcal{F} = \{F_1, \cdots, F_n\}$ *of n itemsets and l, where $l \in \mathbb{N}$ and $l \leq n$, and a distance measure* $Jaccard(F_i, F_j)$: $F_i, F_j \in \mathcal{F}$ *between items F_i and F_j, the l-diversity problem is to select the set* $F : F \subseteq \mathcal{F}$*, such that*

$$F^* = \max_{\substack{F \subseteq \mathcal{F} \\ |F|=l}} f(F), \text{ where } f(F) = \sum_{i=1}^{l} \min_{1 \leq j \leq l, i \neq j} Jaccard(F_i, F_j), F_i, F_j \in F \text{ [20][16]}$$

(1)

p-dispersion-sum maximizes the distance between all members of the selected subset.

Definition 2 *p-**dispersion-sum**. Given a set* $\mathcal{F} = \{F_1, \cdots, F_n\}$ *of n itemsets and l, where $l \in \mathbb{N}$ and $l \leq n$, and a distance measure* $Jaccard(F_i, F_j) : F_i, F_j \in \mathcal{F}$ *between itemsets F_i and F_j, the l-diversity problem is to select the set $F : F \subseteq \mathcal{F}$, such that*

$$F^* = \max_{\substack{F \subseteq \mathcal{F} \\ |F|=l}} f(S), \text{ where } f(F) = \frac{1}{l(l-1)} \sum_{i=1}^{l} \sum_{j>1}^{l} Jaccard(F_i, F_j) \text{ [20,12]}$$

(2)

---

[1] The canonical names from the literature, p-dispersion-min-sum and p-dispersion-sum, are maintained here, even though in this context, they should be called "l-dispersion-min-sum" and "l-dispersion-sum."

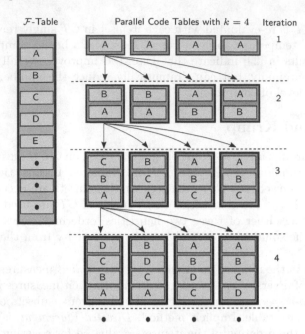

Fig. 3. At each iteration, the next $F \in \mathcal{F}$ is inserted into each of the parallel code tables at a depth $\frac{i}{l}$. Refinement is shown only for one table in each iteration.

p-dispersion-sum has the side-effect of pushing the selected members to the boundaries of the original set. This results in selected sets that are less diverse and representative of the dataset than those that are selected by p-dispersion-min-sum [20].

The Directed Placement diversity heuristic functions by inserting the next candidate itemset, $F \in \mathcal{F}$ at a position with different fractional depths into l parallel instances of CT. The depth inserted into CT is a function of l, where the depth is $\frac{i}{l}|CT| : i = 1, \ldots, l$. Because the role of each itemset in the covering algorithm is dependent on its position in CT, positioning F at different depths explores diverse solution paths. This method of diversity is implicit, because the diversity between different CT tables is not measured directly. See Figure 3.

An additional heuristic ordering of $\mathcal{F}$ called *Reverse Standard Candidate Order* (RSCO) is introduced here. It orders the candidate itemsets primarily by cardinality in ascending order, secondarily by support in descending order, and tertiarily by lexicographical order as a tie-breaker. In combination with the Standard Cover Order heuristic for covering transactions, RSCO attempts to mimic the PRUNING subalgorithm; candidate itemsets with larger cardinality are examined later but are inserted before the smaller itemsets already in CT. With Standard Cover Order, small itemsets whose potential ability to efficiently cover transactions are "shadowed" by larger itemsets and have a lower relative frequency used for the compression calculation. In contrast, using RSCO, smaller itemsets that may have a beneficial effect, yet show up too late in the list to be considered with Standard Candidate Order, can still be evaluated.

5   Experimental Results

Compression of $\mathcal{D}$ with CT, $L(\mathcal{D}, CT)$, is compared to a baseline compression of $\mathcal{D}$ encoded using only ST, $L(\mathcal{D}, ST)$. This paper follows the convention of [24] in using the compression ratio measured in percentages, where lower is better. $L\% = L(\mathcal{D}, CT)/L(\mathcal{D}, ST) * 100$. [24]

The notation $|CT \setminus \mathcal{I}|$ indicates the number of non-singleton itemsets used. For a given compression level, a smaller number of itemsets is considered more interesting. KRIMP optimizes for both $L\%$ and $|CT \setminus \mathcal{I}|$ by evaluating $L\%$ first, and then bettering $|CT \setminus \mathcal{I}|$ with the PRUNING subalgorithm.

All experiments were conducted in KNIME [6] and used APRIORI [1,8] with a *minsupport* of 1 to generate the set of *closed itemsets*. The datasets used were the LUCS-KDD-DN [10] discretized versions of the Breast Cancer Wisconsin (Original) [25] (*Breast*) and Pima Indians Diabetes Data Set [22] (*Pima*) datasets available from the UCI-ML Data Repository [5].

Evaluations in Sections 5.1 and 5.2 compare three methods, KRIMP$_{Greedy}$, KRIMP$_{RSCO}$ and KRIMP$_{Diverse}$. KRIMP$_{Greedy}$ refers to the baseline "standard" KRIMP implementation used within KNIME; KRIMP$_{RSCO}$ refers to the implementation in KNIME, using RSCO for ordering $\mathcal{F}$ rather than Standard Candidate Order, because the results with RSCO for the *Breast* and *Pima* datasets were actually better than Standard Candidate Order;[2] and KRIMP$_{Diverse}$ refers to KRIMP with a method of diversity being evaluated. KRIMP has two performance metrics for a model; solution pairs are shown in the form of $\langle L\%, |CT \setminus \mathcal{I}| \rangle$. Results are shown with and without PRUNING for KRIMP$_{Greedy}$ and KRIMP$_{RSCO}$. All experiments with KRIMP$_{Diverse}$ were performed without PRUNING because we felt it would introduce another variable of diversity for which we were not controlling.

Experimental solution pairs are also shown at the position found $\langle l, k \rangle$, where l is the number of refined models and k is the number of models selected according to compression performance.

5.1   Diverse Candidate Selection

Both of these methods of subset selection are performed with replacement, because early experiments without replacement on $\mathcal{F}$ candidate tables generated from closed itemsets demonstrated that there were simply not enough candidate itemsets for evaluation to generate solutions sets of reasonable performance for larger values of k. With replacement, in order for the algorithm to come to completion, the first element in $\mathcal{F}$ is removed after each iteration, ensuring the algorithm's completion after $|\mathcal{F}|$ iterations. This method naturally entails a dependency on the initial ordering of $\mathcal{F}$.

A summary of the results for the *Breast* dataset using the p-dispersion-min-sum diversity metric for $\mathcal{F}$ candidate selection can be found in Table 1. The experiments

---

[2] This is just an artifact for these two datasets. Preliminary results not shown here demonstrate that RSCO does indeed perform better than SCO for some datasets, albeit not consistently across all datasets tested.

Table 1. *Breast* Dataset Results Summary

	Heuristic		Pruning	$L\%$	$\|CT \setminus \mathcal{I}\|$
	$\mathcal{F}$ (Candidate Table)	CT (Code Table)			
KRIMP$_{Greedy}$	Standard Candidate	Standard Cover	no	18.11	29
	Order	Order	yes	17.61	28
KRIMP$_{RSCO}$	RSCO	Standard Cover	no	17.86	28
		Order	yes	17.82	26
KRIMP$_{Diverse}$	p-dispersion-min-sum + RSCO	Std. Cover Order	no	17.97	28
	p-dispersion-sum + RSCO	Std. Cover Order		19.42	34
	RSCO	Directed Placement		**17.39**	**26**

Table 2. *Pima* Dataset Results Summary

	Heuristic		Pruning	$L\%$	$\|CT \setminus \mathcal{I}\|$
	$\mathcal{F}$ (Candidate Table)	CT (Code Table)			
KRIMP$_{Greedy}$	Standard Candidate	Standard Cover	no	35.6	66
	Order	Order	yes	34.4	53
KRIMP$_{RSCO}$	RSCO	Standard Cover	no	34.3	63
		Order	yes	33.7	**49**
KRIMP$_{Diverse}$	RSCO	Directed Placement	no	**32.9**	56

were run with all combinations of $l \in \{5, 10, 20, 30, 40, 50\}$ and $k \in \{1, 5, 10, 15\}$. The best solution for KRIMP$_{Diverse}$ with p-dispersion-min-sum, $\langle 17.97\%, 28 \rangle$ was found at $\langle l, k \rangle = \langle 50, 10 \rangle$, which was better than KRIMP$_{Greedy}$ without PRUNING $\langle 18.11\%, 29 \rangle$, but not better than KRIMP$_{Greedy}$ with PRUNING $\langle 17.61\%, 28 \rangle$. KRIMP$_{RSCO}$ with PRUNING performed even better at $\langle 17.82\%, 26 \rangle$.

KRIMP$_{Diverse}$ with p-dispersion-sum was run with all combinations of $l \in \{5, 10, 20, 30, 40, 50\}$ and $k \in \{1, 5, 10, 20, 50\}$ and as expected, the best solution pair $\langle 19.42\%, 34 \rangle$ was not nearly as good as that with p-dispersion-min-sum, and was found in an even larger search space of $\langle l, k \rangle = \langle 50, 30 \rangle$.

The experiments with KRIMP$_{Diverse}$ with p-dispersion-min-sum were run over a smaller search space when compared to KRIMP$_{Diverse}$ with p-dispersion-sum after recognizing that the results had already reached the goal of beating one of the KRIMP$_{Greedy}$ scores.

Due to run-time constraints (See Section 6.) experiments were not performed with diversity-based candidate selection on the *Pima* dataset.

5.2 Diverse Cover Order

The results of the Directed Placement heuristic with RSCO as the Candidate Selection heuristic are also summarized in Tables 1 and 2. Experiments were performed on the *Breast* dataset with all combinations of $l \in \{5, 10, 20, 30, 40, 50\}$ and $k \in \{1, 5, 10, 15\}$. The heuristic found a solution $\langle 17.39\%, 26 \rangle$ outperforming the best KRIMP variant. Additionally, the solution was found in a much smaller search space, when compared to Diverse Candidate Selection, with the best solution found first at $\langle l, k \rangle = \langle 10, 10 \rangle$.

The *Pima* dataset showed an interesting property in that paths to better solutions were sensitive to increasing k, the number of selected models to be refined in the next step, but not in l, the number of refinements in each step. Better results, $\langle 33.2\%, 61 \rangle$, than KRIMP_{Greedy} and KRIMP_{RSCO} both without PRUNING were found immediately at $\langle l, k \rangle = \langle 5, 5 \rangle$. In fact, all solutions found with the Directed Placement heuristic with RSCO showed better $L\%$ than either KRIMP_{Greedy} or KRIMP_{RSCO} with or without PRUNING. The best solution found for the *Pima* dataset was $\langle 32.9\%, 56 \rangle$ at $\langle l, k \rangle = \langle 5, 50 \rangle$. This result has significantly better compression and yields nearly the number of itemsets as KRIMP_{Greedy}, but not nearly as good as KRIMP_{RSCO}.

6   Discussion and Future Work

In general, absolute timing values are not necessary for timing comparisons. To a first order of approximation, KRIMP runs in $O(|\mathcal{F}| \times |\mathcal{D}| \times \theta)$ where θ is a factor describing the average length of CT during the entire execution of the algorithm. (It should be noted that the authors of [24] saw a performance improvement in execution speed after implementing the PRUNING subalgorithm, because of a smaller value of θ.) Accounting for application of a diversity measure and the use of a performance measurement for selection, KRIMP runs in $O((|\mathcal{F}| + \Delta + \Psi) \times |\mathcal{D}| \times \theta)$, where $O(\Delta)$ is the measure of the complexity of the diversity heuristic, and $O(\Psi)$ is a measure of the complexity of the performance measurement.

Although p-dispersion-min-sum was able to find comparable results to the standard KRIMP implementation (better than KRIMP_{Greedy} without PRUNING), the computational cost is significant. Selecting a subset of p diverse elements from a larger set is a variation of the p-dispersion problem and is $\mathcal{NP}$-hard [13]. Moreover, a comparison of this metric to p-dispersion-sum demonstrates what could be a pitfall for applying p-dispersion-sum: a much wider solution space had to be searched, $\langle l, k \rangle = \langle 50, 10 \rangle$ versus $\langle l, k \rangle = \langle 50, 30 \rangle$. Although at least one of these diversity measures fulfills the desire to show that widened data mining can find better solutions than the traditional greedy algorithm, it is insufficient for a requirement of finding better solutions in the same or less time than the traditional greedy algorithm, which is the ultimate goal of Widening.

Directed Placement, however, was able to significantly improve on the solution found by standard KRIMP in [24]. For the *Breast* dataset, the results were even better than the results found with KRIMP_{RSCO}. Directed Placement also showed a partially better solution with the *Pima* dataset. In comparison to the other diversity metrics presented here, Directed Placement has a much smaller overhead for generating diverse solution paths. It must be noted, however, that the claim of "better solutions in the same or faster time" in this case is not strictly accurate. For large values of $|\mathcal{F}|$ and $|\mathcal{D}|$, the influence of $O(\Delta)$ for the Directed Placement diversity heuristic becomes negligible. The evaluation of the models for selection, $O(\Psi)$ is also negligible for KRIMP, because it is merely a comparison of the best $L\%$. Additionally, Directed Placement provided the best Widened result in a significantly smaller search region than the other diversity heuristics.

Ideally, a Widened algorithm is able to find a better or the best solution in the same time as the traditional greedy algorithm. The pitfall of the methods described here is that both a performance evaluation (Ψ) and a synchronized comparison of results from the parallel workers are required. This would be avoided with a *communication-less* [15] approach where the parallel workers would be able to refine and select without requiring a synchronized comparison step. Additionally, although the better solutions found by Widened KRIMP meet the definition of "better," further research into how well the smaller sets perform as classifiers or in other KRIMP applications is necessary. The effects of including the PRUNING subalgorithm on the dataset compression, and the corresponding solution space paths also require further investigation, as does the magnitude and interplay between l and k for different datasets.

7   Conclusion

In this paper we have validated Widening for the first time using a state-of-the-art algorithm for itemset mining, KRIMP, and shown that it is possible to use the novel approach of Widening to find significantly better solutions than that of the traditional greedy algorithm by searching diverse regions of a solution space in parallel. We have also introduced RSCO, a new Candidate Table ordering heuristic for KRIMP that can provide even better results for some datasets.

References

1. Agrawal, R., Srikant, R.: Fast algorithms for mining association rules. In: Proceedings of the 20th International Conference on Very Large Data Bases, vol. 1215, pp. 487–499 (1994)
2. Akbar, Z., Ivanova, V.N., Berthold, M.R.: Parallel data mining revisited. Better, not faster. In: Hollmén, J., Klawonn, F., Tucker, A. (eds.) IDA 2012. LNCS, vol. 7619, pp. 23–34. Springer, Heidelberg (2012)
3. Akl, S.G.: Parallel real-time computation: Sometimes quantity means quality. In: Proceedings of the International Symposium on Parallel Architectures, Algorithms and Networks, I-SPAN 2000, pp. 2–11. IEEE (2000)
4. Arlia, D., Coppola, M.: Experiments in parallel clustering with DBSCAN. In: Sakellariou, R., Keane, J.A., Gurd, J.R., Freeman, L. (eds.) Euro-Par 2001. LNCS, vol. 2150, pp. 326–331. Springer, Heidelberg (2001)
5. Bache, K., Lichman, M.: UCI Machine Learning Repository (2013)
6. Berthold, M.R., Cebron, N., Dill, F., Gabriel, T.R., Kötter, T., Meinl, T., Ohl, P., Sieb, C., Thiel, K., Wiswedel, B.: KNIME: The Konstanz Information Miner. In: Preisach, C., Burkhardt, H., Schmidt-Thieme, L., Decker, R. (eds.) Data Analysis, Machine Learning and Applications - Proceedings of the 31st Annual Conference of the Gesellschaft für Klassifikation e.V (GfKL 2007), Berlin, Germany. Studies in Classification, Data Analysis, and Knowledge Organization, pp. 319–326 (2007)
7. Böhm, C., Noll, R., Plant, C., Wackersreuther, B., Zherdin, A.: Data mining using graphics processing units. In: Hameurlain, A., Küng, J., Wagner, R. (eds.) Transactions on Large-Scale Data- and Knowledge-Centered Systems I. LNCS, vol. 5740, pp. 63–90. Springer, Heidelberg (2009)

8. Borgelt, C., Kruse, R.: Induction of association rules: Apriori implementation. In: Compstat, pp. 395–400. Springer (2002)
9. Chan, P., Stolfo, S.J.: Experiments on multistrategy learning by meta-learning. In: Proceedings of the Second International Conference on Information and Knowledge Management, pp. 314–323 (1993)
10. Coenen, F.: LUCS-KDD DN software (2003)
11. Dhillon, I.S., Modha, D.S.: A data-clustering algorithm on distributed memory multiprocessors. In: Zaki, M.J., Ho, C.-T. (eds.) KDD 1999. LNCS (LNAI), vol. 1759, pp. 245–260. Springer, Heidelberg (2000)
12. Drosou, M., Pitoura, E.: Comparing diversity heuristics. Technical report, Technical Report 2009-05. Computer Science Department, University of Ioannina (2009)
13. Erkut, E.: The discrete p-dispersion problem. European Journal of Operational Research 46(1), 48–60 (1990)
14. Farivar, R., Rebolledo, D., Chan, E., Campbell, R.: A parallel implementation of k-means clustering on GPUs. In: Proceedings of International Conference on Parallel and Distributed Processing Techniques and Applications (PDPTA), pp. 340–345 (2008)
15. Ivanova, V.N., Berthold, M.R.: Diversity-driven widening. In: Proceedings of the 12th International Symposium on Intelligent Data Analysis (IDA 2013) (2013)
16. Jaccard, P.: Étude comparative de la distribution florale dans une portion des Alpes et des Jura. Bulletin del la Société Vaudoise des Sciences Naturelles (1901)
17. Kantabutra, S., Couch, A.L.: Parallel k-means clustering algorithm on nows. NECTEC Technical Journal 1(6), 243–247 (2000)
18. Liu, G., Lu, H., Yu, J.X., Wei, W., Xiao, X.: AFOPT: An efficient implementation of pattern growth approach. In: Proceedings of the ICDM Workshop on Frequent Itemset Mining Implementations (2003)
19. Lowerre, B.T.: The HARPY speech recognition system. PhD thesis, Carnegie Mellon University, Pittsburgh, PA, USA (1976)
20. Meinl, T.: Maximum-Score Diversity Selection. PhD thesis, University of Konstanz (July 2010)
21. Rissanen, J.: Modeling by shortest data description. Automatica 14(5), 465–471 (1978)
22. Smith, J.W., Everhart, J.E., Dickson, W.C., Knowler, W.C., Johannes, R.S.: Using the adap learning algorithm to forecast the onset of diabetes mellitus. In: Proceedings of the Symposium on Computer Applications and Medical Care, vol. 261, p. 265 (1988)
23. Stoffel, K., Belkoniene, A.: Parallel k/h-means clustering for large data sets. In: Amestoy, P.R., Berger, P., Daydé, M., Duff, I.S., Frayssé, V., Giraud, L., Ruiz, D. (eds.) Euro-Par 1999. LNCS, vol. 1685, pp. 1451–1454. Springer, Heidelberg (1999)
24. Vreeken, J., van Leeuwen, M., Siebes, A.: Krimp: Mining itemsets that compress. Data Mining and Knowledge Discovery 23(1), 169–214 (2011)
25. Wolberg, W.H., Mangasarian, O.L.: Multisurface method of pattern separation for medical diagnosis applied to breast cytology. Proceedings of the National Academy of Sciences 87(23), 9193–9196 (1990)
26. Zhao, W., Ma, H., He, Q.: Parallel k-Means Clustering Based on MapReduce. In: Jaatun, M.G., Zhao, G., Rong, C. (eds.) Cloud Computing. LNCS, vol. 5931, pp. 674–679. Springer, Heidelberg (2009)

Finding the Intrinsic Patterns
in a Collection of Time Series

Anke Schweier and Frank Höppner

Ostfalia University of Applied Sciences
Dept. of Computer Science, D-38302 Wolfenbüttel, Germany

Abstract. With most approaches to pattern discovery in time series
the notion of a pattern is defined a priori and then an algorithm for
the efficient discovery of patterns is proposed. But finding the *intrinsic patterns* in a collection of time series may require a search for the
best pattern representation, too. For one dataset it may be important
to consider absolute points in time, for other datasets only the shapes
may be of interest. With some datasets reoccurring subseries match the
pattern closely and with others only loosely. We propose an MDL-based
approach to search not only for patterns, but for the *intrinsic pattern
representation*. The preliminary results of this unsupervised method are
promising, because in the examined (supervised) datasets the identified
representations led to patterns that discriminate between classes.

1 Introduction

The ubiquity of sensor technologies (e.g. in mobile phones) and the affordability
of storage capacities attract more and more companies to continuously record
and store data. Prominent examples are 'open microphones' in new Android
phones to continuously identify user commands or to create a play-list of all
songs you (incidentally) came across today (www.shazam.com). Many daily activities (like driving a car) are potentially interesting (for the car manufacturer
or insurance company) such that one may suspect that your car "likely has a
black box spying on your already" [3]. Without necessarily sharing the visions
behind these applications, the examples demonstrate that temporal data (such
as time series) become increasingly popular and common.

To explore a collection of time series, it is helpful to summarise the series
somehow, that is, to identify subseries that repeat often (within the same series
or across different series). Such patterns can, however, be perceived in many
different ways: using absolute time points ("driving to work at 6:30 in the morn-
ing"), shapes ("sharp increase followed by sudden drop"), etc. And to identify
reoccurring patterns another important aspect is the accuracy of the matching
step: do we have to carve out patterns exactly or only vaguely in order to find
repeating occurrences? Most approaches from the literature define the type of
patterns they are going to discover, but do not consider a search for the best
pattern type. In this preliminary work, we investigate the possibility of identifying the *intrinsic patterns* in a collection of time series, that is not only the

H. Blockeel et al. (Eds.): IDA 2014, LNCS 8819, pp. 286–297, 2014.
© Springer International Publishing Switzerland 2014

patterns but the conditions under which they expose best, with the help of the minimum description length (MDL) principle.

The paper is organised as follows: In the next section we review MDL and related approaches from the literature. Then, Section 3 gives an overview of the approach and the considered pattern representations. The algorithmic approach is covered in Section 4. Using different datasets from the UCR time series repository [6] the experimental evaluation is presented in Section 5. Finally, Section 6 concludes the paper.

2 Definitions and Related Work

A time series T of length m is an ordered sequence of real values $T = (x_i)_{i=1\ldots m} \in \mathbb{R}^m$. In principle, any x_i may have arbitrary precision, but we assume that number are discretized to k different values (as suggested in [4]).

Minimum Description Length Principle. The description length $DL(T)$ of series T is the number of bits required to encode T. Depending on the type of encoding $DL(T)$ will vary. Assuming $k = 16$, a naive, direct calculation of $DL(T)$ requires 4 bits per value, amounting to $DL(T) = \log_2(k) \cdot m$. If we use different code lengths l_x to encode value x, we arrive at $DL(T) = \sum_{i=1}^{m} l_{x_i}$. We may obtain such a variable-length code from Huffman coding [5], which assigns shorter codes to more frequently occurring values, thereby minimising the overall code length.

Rather than a direct encoding of all values in T, a compact representation or approximation of T may help to further reduce the description length. Instead of the original time series T, a model M may be encoded using $DL(M)$ bits (depending on the type of model used). As the intention of the model is to approximate the original data, there is a loss of precision when considering model M instead of the original series T. To allow a lossless reconstruction of T, we have to encode the differences between model M and original series T, too. In total, the description of T via model M requires $DL(M)$ bits for encoding the model plus $DL(T|M)$ for a full reconstruction of T given the model:

$$DL(T, M) = DL(M) + DL(T|M)$$

If a model M captures the main characteristics of the series T well, using model M as an intermediate step may pay off in terms of the total description length, that is, we may observe $DL(T, M) < DL(T)$. Figure 1 illustrates such a situation. Encoding the series T of length 30 directly (4 bits per value) amounts to $30 \cdot 4 = 120$ bits. The Huffman code assigns codes of length 3 to the more frequent values and 4 to the less frequent values, amounting to 97 bits in total. An adaptive piecewise constant approximation (APCA) of T is shown in Figure 1 (red line). This model may be represented by a series of pairs denoting the point in time at which the segment starts and the value that holds within the segment:

$$M = ((1, 3), (17, 15), (25, 8)) \tag{1}$$

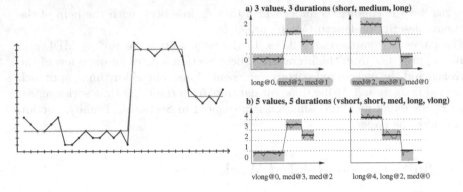

Fig. 1. An example time series

Fig. 2. Depending on the discretization, a pattern may or may not show up

With a direct encoding (still assuming $k = 16$ different values and 24 points in time) we get (when ignoring the first time point as it is fixed to 1): $DL(M) = 3 \cdot \log_2(16) + 2 \cdot \log_2(24) \approx 22$ bits. To reconstruct the original series T, we still need the deviations from the model:

$$\Delta_M(T) = (2, 1, 0, 0, 1, 2, -2, -2, -1, 0, -1, -1, 0, -1, 0, -2, 1, 0, 0, -1, 0, 0, 1, -1, \ldots$$

In this example, the delta series consists of five values only $(-2, -1, 0, 1, 2)$ and may be (naively) encoded by 3 bits per value ($DL(T|M) = 90$) or less when using Huffman coding ($DL(T|M) = 65$). In either case, the total description length $DL(M) + DL(T|M)$ became smaller, because the model M captures T very well, such that the deviations can be encoded more efficiently. To save even more bits, we may choose a coarser granularity k_M for the model than the granularity k_T used for T and $\Delta(T)$.

To find the inherent structure of T, the MDL principle advocates to search for the best encoding [2]. In [4] a range of possible models is constructed[1] for a given time series T. The model with the minimal description length successfully identified the best-suited model for T.

Sequitur. To identify chunks of repeating segments we will use Sequitur [9], which is a string compression algorithm that constructs a context-free grammar from a text string and a compressed representation of the input string using non-terminal symbols of the grammar. For instance, the input sequence *aabaab* would be compressed to XX with two rules $X \to Yb$ and $Y \to aa$ (using capital letters for non-terminals). Sequitur has been used in [7] to derive patterns (grammar rules) from a symbolic approximation (SAX [8]) of time series. The greedy algorithm has several nice properties (e.g. a new non-terminal is introduced only if it can be

---

[1] e.g. adaptive piecewise constant approximations (APCA), piecewise linear approximation (PLA), discrete Fourier decomposition (DFT), etc.

used at least twice for the compression of the input sequence) and linear runtime complexity. As in [7], we will use Sequitur to compress segment series.

Time Series Patterns. Usually approaches to time series pattern discovery define some distance measure between subseries and apply a sliding window approach to compare subseries within the same or between different series (cf. [1,7] and ref. therein). Such approaches assume, for instance, that the position of a pattern in time is not relevant and/or that the pattern does not exceed the window length. It is also common to perform a z-score normalisation of time series, thereby losing the capability of focusing patterns on exact slopes or exact values. In this work, we want to avoid such assumptions and investigate if the MDL principle can reveal the circumstances under which patterns show up prominently.

3   Outline of the Idea

We assume a set $\mathcal{T}$ of time series is given, not necessarily all of the same length. We want to investigate, how the idea of finding a best representation of a single series by MDL from [4] can be successfully extended to the problem of finding a set of rules or patterns for a whole set of time series. While in [4] the *raw data* were the time series and the *models* were the APCA representations (amongst others), we start with APCA-transformed series, which take the role of *raw data* now. We use the term *segment series* for a given APCA model to emphasise that they become the raw data and to avoid confusion with the patterns that will serve as models hereafter.

A model is a condensed, lossy representation of the original segment series. Since we encode *all* segment series rather than just one, we hope to benefit from similar subsequences within the same and across different segment series. Once such re-occurring subsequences have been identified, we encode them as part of our model and refer to them rather than encoding them multiple times. The search for the best representation involves two aspects: (1) Similar to the different types of time series approximation (piecewise constant, piecewise linear, etc.) we can think of different segment representations that lead to different *types* of patterns (see below). (2) Secondly, for any kind of representation, we may consider two segments as being sufficiently similar (to match each other) if they become identical under some discretization.

Figure 3 shows two segment series (dashed blue line and dotted red line). We consider a number of possible representations for a given segment:

ATAV: The most direct representation of a segment series is $(t_i, x_i)_{i=1...m}$ where t_i is the starting point in time of the ith segment and x_i its value (ATAV: absolute time, absolute value) as used in (1). Then, two segment series share a common subsequence only if they are aligned in both dimensions simultaneously. The dotted red and dashed blue series in Fig. 3(top left) share the segment series drawn in black.

RTAV: Rather than encoding absolute time points, the tuples may store only relative time (duration), together with the absolute value (RTAV: relative

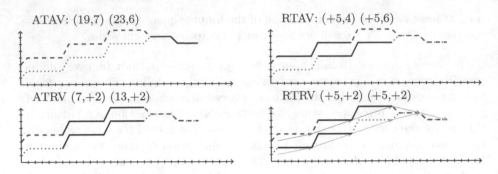

Fig. 3. Two segment series and (some) shared segments in various representations

time, absolute value). Identical subsequences may then appear at different positions in time but have identical values (top right).

ATRV: Similarly, relative values of the time series may be stored (keeping absolute time points). Identical subsequences then occur at the same point in time and change by the same amount in their value (lower left).

RTRV: Finally, both values may be encoded relatively to the previous segment. The segment itself is then interpreted as a vector (additionally drawn in the lower right image). Identical subsequences occur at different points in time and at different values, but keep the same shape.

We consider all pattern types as potentially useful. If the time series at hand are speed profiles recorded from car drivers (speed at time t, $t = 0$ at start of journey), we expect characteristic speed levels to reoccur in patterns (speed limits). To discover driving patterns, RTAV may thus be the best representation. Air pressure time series recorded at similar weather conditions (e.g. stormy) may occur at different levels of air pressure but usually exhibit similar slopes, so RTRV might be the best segment representation. When examining the effect of marketing effort on product orders over time ($t=0$ for start of sales promotion), we may observe characteristic lags between an increase in sales figures depending on the involved marketing channels, so ATRV may be a the appropriate representation. ATAV may be considered as the least interesting representation, as it corresponds to a 1:1 match of subseries. We do not consider it in this paper due to lack of space.

Regarding the matching of patterns, we consider two segments as *matching* if their discretized versions become identical. This is illustrated in Figure 2 for RTAV: Using a discretization into three values and three durations the sequence "*(medium length, value 2), (medium length, value 1)*" occurs in both examples (top row), but we observe no repetition at a finer granularity.

We want the MDL principle to identify the representation that characterizes a given dataset best. We will employ Sequitur to identify the patterns, so the Sequitur outcome (grammar and compressed sequence) corresponds to the model and defines $DL(M)$. The difference between the Sequitur model and the set $\mathcal{T}$ determines $DL(\mathcal{T}|M)$.

Algorithm 1.

Require: $\mathcal{T}$ set of time series, maximal temporal granularity k_T and value gr. k_V
Ensure: Best pattern representation (e.g. RTRV) and time/value discretization

```
 1: best = ∞
 2: S = {S | T ∈ T, construct segment series S from an APCA representation of T}
 3: for all segment representations R (ATRV, RTAV, ...) do
 4:     Let S' be the set of segment series S in representation R
 5:     for all considered segment discretizations (k₁, k₂) ∈ ℕ≤kT × ℕ≤kV do
 6:         D = {D | S ∈ S', D is the (k₁, k₂)-discretized segment series S}
 7:         merge consecutive segments in segment series of D ∈ D
 8:         find Sequitur model M (grammar and cseq) from craw
 9:         if DL(M) + DL(T|M) < best then
10:             best = DL(M) + DL(T|M), store R, k₁, k₂
11:         end if
12:     end for
13: end for
14: return best representation (stored R, k₁, k₂)
```

4   Algorithmic Approach

We propose a simple approach (cf. Algorithm 1) to find the best patterns from a set of time series. A pattern is a subsequence of segments (encoded in alternative ways) at a given resolution (alternative discretizations of segments). The hypothesis is that the inherent properties of the series can be best exploited by the pattern representation that leads to a minimal description length for the whole set of series.

4.1   Preprocessing the Series

As already mentioned, $\mathcal{T}$ denotes a set of time series $(x_1, \ldots, x_n)$ of varying lengths. Each time series is transformed into a piecewise constant approximation (using e.g. [4]), where the length of each segment may vary (line 2 of Algorithm 1). All further processing is done upon these APCA representations. As this approximation is performed for each time series individually, each series may come up with a different set of discrete values to approximate the original series. From the approximations, a segment series $S' = ((a_1, b_1), \ldots, (a_m, b_m))$ is constructed for each series S, where a_i denotes absolute or relative temporal information and b_i absolute or relative time series values, depending on the currently chosen segment representation (see page 289).

The next step (line 4) performs a discretization of the segments, which involves the discretization of both tuple values into k_1 and k_2 discretized values, resp., such that we deal with at most $k := k_1 \cdot k_2$ different discretized segments. For any choice of k_i we divide the range of values into k_i equally sized intervals. We refer to a discretized value of x or t by putting it into squared brackets $[x]$ or $[t]$. It may happen that two consecutive segments become similar after discretization such that they are better represented by a single segment (likely to happen if the APCA granularity is high but the current segment representation is coarse). If and how segments are merged depends on the chosen representation:

RTAV: Two consecutive segments $(\Delta t_1, v_1)$ and $(\Delta t_2, v_2)$ may refer to the same discretized value $v = [v_1] = [v_2]$ and are merged into one segment $(\Delta t_1 + \Delta t_2, v)$.

ATRV: Two consecutive segments $(t_1, \Delta v_1)$ and $(t_2, \Delta v_2)$ having the same discretized time point $t = [t_1] = [t_2]$ are merged into a segment $(t, \Delta v_1 + \Delta v_2)$.

RTRV: We interpret a segment $(\Delta t, \Delta v)$ as having a slope of $\frac{\Delta v}{\Delta t}$ for Δt time units; so we merge consecutive segments $(\Delta v_1, \Delta t_1)$ and $(\Delta v_2, \Delta t_2)$ to $(\Delta v_1 + \Delta v_2, \Delta t_1 + \Delta t_2)$ if both segments encode the same (discretized) slope.

Finally any segment series $S = ((a_1, b_1), \ldots, (a_m, b_m)) \in \mathcal{S}'$ has been transformed to a discretized series $D = (d_1, d_2, \ldots, d_{m'}) \in \mathcal{D}$ with $|\mathcal{D}| = k_1 \cdot k_2 =: k$, $m' \leq m$. For simplicity, we use numbers $1, 2, \ldots, k$ to refer to the available types of discretized segments. Two subseries with similar but different APCA representations may now, depending on the discretization parameters, appear identical after discretization and merging (cf. Figure 2).

4.2   Finding the Model

Sequitur shall be used to identify re-occurring subsequences of segments. As the discretized segments are represented by numbers, patterns correspond to sequences in $\{1, \ldots, k\}$. Let us denote a *tuple concatenation operator* by $\bullet$, i.e., $(a, b, c) \bullet (d, e) = (a, b, c, d, e)$. With d_i denoting the ith discretized series, the full set $\mathcal{D}$ of m series is encoded into a single sequence (over the alphabet $\{-m, \ldots, -1, 1, \ldots, k\}$):

$$c_{\text{raw}} := d_1 \bullet (-1) \bullet d_2 \bullet (-2) \bullet d_3 \bullet (-3) \cdots d_n$$

The negative numbers serve as separators between the encoded segment series. As Sequitur requires a symbol to occur at least twice before introducing a rule, these separators effectively prevent Sequitur from elaborating rules that connect symbols from the end of series d_i and the beginning of series d_{i+1}. Such rules would depend on the (arbitrary) order of series d_i and are therefore undesired.

The code c_{raw} represents the input to Sequitur. Sequitur delivers a grammar based on a set of new (non-terminal) symbols, which we encode also by numbers (starting at $k + 1$). A rule of the grammar thus reads like $A \to BC$ where A, B and C are numbers. A represents a non-terminal (thus $A > k$) and B (as well as C) may refer to a terminal symbol (discretized segment; $1 \leq B \leq k$) or another non-terminal symbol ($B > k$). Sequitur also delivers a compressed sequence c_{seq} from which the original sequence c_{raw} can be reconstructed by replacing non-terminals with the resp. right-hand side of its rule. All separators $s < 0$ (e.g. (-1), (-2) from c_{raw} above) in c_{seq} may be removed completely or replaced by a single separator symbol (0) to preserve the separation of the series (but different symbols were only necessary to prevent Sequitur from deriving rules that link different series). Thus, if the grammar consists of r rules, the output is a sequence over the alphabet $\mathcal{A} = \{0, 1, \ldots, k + r\}$ ($\leq k$: discretized segments, $> k$: non-terminals).

For instance, we obtain $c_{\text{raw}} = (1, 3, 2, 4, -1, 1, 2, 4, 3, -2, 3, 1, 2, 4)$ from three discretized sequences $d_1 = (1, 3, 2, 4)$, $d_2 = (1, 2, 4, 3)$ and $d_3 = (3, 1, 2, 4)$. Sequitur may deliver two rules $5 \to 24$ and $6 \to 15$ (with new non-terminals 5 and 6), leading to $c_{\text{seq}} = (1, 3, 5, 0, 6, 3, 0, 3, 6)$. This procedure is carried out for all considered discretizations and all considered pattern representations in line 8 of Algorithm 1.

4.3   Calculating DL

The description length of the model consists of the Sequitur grammar and the compressed collection of sequences. We encode this as follows: (1) the number r of rules, (2) the right-hand side of all rules (no separation necessary because the right-hand side of a Sequitur rule has always 2 symbols), (3) the compressed sequence c_{seq}. We use a Huffman code to obtain minimal coding costs.

Secondly, we have to determine the difference between the original segment series S and the model. Let us ignore the merging step of line 7 for the moment. The Sequitur compression can be reversed to arrive at the original sequence c_{raw}, so we have to encode the differences between the discretized segments of c_{raw} and the original segments of S'. We apply the same pointwise differencing as described for the case of time series in section 2: If the original segment series is $S = ((1, 4), (7, 2), (17, 15), (25, 8))$ and the time points and values are discretized to $\{1, 5, 10, 15, 20, 25, 30\}$ and $\{2, 6, 10, 14\}$, resp., we obtain

$$D = ((1, 6), (5, 1), (15, 14), (25, 10))$$
$$\text{and } \Delta_D(S) = ((0, -2), (2, 1), (2, 1), (0, -2)).$$

The better the discretization adopts to the segments occurring in S, the shorter $DL(S|M)$. Again, a Huffman code is used to encode $\Delta_D(S)$.

The merging of segments in line 7 is necessary to join segments that were considered different in their APCA representation but become identical under the currently applied segment discretization. Without such a merging step, we have as many original as discretized segments (1:1 relationship), but merging may reduce the number of discretized segments. Thus, from a single encoded segments in c_{raw} we may have to reconstruct multiple original segments in S'. For example, if $S = ((1, 4), (7, 2), (15, 14), (17, 15), (25, 8))$ is discretized to $((1, 6), (5, 1), (15, 14), (15, 14), (25, 10))$ and subsequently merged to $((1, 6), (5, 1), (15, 14), (25, 10))$, we have to keep in mind which segments were merged in order to calculate $\Delta_D(S)$ correctly:

$$
\begin{aligned}
S &= \quad (1, 4) \quad (7, 2) \ (15, 14) \ (17, 15) \quad (25, 8) \\
D &= \quad (1, 6) \quad (5, 1) \qquad (15, 14) \qquad (25, 10) \\
\Delta &= (0, -2) \ (2, 1) \quad (0, 0) \ (2, 1) \quad (0, -2)
\end{aligned}
$$

There are multiple ways to encode how many segments of Δ belong to a single segment of D: either we include counts for the number of Δ-segments belonging to the next D-segment or we insert a special "glue" symbol (g) between Δ-segments belonging to a merged D-segment:

$$\text{counts:} \quad 1(0,-2) \quad 1(2,1) \quad 2(0,0)(2,1) \quad 1(0,-2)$$
$$\text{glue symbol:} \quad (0,-2) \quad (2,1) \quad (0,0)g(2,1) \quad (0,-2)$$

5 Experimental Evaluation

We examine three datasets to evaluate whether the identification of *intrinsic patterns* via MDL is viable: (1) a (one-dimensional) random walk dataset, (2) the CBF dataset consisting of short time series from three different classes (describing a cylinder, a bell and a funnel) and (3) the symbols dataset consisting of 6 different hand-drawn symbols on a touchscreen. The latter two datasets are taken from [6]. These datasets have been chosen for the first experiments because the random walk should not contain any particular patterns (by construction) while the other two datasets are known to contain patterns (cf. [6]).

By the term *configuration* we refer to both, the chosen pattern type (e.g. RTRV) and the granularity used for time and value discretization. Once Algorithm 1 has identified the best configuration, how do we know if this representation succeeded in capturing the patterns inherent in the time series collection? The CBF and symbols dataset have class labels, so we investigate if the discovered rules correlate with class labels. The algorithm is not aware of the class labels, but we expect intrinsic patterns to correspond to class-specific subseries. For rules of the grammar that apply to at least 5% of the series, we qualitatively examine their entropy. For each class, we report the rule with the lowest entropy.

Table 1 shows the results for all three datasets. The smallest total description lengths (per pattern type) are shown in column *total*. In all three datasets, the configuration with the smallest total length is of type RTRV. However, for the random walk and CBF dataset, only a few patterns (meeting the 5% usage threshold) were found for RTRV configurations. The minimal cost configuration does not seem to be a good indicator to identify the intrinsic pattern representation. We attribute this observation to the fact, that the RTRV representation benefits from the smoothness of the considered time series: consecutive segments are only a few time indices apart and (thanks to the smoothness) the deviation between consecutive values is also comparatively small. Many small differences (obtained from relative values) are encoded much more efficiently than absolute values that distribute more uniformly. This gives RTRV an advantage over RTAV and ATRV, because two values are encoded relatively rather than just one.

But this affects the encoding of the differences $\Delta_D(S)$ only (column *delta*). The description length of the (compressed) model involves only c_{raw}, which is just a sequence over an alphabet of terminal and non-terminal symbols. To evaluate how well a configuration supports the 'compressability' of c_{raw}, we report the the size of the model as a fraction of the size of an uncompressed model (as if Sequitur delivered an empty grammar for c_{raw}) in column *reduced size*. A smaller percentage indicates a better compressability of the model and is thus considered to be a better indicator for the best pattern representation.

Table 1. Results for the three data sets

a)

random walk	best granularity		description length			reduced
	time	value	model	delta	total	size
RTAV	2	9	6191	69400	75591	40.6%
ATRV	4	2	3352	76603	79955	31.2%
RTRV	6	3	3352	69400	72752	85.4%

b)

CBF	best granularity		description length			reduced	entropy of best rule		
	time	value	model	delta	total	size	cylinder	bell	funnel
RTAV	3	4	6594	40391	47525	37.5%	0.00	0.00	0.00
ATRV	3	2	3898	38063	41961	39.3%	0.49	1.53	0.00
RTRV	2	2	2554	35020	37574	91.1%	–	0.98	–

c)

symbols	best granularity		description length			reduced	entropy of best rule				
	time	value	model	delta	total	size	1	2	3	4|5	6
RTAV	3	7	30k	379k	409k	23.8%	0.53	0.96	0.00	0.61	0.00
ATRV	13	2	23k	350k	374k	21.4%	0.95	2.00	0.00	0.11	0.00
RTRV	11	52	67k	259k	327k	45.9%	1.63	0.55	0.57	0.89	0.97

The 500 series in the random walk dataset (without class labels) consist of 250 values and start at $x_1 = 0$. Although there are no patterns imputed in the dataset there may nevertheless be incidental repetitions to be discovered by Sequitur. RTRV patterns are sequences of "increase by Δv_i within Δt_i time units". This representation is useful to approximate the up's and down's of the random walk locally, but due to the random nature of the dataset, a longer series of certain up's and down's is unlikely to repeat itself, so the representation is of limited use to identify reoccurring patterns (size remains 85.4% of uncompressed model, cf. Figure 1a). With ATRV we achieve the best model compression: The temporal granularity of 4 subdivides the time axis into 4 intervals and Δv takes only two values (increasing, decreasing). At this configuration, any random walk consists of a series of length 4 only, a particular series may be described as, e.g., 'values increase in the first and last quarter, but decrease in the second and third'. Only 2^4 different sequences exist, which is exploited by Sequitur. Thus, the best configuration takes a rather global perspective on the series, which is a reasonable result for random walk data.

The results for the CBF dataset (900 series, 3 classes) are shown in Table 1b. Only a few short rules are discovered by Sequitur for RTRV with little connection to the classes. The best model compression was achieved for the RTAV representation – and it is also the RTAV model which delivered patterns that best correspond to classes. Relative times are reasonable for CBF, because the imputed patterns are randomly displaced in time, as well as absolute values, because all CBF series jumps and linearly interpolates between two values only.

However, the APCA approximation of the CBF series are quite short (only 3-5 segments remain per series), which limits the length of discoverable patterns. This is quite different for the symbols dataset (995 series, 6 classes, cf. Figure 4);

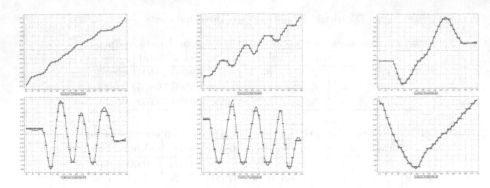

Fig. 4. Examples from the symbols dataset (note the similarity of classes 4 and 5)

results are shown in Table 1c. This time the optimal RTRV granularity is much higher and the discovered rules are much longer, non-terminals of the grammar represent sequences of up to 13 segments. The RTRV model compression is much more competitive compared to the CBF and random walk datasets, the best rule (in terms of entropy) for class #2 is of type RTRV. This is due to the fact that the time series consist of similar shapes that repeat across different classes and also within series of the same class. The highest model compression is achieved with ATRV (down to 21.4%). Series from multiple classes have long up/downward trends, which are also exploited by patterns of type RTAV and RTRV, but classes 1∪2, 3 and 6 can be easily distinguished if we know where these trends occur in time. Again, the best configuration provides those patterns that are most meaningful with respect to class labels.

6   Conclusions

Patterns may disguise themselves in time series in quite different ways. To identify similar subsequences, the shape of the subsequence may be important, the position in the time series, their absolute value, etc. No repetition will be exactly identical, but it is not a priori clear under which resolution patterns will show up. In this preliminary work we explored if the MDL principle can successfully be applied to identify the best pattern representation in terms of pattern types (absolute/relative values) and degree of similarity (discretization granularity). The preliminary results are promising: Despite the simplistic approach, the configuration that led to the highest model compression always delivered patterns that correspond best to class labels (which were unknown to the MDL approach).

References

1. Fu, T.-C.: A review on time series data mining. Engineering Applications of Artificial Intelligence 24(1), 164–181 (2011)

2. Grunwald, P.D.: The Minimum Description Length Principle. University Press Group Ltd. (2007)
3. Hill, K.: Hate To Break It To You, But Your Car Likely Has A Black Box 'Spying' On You Already (2012), http://onforb.es/I7BRLJ
4. Hu, B., Rakthanmanon, T., Hao, Y., Evans, S., Lonardi, S., Keogh, E.: Discovering the Intrinsic Cardinality and Dimensionality of Time Series Using MDL. In: Proc. 11th Int. Conf. on Data Mining (ICDM), pp. 1086–1091 (2011)
5. Huffman, D.: A Method for the Construction of Minimum-Redundancy Codes. Proceedings of the IRE 40(9), 1098–1101 (1952)
6. Keogh, E., Zhu, Q., Hu, B., Hao, Y., Xi, X., Wei, L., Ratanamahatana, C.A.: The UCR Time Series Classification/Clustering Homepage (2011)
7. Li, Y., Lin, J.: Approximate variable-length time series motif discovery using grammar inference. In: Proceedings of the Tenth International Workshop on Multimedia Data Mining - MDMKDD 2010, pp. 1–9 (2010)
8. Lin, J., Keogh, E., Wei, L., Lonardi, S.: Experiencing SAX: A novel symbolic representation of time series. Data Mining and Knowledge Discovery 15(2), 107–144 (2007)
9. Nevill-Manning, C.G., Witten, I.W.: Identifying Hierarchical Structure in Sequences: A linear-time algorithm. Journal of Artificial Intelligence Research 7, 67–82 (1997)

A Spatio-temporal Bayesian Network Approach for Revealing Functional Ecological Networks in Fisheries

Neda Trifonova[1], Daniel Duplisea[2], Andrew Kenny[3], and Allan Tucker[1]

[1] Department of Computer Science, Brunel University, London, UK
[2] Fisheries and Oceans, Canada
[3] Centre for Environment, Fisheries and Aquaculture Science, Lowestoft, UK

Abstract. Ecosystems consist of complex dynamic interactions among species and the environment, the understanding of which has implications for predicting the environmental response to changes in climate and biodiversity. Machine learning techniques can allow such complex, spatially varying interactions to be recovered from collected field data. In this study, we apply structure learning techniques to identify functional relationships between trophic groups of species that vary across space and time. Specifically, Bayesian networks are created on a *window of data* for each of the 20 *geographically* different and *temporally* varied sub-regions within an oceanic area. In addition, we explored the *spatial* and *temporal* variation of *pre-defined functions* (like predation, competition) that are generalisable by experts' knowledge. We were able to discover meaningful ecological networks that were more precisely *spatially-specific* rather than temporally, as previously suggested for this region. To validate the discovered networks, we predict the biomass of the trophic groups by using dynamic Bayesian networks, and correcting for spatial autocorrelation by including a *spatial node* in our models.

1 Introduction

In recent decades it has become clear that ecosystem structure and function can change over relatively short time [13]. Functional changes can significantly affect the abundance and distribution of fish populations, either directly or by affecting prey or predator populations [11]. The effect of predators has been shown to influence prey populations and vice versa and has been described to be of the same or greater magnitude than fishing alone [11]. Different species may have similar functional roles (the functional status of an organism) within a system depending on the region. For example, one species may act as a predator of another which regulates a population in one location, but another species may perform an almost identical role in another location. If we can model the function of the interaction rather than the species itself, data from different regions can be used to confirm key functional relationships, to generalise over systems and to predict impacts of forces such as fishing and climate change.

H. Blockeel et al. (Eds.): IDA 2014, LNCS 8819, pp. 298–308, 2014.

One way to understand community structure and stability is examination of the functional relationships (such as prey-predator) between species in their potential habitat (space) and across time. In this way, learning functional relationships can provide a metric for assessing community structure and resilience in response to natural and anthropogenic influences [8]. In this study, we aggregate individual species into trophic species (functional groups of taxa that share the same set of predators and prey within a food web): invertebrates, pelagics (pelagic fish that live in the pelagic zone of ocean waters - being neither close to the bottom nor near the shore), small piscivorous (fish-eating species) and large piscivorous and top predators from the northern Gulf of St. Lawrence groundfish and shrimp summer survey since 1990 and examined how the learned functional relationships between the trophic groups varied in time and space.

Interactions among species make it difficult to predict how ecological communities will respond to environmental degradation, yet to do so we must understand the functional networks that form the systems [4]. The functional network approach to understand community structure and resilience is an on-going approach combining known topological features of food webs with quantitative variation in species interactions to predict community stability. Recently, an approach has arisen in biology that is capable of inferring network structures, capturing nonlinear, stochastic and arbitrary combinatorial relationships: Bayesian Networks (BNs) [10]. Formally, a BN exploits the conditional independence relationships over a set of variables, represented by directed acyclic graphs (DAG) [6]. Each node in the DAG is characterised by a state which can change depending on the state of other nodes and information about those states propagated through the DAG. By using this kind of inference, one can change the state or introduce new data or evidence into the network, apply inference and inspect the posterior distribution. Structure learning of these models from data is an NP-hard problem and many studies have been conducted on this subject, leading to three different approaches: constraint-based methods, score-based and hybrid methods [2]. We focus in this paper on BN structure learning using score-based method, specifically learning a distinct network for each sub-region of the Gulf of St. Lawrence oceanic area.

In this paper, we examine how aggregated species interact at different *spatial* scales and over *time* to understand what mechanisms are involved in shaping the ecological networks and functional dynamics of food webs. Specifically, we explore how *pre-defined functional relationships* vary in *time* and *space* in order to better understand community structure and resilience. At larger spatial scales, although fishing can still be the dominant driver of functional changes, the consequences of fishing are not predictable without understanding the food web dynamics [11].

2 Methods

2.1 Species Collection

We analysed data from the northern Gulf of St. Lawrence (48.00°N, 61.50°W, Fig.1a) groundfish and shrimp summer survey (1990-2013). The survey utilises a stratified random sampling design [3] with a standard tow using a benthic otter trawl. For each tow, all the fish were weighed and a subsample (200 individuals per species) was taken for computing length-frequency distributions. These length-frequency distributions were the basis of the data used here.

2.2 Data Preparation

K-means [9] was applied to limit the number of variables and cluster the number of sampling stations (originally over 200 sampling stations per year, Fig.1b) on the mean latitude and longitude, resulting in 20 spatial clusters (or subregions, Fig.1c). Note that differences in density of the clustered stations could explain the slight spatial contrast between Fig.1b and Fig.1c. The number of stations varied within each cluster so the biomass (the total quantity or weight of organisms in a given area or volume) was averaged over the same species and within the same year. Then, fish and invertebrate species were aggregated into the relevant trophic group by summing up the biomass. The nature of individual species summed into the trophic guilds varied between the spatial clusters but this was not of importance since they were always aggregated into the correct trophic group. This was performed for each of the 20 clusters and for each year in the time window: 1990-2013, ending up with four variables for each spatial cluster across the time window.

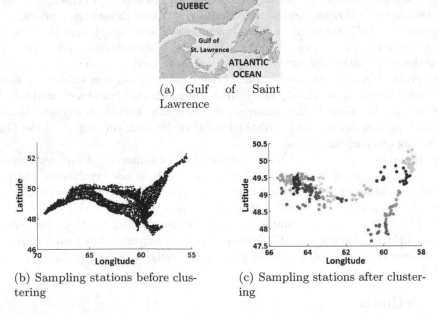

(a) Gulf   of   Saint Lawrence

(b) Sampling stations before clustering

(c) Sampling stations after clustering

Fig. 1. Locations of the oceanic region of St. Lawrence (a) and the sampling stations before clustering (b) and after clustering (c)

2.3 Structure Learning of BNs

Our model is a BN in which nodes represent trophic groups and edges (connections between nodes) represent potential species interactions. Note we infer static BNs from temporal data for each of the 20 spatial clusters. Hill-climbing procedure was applied for learning the static BN structure. The search begins with an empty network. In each stage of the search, networks in the current neighbourhood are found by applying a single change to a link in the current network such as *add arc* or *delete arc* and choose the one change that improves the score the most. We used the Bayesian Information Criterion (*BIC*) for scoring candidate networks [14]. The *BIC* function is a combination of the model log-likelihood and a penalty term that favours less complex models- as such it is similar to the minimum description length: $BIC = \log P(\Theta) + \log P(\Theta|D)$ - 0.5 $k \log(n)$ where Θ represents the model, D is the data, n is the number of observations (sample size) and k is the number of parameters. $\log P(\Theta)$ is the prior probability of the network model Θ, $\log P(\Theta|D)$ is the log-likelihood while the term $k \log(n)$ is a penalty term, which helps to prevent over-fitting by biasing towards simpler, less complex models.

The hill-climb structure learning approach was conducted with 10 random restarts. In this approach, we apply the search until we hit a local maximum. Then, we randomly perturb the network structure and repeat the process for some number of iterations, in the case of the network analysis for individual clusters alone (20 clusters, each matrix with the size of *4x24*), we apply the learning procedure for 500 iterations. In addition, to learn the model structure for each year in the time window, the hill-climbing was conducted on a window of data (size of window= 10). In this way, we would be able to capture any significant functional interactions over the previous 10 years.

Spatial autocorrelation, the phenomenon that observations at spatially closer locations are more similar than observations at more distant observations, is nearly ubiquitous in ecology and can have a strong impact on statistical inference [1]. To incorporate potential spatial autocorrelation in our model, we connect each node in the network to an enforced parent node that represents the average biomass from the spatial neighbourhood (the four nearest neighbours) of the current geographic location (or cluster) [1]. By applying the windowing approach, we produced two variants of our BN model: one that excludes a spatial node and one including the spatial node.

2.4 Detection of Pre-defined Functions

A library of simple BNs, representing species interactions or functional relationships, based on expertise knowledge (Table 1, **I**-invertebrates, **P**-pelagics, **SP**-small piscivorous and **LP**-large piscivorous and top predators) was created. Then, the experiment was conducted, in which each cluster was individually analysed to identify how the known functional relationships vary across time, but also to discover relationships between trophic groups, producing structures for 20 static BN models, equivalent to each one of the sub-regions in the Gulf of St. Lawrence oceanic area. Note that we detect the equivalence classes of each functional relationship and score the confidence of each relationship being in the network over space and time. Our model adopting random restarts was preferably chosen compared to conditional independence tests for example as we wanted to learn the confidence of each functional relationship being in the network and not just examine the dependency relationships. We defined functional relationships of high confidence as those in which we have the greatest confidence of being in the network (threshold ≥ 0.3).

Table 1. Pre-defined Functional Relationships

Pre-defined Functional Relationships and Descriptions	
1. $I->SP<-P$	Competition
2. $P<-I->SP$	Predation
3. $P<-I->SP, I->LP$	Predation
4. $P<-I->SP, P->LP$	Predation
5. $P<-I->SP->LP, P->LP$	Predation
6. $P<-I->SP, LP<-SP->P$	Intraguild Predation
7. $LP<-I->P->SP->LP$	Omnivory
8. $P<-I->SP->LP$	Predation

2.5 Dynamic Bayesian Networks and Prediction

As well as learning functional relationships over space and time, we also explore network predictions over time. We choose to validate the networks through prediction by inferring dynamic Bayesian networks (DBNs) for each cluster and comparing the predicted biomass by either including or removing the spatial node from the model. Modelling time series is achieved by the DBN where nodes represent variables at particular time slices [6]. More precisely, a DBN defines the probability distribution over $\mathbf{X}[t]$ where $\mathbf{X}=X_1...X_n$ are the n variables observed along time t. To predict the biomass of each trophic group, we first infer the biomass at time t by using the observed evidence from time t-1. Two sets of experiments were then conducted: one that excludes the spatial node (DBN) and in the other, spatial node was included in the model (DBN+ spatial) to see if the node improves prediction. Non-parametric bootstrap analysis [6] was applied 250 times for each variant of the model (resulting in two model variants for each of the clusters) to obtain statistical validation in the predictions.

3   Results and Discussion

We were able to discover meaningful networks of functional relationships from ecological data, giving us confidence in the novel methods and results presented here. While the precise explanation behind the varying spatio-temporal confidence of some of the discovered relationships is not known, we expect them to be reflective of the underlying interactions within the community, thus suggesting similarity to the majority of the weak and some strong interactions expected of stable systems [12].

3.1 Functional Relationships Revealed by Hill-Climbing

We now examine how learned by the model relationships amongst trophic groups of species vary across time and space. The relationship between invertebrates and pelagics (I-P) was found to be strongly significant (range: 0.3-1) and consistent in time and space (Fig.2a,b). Cluster 7 was the only cluster in which the relationship was found throughout the entire time series and in cluster 5 the relationship was found to be with highest confidence throughout time. Temporally, the confidence for the I-P relationship in majority of the clusters was found to be generally increasing with a small decline over recent years. The invertebrates-small piscivorous fish (I-SP) relationship had the highest confidence throughout time in cluster 4. The relationship was relatively consistent in time for individual clusters. For cluster 19, the invertebrates- large predators (I-LP) relationship was identified throughout the entire time series but the most highly significant confidence was found for cluster 17 (range: 0.3-0.8). Temporally, both relationships: I-SP and I-LP, for majority of clusters were relatively stable but with declining trend at end of the time series.

We now consider the pelagics- small piscivorous fish (P-SP) relationship (Fig.2c,d). As with the I-P relationship, here P-SP was also the most highly confident for cluster 5 (range: 0.3-1). This P-SP relationship was highly consistent in time for clusters 4 and 16 in which the relationship was found throughout the entire time series. Compared to P-SP, for the pelagics- large predators (P-LP) relationship, cluster 10 was the one in which the relationship was highly confident (range: 0.3-1). However, cluster 5 was the one in which the relationship was consistent throughout time. Across time, both relationships varied for the different clusters and it was difficult to find any temporal trends. However, some clusters declined around 2007 to 2010 (for example 7, 15) whilst clusters 11 and 19 increased around the same time and in most recent years.

The most highly confident small piscivorous- large predators (SP-LP) relationship (Fig.2e,f) in time that was also consistent in the series was found for cluster 20 (range: 0.3-1). The relationship was also consistent in time for cluster 9. Across time, similarly to the previous relationship, some clusters were relatively stable but some decline occurred around 2007 to 2008 (clusters: 15, 5 and 7), whilst in other clusters increase in confidence was found for more recent years (for example clusters 1, 9).

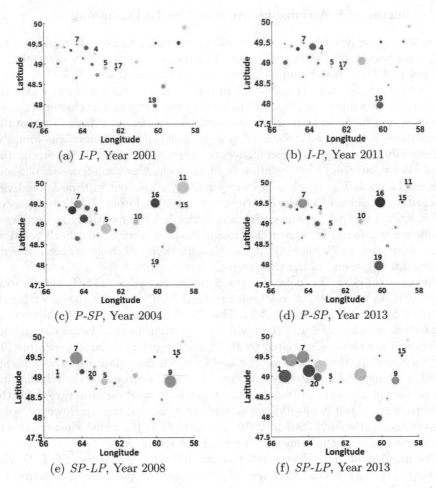

Fig. 2. The learned *I-P*, *P-SP* and *SP-LP* relationships for all 20 spatial clusters (size of scattered bubbles is equivalent to the estimated confidence by the hill-climb). The clusters mentioned in 3.1 are numbered.

Overall, the identified functional relationships were found to be consistently confident in time however we notice the spatially-specific differentiation. Such spatial heterogeneity could result from habitat fragmentation leading to decreased dispersal or the optimal habitat being located in a more restricted area, leading to increased aggregation [7]. Individual year effects are very strong for this area as time increases, as already suggested by [5] which makes it difficult to determine temporal trends. However, some of the clusters' temporal increase in early to mid-2000 (specifically for *I-SP* (cluster 5), *P-LP* (cluster 7) and *SP-LP* (cluster 5), Fig.3a,b,c) could be owed to the fisheries moratorium in the area placed in 1994. In addition, our findings of recent temporal decline for some of the clusters' relationships (*P-LP* (cluster 5), *SP-LP* (cluster 19), Fig.3b,c) we suggest to be due to predation release of small abundant species by the selective fishing of larger predators [7]. Note again the temporal variation of the systems was set apart in geographically-specific order, possibly due to site-specific fisheries exploitation targeting particular species.

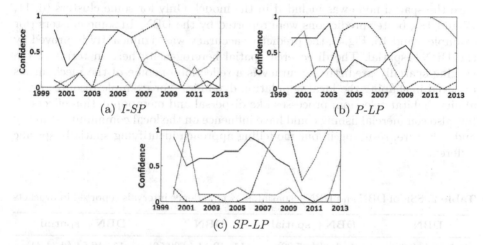

(a) *I-SP*   (b) *P-LP*

(c) *SP-LP*

Fig. 3. The learned *I-SP*, *P-LP* and *SP-LP* relationships for clusters 5, 7 and 19 (represented by solid, dash and dot line respectively) for the time window: 2000-2013

3.2   Summary of Discovered Functional Relationships

Next, we consider the variation of the pre-defined known functional relationships (Table 1) temporally and spatially. First, *function 1* and *2* were identified in all clusters. However, the significance of both functions varied across time with some consistency in terms of spatial clusters. We find the emergence of "characteristic scales" of functional relationships, identified at spatially-specific geographic scales. Temporally, there was some decline in the significance of *function 1* and *function 2*, specifically in more recent years: 2010 to 2013 in all clusters. At the same time clusters like 9, 5 and 20 were found to be with relatively strong significance throughout time, outlining the importance of habitat quality at specific locations implying that in some regions prey are more affected by predators than in others. *Function 3* and *4, 5, 6, 7* and *8* were not identified for all clusters and were only found in some years. However, again there was some spatial consistency in terms of different functions identified outlining only some clusters, highlighting the fact that relationships are scale dependant but also the importance of functional relationships for the local food web dynamics and structure. Other possible explanations include species abundance and distributional changes but in either case fishing could have had an important role.

3.3   DBNs and Prediction

We now turn to the generated predictions by the DBNs for each spatial cluster. To recall, two variants of each model were produced: DBN excluding the spatial node and DBN+ spatial in which the spatial node was enforced and connected to each one of the other variables. Predictive performance between the two model variants was compared (Table 2). In general, predictive accuracy was improved once the spatial node was included in the model. Only for some clusters (6, 11, 17 and 18), better predictions were reported by the DBN. In some clusters (for example 5 and 15, Fig.4), the predictive accuracy was significantly improved by the DBN+ spatial. The discovered spatial heterogeneity here in terms of the varying spatially predictive accuracy is a reflection of some of the mechanisms involved in shaping the local population dynamics. For example resource availability, habitat selection, processes like dispersal and metapopulation effects [7] but also commercial fishing could have influence on the local community stability and structure, resulting in our modelling approach identifying spatially-specific differences.

Table 2. SSE of DBN and DBN+ spatial. 95% confidence intervals reported in brackets

DBN	DBN+ spatial	DBN	DBN+ spatial
1. 5.58 (±9.29)	**1.** 4.38 (±7.08)	**11.** 12.44 (±20.56)	**11.** 16.54 (±34.34)
2. 0.24 (±0.36)	**2.** 0.14 (±0.12)	**12.** 69.90 (±308.02)	**12.** 30.55 (±64.30)
3. 16.20 (±29.92)	**3.** 10.76 (±17.16)	**13.** 12.68 (±16.63)	**13.** 12.06 (±9.63)
4. 10.09 (±14.70)	**4.** 9.68 (±12.58)	**14.** 196.11 (±271.68)	**14.** 109.37 (±102.42)
5. 44.20 (±51.17)	**5.** 11.27 (±12.47)	**15.** 77.45 (±605.26)	**15.** 23.62 (±47.80)
6. 20.20 (±40.42)	**6.** 20.22 (±34.29)	**16.** 17.15 (±18.40)	**16.** 14.86 (±13.46)
7. 25.29 (±55.38)	**7.** 19.47 (±26.86)	**17.** 5.88 (±8.78)	**17.** 6.12 (±6.67)
8. 38.72 (±46.22)	**8.** 19.59 (±11.78)	**18.** 2.68 (±3.94)	**18.** 3.43 (±3.72)
9. 125.19 (±240.49)	**9.** 92.14 (±111.67)	**19.** 80.32 (±112.90)	**19.** 77.19 (±72.43)
10. 104.31 (±167.02)	**10.** 60.62 (±62.38)	**20.** 13.20 (±22.08)	**20.** 10.70 (±13.64)

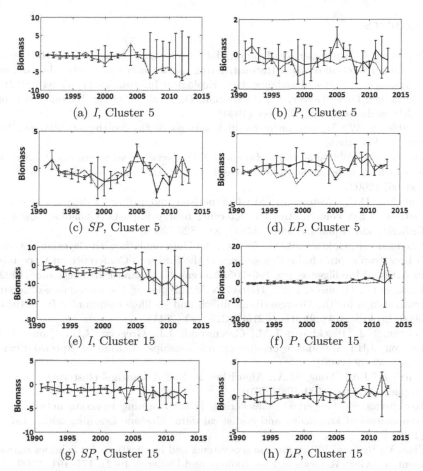

Fig. 4. Biomass predictions generated by DBN+ spatial for clusters 5 and 15 for the four trophic groups: *I*, *P*, *SP* and *LP*. Solid line indicates predictions and dash-dot line indicates standardised observed biomass. 95% confidence intervals report bootstrap predictions' mean and standard deviation.

4   Conclusion

In this paper we have exploited the use of BNs with *spatial nodes* in order to identify patterns of functional relationships which proved significant in terms of predictive accuracy of our models, further concluding the spatial heterogeneity in this oceanic region. We have also used knowledge of functional interactions between species to identify changes over time. Our results show highly confident but *spatially* and *temporally* differentiated ecological networks that indicate spatial relationship of species and habitat with the particular mechanisms varying from facilitation through trophic interactions. Future work will involve detailed analysis of each individual cluster with expansion on the functional networks.

References

1. Aderhold, A., Husmeier, D., Lennon, J.J., Beale, C.M., Smith, V.A.: Hierarchical Bayesian models in ecology: Reconstructing species interaction networks from non-homogeneous species abundance data. Ecological Informatics 11, 55–64 (2012)
2. Chickering, D.M., Geiger, D., Heckerman, D., et al.: Learning Bayesian networks is NP-hard. Tech. rep., Citeseer (1994)
3. Doubleday, W.: Manual on groundfish surveys in the Northwest Atlantic. Tech. rep., NAFO (1981)
4. Dunne, J.A., Williams, R.J., Martinez, N.D.: Network structure and biodiversity loss in food webs: robustness increases with connectance. Ecology Letters 5(4), 558–567 (2002)
5. Duplisea, D.E., Castonguay, M.: Comparison and utility of different size-based metrics of fish communities for detecting fishery impacts. Canadian Journal of Fisheries and Aquatic Sciences 63(4), 810–820 (2006)
6. Friedman, N., Goldszmidt, M., Wyner, A.: Data analysis with Bayesian networks: A bootstrap approach. In: Proceedings of the Fifteenth Conference on Uncertainty in Artificial Intelligence, pp. 196–205. Morgan Kaufmann Publishers Inc. (1999)
7. Frisk, M.G., Duplisea, D.E., Trenkel, V.M.: Exploring the abundance-occupancy relationships for the Georges Bank finfish and shellfish community from 1963 to 2006. Ecological Applications 21(1), 227–240 (2011)
8. Gaston, K.J., Blackburn, T.M., Greenwood, J.J., Gregory, R.D., Quinn, R.M., Lawton, J.H.: Abundance–occupancy relationships. Journal of Applied Ecology 37(s1), 39–59 (2000)
9. Hartigan, J.A., Wong, M.A.: Algorithm as 136: A k-means clustering algorithm. Applied Statistics, 100–108 (1979)
10. Heckerman, D., Geiger, D., Chickering, D.M.: Learning Bayesian networks: The combination of knowledge and statistical data. Machine Learning 20(3), 197–243 (1995)
11. Jiao, Y.: Regime shift in marine ecosystems and implications for fisheries management, a review. Reviews in Fish Biology and Fisheries 19(2), 177–191 (2009)
12. Milns, I., Beale, C.M., Smith, V.A.: Revealing ecological networks using Bayesian network inference algorithms. Ecology 91(7), 1892–1899 (2010)
13. Scheffer, M., Carpenter, S., Foley, J.A., Folke, C., Walker, B.: Catastrophic shifts in ecosystems. Nature 413(6856), 591–596 (2001)
14. Schwarz, G., et al.: Estimating the dimension of a model. The Annals of Statistics 6(2), 461–464 (1978)

Extracting Predictive Models from Marked-Up Free-Text Documents at the Royal Botanic Gardens, Kew, London

Allan Tucker[1] and Don Kirkup[2]

[1] Department of Computer Science, Brunel University, UK
[2] Royal Botanical Gardens at Kew, UK
allan.tucker@brunel.ac.uk

Abstract. In this paper we explore the combination of text-mining, un-supervised and supervised learning to extract predictive models from a corpus of digitised historical floras. These documents deal with the nomenclature, geographical distribution, ecology and comparative morphology of the species of a region. Here we exploit the fact that portions of text in the floras are marked up as different types of trait and habitat. We infer models from these different texts that can predict different habitat-types based upon the traits of plant species. We also integrate plant taxonomy data in order to assist in the validation of our models. We have shown that by clustering text describing the habitat of different floras we can identify a number of important and distinct habitats that are associated with particular families of species along with statistical significance scores. We have also shown that by using these discovered habitat-types as labels for supervised learning we can predict them based upon a subset of traits, identified using wrapper feature selection.

1 Introduction

In the last two decades, there has been a surge in data related to biodiversity of plants through, for example, on-line publications, DNA-sequences, images and metadata of specimens. Much of the new data is characterised by its semi-structured, temporal, spatial and 'noisy' nature arising from disparate sources. Here, we focus on the use of textual data in floras. These are the traditional taxonomic research outputs from organisations such as the Royal Botanical Gardens at Kew, London, and deal with the nomenclature, geographical distribution, ecology and comparative morphology of the species of a region, explicitly linked to defined taxonomic concepts. We exploit the use of data mining (and in particular text mining) in combination with machine learning classifiers in order to build predictive models of habitat based upon plant traits.

Text mining has grown in popularity with the digitisation of historical texts and publication [12]. In particular, the use of text mining for bioinformatics data has led to a number of different approaches. For example, medline abstracts have been mined for association between genes, proteins and disease outcome [11]. These can

H. Blockeel et al. (Eds.): IDA 2014, LNCS 8819, pp. 309–320, 2014.

vary from simple statistical approaches to more complex *concept profiles* as developed in [8] where a measure of association between a pair of genes is calculated based not only on the co-occurrence of entities in the same document, but also on indirect relations, where genes are linked via a number of documents. An association matrix for gene-pairs can be generated, where each entry represents the strength of the relationship between genes, based on a database of scientific literature. Business Intelligence is another area where text mining has proved popular in relation to tweet messages and sentiment analysis [5]. In ecology, the use of text mining is a little less explored though there is a growing interest in the use of these approaches to extract knowledge [14].

There is a growing effort to taking a predictive approach to ecology [3] with the availability of larger and more diverse datasets. If we can build models that can predict biodiversity or species distribution, for example, then we will have greater confidence that the models capture important underlying characteristics. A related discipline, 'systems ecology', encourages a focus on holistic models of ecosystems [10]. This follows the success of similar approaches in molecular biological applications Many novel techniques developed from bioinformatics can be translated to the ecological domain [15]. Indeed, here we make use of a statistic that was previously developed for validating clusters in microarray data.

In this paper, we explore the use of text-mining where we exploit the fact that the flora that we analyse are marked-up to distinguish between descriptions of different plant traits and habitats. We cluster habitat texts and use a statistic (originally designed to validate clusters of genes from microarray experiments) to validate the discovered habitat-types against the plant taxonomy. We then exploit the trait texts to build probabilistic classifiers [6] for predicting the habitat clusters. The motivation for this research is to permit exploration of taxonomic and functional trait diversity. This will lead to better plant functional type classifications for input to vegetation models under differing climate change scenarios. From this, we can gain a better understanding of plant species distribution, vital for effective species and habitat conservation. In the next section we describe the general pipeline that we have developed to build these predictive models from the marked up text and plant taxonomy. We also describe the probabilistic models and statistics that we use to assess our results. In the results section we document the results from the different stages of the pipeline with insights from plant ecology before concluding.

2   Methods

2.1   Data

The Flora of Tropical East Africa (FTEA) is one of the largest regional tropical Floras ever completed, covering 12,500 wild plant species from Uganda, Kenya and Tanzania. Together with Flora Zambesiaca and Floras Somalia, these floras cover equatorial, tropical and subtropical biomes of [16] and major phytochoria of [17]. Virtually all the main vegetation types are represented. These floras have been digitised to create the EFLORAS database - a unique data source

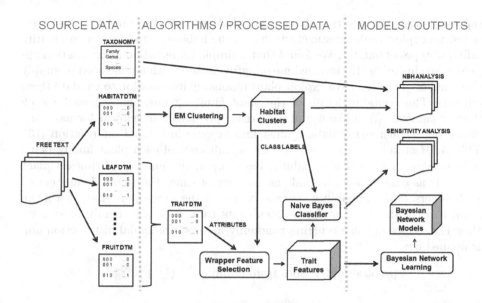

Fig. 1. Pipeline for Converting Free Marked-Up Flora Text into Predictive Models of Ecosystems

of tropical plant species distribution, ecology and morphology, together with historical data on plant collectors [9] in the EFLORAS corpus. Each document represents a *taxon* (in this case a species) which is identified by a unique ID and contains a digitised paragraph, tagged as to whether it describes a number of different characteristics: *habitat, habit, leaf, fruit, or seed* (flower features were not available for this study). In total there are 8252 documents (i.e. species), containing each paragraph. Standard text mining procedures were employed for each paragraph type in order to remove stop-words, white spaces, punctuation and numbers, and to stem all necessary words [4]. This results in an n (terms) by m (documents) matrix for each paragraph, where cells contain the number of times a term has appeared in the corresponding document. Terms can include anything such as 'bilobate', 'golden' and 'elongated' reflecting traits but also other terms such as 'beautifully' and 'actually' reflecting a particular author's writing style.

2.2   Experiments

We exploit the text concerning plant traits to predict habitat. Therefore, the matrices for all types of trait are combined into a single document term matrix, the *trait matrix*, for all types except habitat. Clearly, the combined trait matrix and the *habitat matrix* are sparse and any terms that appear in less than 10% of all documents are removed from both. This reduces the size of the trait matrix to 759 terms, and the habitat matrix to 106. We use the *tm* package in R for all of this processing [4]. Having processed the text into two matrices: one that

represents the plant traits and the other that represents the habitat characteristics, we exploit probabilistic clustering to the habitat data in order to identify
different types of habitat. We found that a simple Expectation Maximisation approach to clustering [1] identified meaningful clusters without the need to supply
the number of clusters. We exploit plant taxonomy information to validate these
clusters. This contains details of the plant family, genus, and species for each
flora document. We make use of a statistic previously developed for assessing
clustering in microarray data against known gene functional information [13].
This *NBH* statistic is used to score the significance of each plant family being
associated with a particular habitat based upon the number of times a plant
family is associated with it, and the number of times the family is associated
with others. This probability score is based on the hypothesis that, if a given
habitat, i, of size s_i, contains x documents from a defined family of size k_j,
then the chance of this occurring randomly follows a binomial distribution and
is defined by:

$$pr(\text{observing } x \text{ docs from family } j) = (\tfrac{k_j}{x})p^x q^{k_j - x}$$

$$\text{where } p = s_i/n,$$
$$q = 1 - p$$

As in [13] we use the normal approximation to the binomial to calculate the
probability where:

$$z = (x - \mu)/\sigma,$$
$$\mu = k_j p,$$
$$\sigma = k_j pq$$

This cluster probability score is used to identify statistically significant families
allocated to each habitat (at the 1% level).

The cluster labels identified through the clustering are then used to identify
predictive features in the trait matrix using a wrapper feature selection approach
[7] to explore combinations of predictive terms. We use the Naïve Bayes Classifier
[6] as the classifier for the wrapper as this was found to be the most predictive.
Whilst we expect there to be interesting interactions between terms, it appears
that the simplicity of the Naïve Bayes is suitable to classifying a large number
of habitats by minimising parameters. What is more, the flexibility of Bayesian
classifiers allow us to use different nodes as predictors so we can use the resultant
models to predict both neighbouring plant traits as well as habitat type.

The Naïve Bayes classifier makes the simplifying assumption that each feature
is independent of each other given the class. This corresponds to the efficient
factorization

$$p(x|c) = \prod_{i=1}^{n} p(x_i|c)$$

Assuming uniform priors, a Bayesian estimate of $p(x_i|c)$ is given by

$$\hat{p}(x_{ip}|c) = \frac{1 + n(x_{ip}|c)}{s + n(c)}$$

where s is the number of discretized states of the gene variable X_i, $n(x_{ip}|c)$ is the number of cases in the dataset where X_i takes on its pth unique state within the samples from class c, and $n(c) = \sum_{p=1}^{s} n(x_{ip}|c)$ is the total number of samples from class c. From $\hat{p}(x|c)$, an estimate of $p(c|x)$ is calculated using Bayes rule and the resulting classification rule assigns the sample x to the class associated to the highest estimated probability.

Having identified the relevant features to predict habitat type, we explore how predictive these features are using a Naïve Bayes classifier under a 10-fold cross-validation regime. Finally, we explore the interactions between features within each habitat type by carrying out 'what if' experiments on a sample of habitats and build Bayesian network structures from data associated with each habitat type to see if any traits / network-of-traits are highlighted for that habitat in particular. The general pipeline is illustrated in Figure 1.

3 Results

3.1 Discovering Habitat Clusters

Having clustered the data into 9 different habitats based upon the document term matrices generated from the habitat corpus, the individual term frequencies were calculated and explored in the context of habitats that they likely represent. The following descriptions could be elicited from experts based upon the terms associated with each habitat cluster.**Habitat 0** - appears to reflect vegetation in wet places that are largely, but not exclusively, upland (For the remainder of the paper we refer to this habitat type as WETLANDS - *WET* when abbreviated). **Habitat 1** reflects a mixture of woody and herbaceous vegetation in drier conditions, including deciduous types (DECIDUOUS BUSHLAND - *BUSHLAND*). **Habitat 2** clearly reflects lowland and upland, wetter forest types (RAINFOREST). **Habitat 3** contains a variety of upland vegetation (MONTANE). **Habitat 4** appears to represent disturbed vegetation and cultivation (DISTURBED) **Habitat 5** contains vegetation in open sites, and margins including cultivation, similar to 4 and not readily separable (OPEN/DISTURBED - *OPEN*). **Habitat 6** is large and contains a combination of open woody and herbaceous vegetation in wetter areas, including evergreen types (WOODLAND + WOODED GRASSLAND - *WOODED*). **Habitat 7** is a mixture of drier lowland forest, scrub and evergreen bush (FOREST + SCRUB + BUSH - *SCRUB*). **Habitat 8** contains mixed habitats including rainforest and dry vegetation (FOREST + SCRUB + BUSH - *SCRUB2*).

The distribution of documents to habitat varied dramatically. In general, the three larger clusters (habitats 1, 6 and 8) were less specific and generally mixed different habitat-types. For this reason, these were omitted from the feature selection and classification analysis though further work will involve exploring finer grain clusters to split these into more detail. The identified habitats were validated by using the *NBH* statistic [13]. The distribution of all plant families occurring in the texts over each specific habitat were explored. Families with an

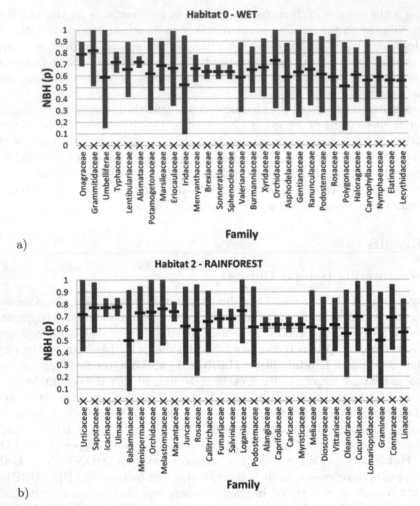

a)

b)

Fig. 2. Significant Plant Families for 2 Selected Habitats Using the NBH Statistic (denoted with an '×') Compared to the Distribution over all Other Habitats (Denoted by Error Bars)

NBH statistic with p values at less than the 1% level were selected and compared to the distribution over the other habitats in order to highlight the specific association between that family and the discovered habitat (shown in Figure 2). These results highlighted some expected families of plants based upon their habitat types: Habitat 0 (WET) is clearly dominated by families of aquatic and marshplants (figure 2a). Habitat 1 (BUSHLAND) contains Burseraceae, Leguminosae, Capparidaceae which are dominant dry bushland components. Portulacaceae is also characteristic of this type of habitat. Habitat 2 (RAINFOREST) contains herbs and understory shrub/treelet families are well represented (including ferns), followed by tree families (figure 2b). Habitat 3 (MONTANE) contains montane ferns. Habitat 4 (DISTURBED) are mostly families with weedy species which make sense for disturbed regions (figure 2c). Habitat 5 (OPEN) includes a mixture of herbaceous and woody families. Habitat 6 (WOODLAND) families are a mixture and this is not surprsing considering the large mixed habitats that were identified earlier. It is intriguing as to why the mistletoe families are so prominent (Loranthaceae, Viscaceae, Santalaceae). Habitat 7 (SCRUB) families are scrub component families and Habitat 8 (SCRUB2) fits with forest herbs shrubs and trees. For all identified families the p-value compared to the distributions of other families and habitats illustrate that they are well separated and significant.

3.2   Plant Trait Feature Selection and Classification

We now turn to the plant trait documents. We wish to use these to predict habitat type. A wrapper feature selection procedure was carried out on the plant traits to identify combinations of traits that characterise the different habitat clusters. A greedy search scored with classification accuracy was used to identify the features. Figure 3 illustrates the identified features and how the expected frequencies of these terms vary for each habitat type. For example, the term *FRUIT_exsert* representing the term 'exsert' in the text describing 'fruit' is identified as relevant and, as can be seen here, has a much higher expected frequency in scrub habitats compared to others. Features marked with an asterisk '*' were those that were expected to be good at discriminating between the habitats.

The results of applying 10-fold cross-validation to predicting the habitat type with Naïve Bayes is shown in Table 1. The predictive accuracy varied depending on the habitat with Habitat 7 (SCRUB) being the most accurately predicted. The table shows the distribution of Areas Under the ROC curves for each habitat. The confusion matrix indicated the typical misclassifications involved mistakenly classifying Habitats 0 (WET) and 2 (RAINFOREST), and 4 (DISTURBED)

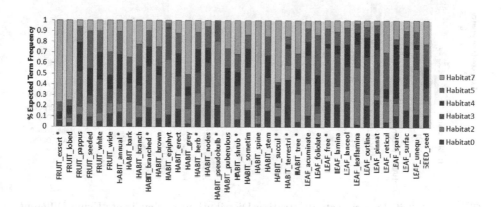

Fig. 3. Identified Features (using Naïve Bayes Wrapper) - Expected Frequencies for each Habitat

Table 1. Classification Results (Confusion Matrix and AUC) of Habitats Given Features as Percentages (10 fold Naïve Bayes Classifier)

-	H0	H2	H3	H4	H5	H7	AUC
Habitat0:	0.06	0.05	0.01	0.03	0.01	0.01	0.70
Habitat2:	0.05	0.15	0.01	0.03	0.01	0.03	0.71
Habitat3:	0.02	0.02	0.02	0.02	0.00	0.01	0.69
Habitat4:	0.03	0.03	0.01	0.08	0.02	0.01	0.70
Habitat5:	0.02	0.03	0.01	0.05	0.02	0.02	0.64
Habitat7:	0.02	0.03	0.00	0.03	0.01	0.07	0.75
Wtd Avg.:	-	-	-	-	-	-	0.70

and 5 (OPEN) which makes sense as vegetation could easily overlap between these.

3.3 'What if?' Experiments

Having identified both habitat type and plant traits relevant to predicting habitat type, we explore the interaction discovered between the different features. This allows us to explore combinations of terms as well as their relationship to different habitats.

Figure 4 illustrates the expected frequencies as inferred from the predictive model for some selected plant traits. The three bars in Figure 4a represent expected frequencies for the other traits when HABIT_annual is set to 0, 1 and 2 respectively. Terms in brackets illustrate the most probable habitat given the observation.

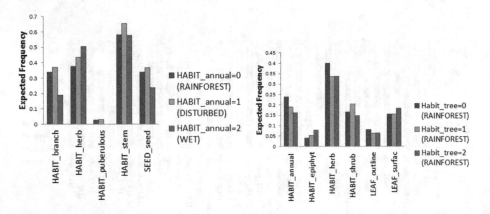

Fig. 4. 'What if' Experiments Illustrating Distributions of Key Traits Using Different Observations on *HABITAT_annual* (left) and *HABITAT_tree* (right)

For the scenario in Figure 4a where *HABIT_annual*=0 is observed, the most likely habitat is RAINFOREST, and the highest expected values are for *HABIT_herb*, *HABIT_stem* and *SEED_seed*. This is somewhat counterintuitive as the features *herb*, *stem* and *seed* are associated with annuals, but here annual has a frequency of zero. For *HABIT_annual*=1, the most likely habitat is DISTURBED, with highest conditional expected frequencies for *HABIT_stem*, *HABIT_herb* and *HABIT_branch*. For *HABIT_annual*=2, the most likely habitat WET, with highest conditional expected frequencies for *HABIT_stem*, *HABIT_herb* and *HABIT_branch*, while *HABIT_puberulous* is 0. This scenario makes sense: As the frequency of *HABIT_annual* is increased, so too are the probabilities of observing the 'annual related' features (stem, herb, branch etc.). There are comparatively few annuals in rainforests but as expected they are a major element of disturbed habitats. In addition, very high numbers of annuals appear to be associated with the wet habitat and these plants apparently are never puberulous (shortly hairy). Aquatic plants are frequently glabrous, that is, without hairs.

Unlike *HABIT_annual*, if we observe *HABIT_tree* as either 0,1 or 2 (see Figure 4b), the most likely habitat is always RAINFOREST. This could be because the habitat is more species diverse. However, the intermediate scenario *HABIT_tree*=1 gives the highest probability for habitat 2. This is because RAINFOREST is a rich habitat which contains both tree and non-tree species. It could be that *HABIT_tree*=2 precludes non-tree species typical of RAINFOREST.

3.4   Networks of Traits

For the final piece of analysis, we explored learning network structures for different habitat-types by splitting the data accordingly and learning networks using the K2 algorithm of [2]. Some sample networks are documented in Figure 5 (detail). Some interesting characteristics emerge when focussing on the 'hub' nodes - those that have higher degree of connectivity. For example, in Habitat 0 (WETLANDS) there are two clear hubs: *HABIT_shrub* and *LEAF_free*. The former links to features of woody plants (expected to be mostly absent from typical habitat 0 plants). The latter contains many aspects of leaf descriptions (lamina, outline, lanceolate, pinnate, surface) but there are some connections which are not immediately clear (connections from *HABIT_terrestrial*, *FRUIT_wide* and *FRUIT_pappus*). The term *LEAF_free* may cover several different situations eg. free stipules, free petiole (all parts of the leaf). In Habitat 2 (RAINFOREST) there is a *HABIT_epiphyte* hub (containing terms *epiphyte* and *pseudobulb*) which could be linked to orchids (an epiphyte is a plant that grows on trees such as orchids). Also *epiphyte* and *tree* are linked which could be related to plants specifically growing on trees. In general, many of the hubs make sense in terms of why they may be connected (often descriptive terms that are related to similar parts of a plant). There are also some interesting relationships that appear to be specific to their habitat such as orchids in rainforests.

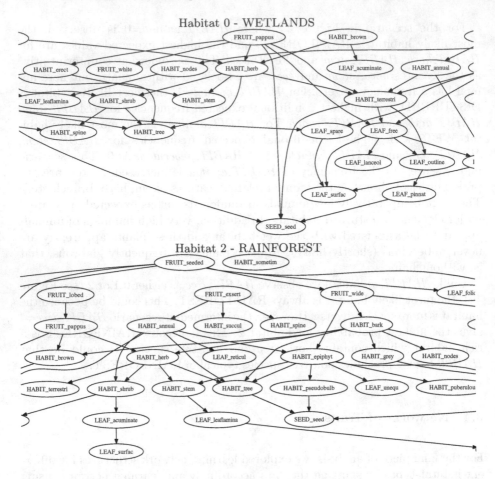

Fig. 5. Portion of Networks Learnt for 2 Sample Habitats with a Focus on Hub Nodes

4   Conclusions

In this paper we have explored a pipeline for converting text documents at the Royal Botanical gardens at Kew, London describing different plant families into models that can predict habitat type and neighbouring plant characteristics, based upon plant traits. The pipeline identifies distinct habitat types and integrates taxonomy data in order to highlight significant plant families within those habitats by exploiting a statistic previously developed for bioinformatics applications. A combination of wrapper feature selection and naive Bayes classification is exploited to identify the discriminative features and build models that can predict both neighbouring plant traits and habitat type. Future work, will involve exploring other predictive capabilities between the text and other data such as the taxonomy. For example, we will explore how well our models can predict families and species directly rather than via the habitat. We will also explore other ways to quantify the value of the 'what if' results.

The paper documents the start of a larger project that explores the hypothesis that a comprehensive understanding of neighbouring species and what a plant looks like will indicate where it grows. Our tools will enable predictions about individual species and their functions in ecosystems of other regions. This will be facilitated through identifying factors (including taxonomic and environmental) that influence biodiversity and stability of ecosystems, vital for effective species and habitat conservation.

References

1. Bilmes, J.: A gentle tutorial on the em algorithm and its application to parameter estimation for gaussian mixture and hidden markov models. Technical Report TR-97-021, ICSI (1997)
2. Cooper, G.F., Herskovitz, E.: A Bayesian method for the induction of probabilistic networks from data. Machine Learning (9), 309–347 (1992)
3. Evans, M.R., Norris, K.J., Benton, T.G.: Introduction: Predictive ecology: systems approaches. Philosophical Transactions of the Royal Society: Part B 367(1586), 163–169 (2012)
4. Feinerer, I., Hornik, K., Meyer, D.: Text mining infrastructure in R. Journal of Statistical Software 25(5), 1–54 (2008)
5. Feldman, R.: Techniques and applications for sentiment analysis. Communications of the ACM 56(4), 82–89 (2013)
6. Friedman, N., Geiger, D., Goldszmidt, M.: Bayesian network classifiers. Machine Learning (29), 131–163 (1997)
7. Inza, I., Larrañaga, P., Blanco, R., Cerrolaza, A.J.: Filter versus wrapper gene selection approaches in dna microarray domains. Artificial Intelligence in Medicine (31), 91–103 (2004)
8. Jelier, R., Schuemie, M.J., Veldhoven, A., Dorssers, L.C.J., Jenster, G., Kors, J.A.: Anni 2.0: A multipurpose text-mining tool for the life sciences. Genome Biology 9(6), R96 (2008)
9. Kirkup, D., Malcolm, P., Christian, G., Paton, A.: Towards a digital african flora. Taxon 54(2) (2005)
10. Purves, D., Scharlemann, J., Harfoot, M., Newbold, T., Tittensor, D.P., Hutton, J., Emmott, S.: Ecosystems: Time to model all life on earth. Nature (493), 295–297 (2013)
11. Steele, E., Tucker, A., Schuemie, M.J.: Literature-based priors for gene regulatory networks. Bioinformatics 25(14), 1768–1774 (2009)
12. Swanson, D.R.: Medical literature as a potential source of new knowledge. Bull. Med. Libr. Assoc. (78), 29–37 (1990)
13. Swift, S., Tucker, A., Vinciotti, V., Martin, N., Orengo, C., Liu, X., Kellam, P.: Consensus clustering and functional interpretation of gene-expression data. Genome Biology 5(11), R94 (2004)
14. Tamames, J., de Lorenzo, V.: Envmine: A text-mining system for the automatic extraction of contextual information. BMC Bioinformatics 11(294) (2010), doi:10.1186/1471-2105-11-294

15. Tucker, A., Duplisea, D.: Bioinformatics tools in predictive ecology: Applications to fisheries. Philosophical Transactions of the Royal Society: Part B 356(1586), 279–290 (2012)
16. Walter, H.: Vegetation of the Earth and Ecological Systems of the Geo-biosphere. Springer (1979)
17. White, F.: The Vegetation of Africa – A descriptive memoir to accompany the Unesco/AETFAT/UNSO vegetation map of Africa. UNESCO (1983)

Detecting Localised Anomalous Behaviour
in a Computer Network

Melissa Turcotte[1,2], Nicholas Heard[3], and Joshua Neil[1]

[1] ACS-PO, Los Alamos National Laboratory, Los Alamos, USA
[2] CNLS, Los Alamos National Laboratory, Los Alamos, USA
[3] Imperial College London and Heilbronn Institute, University of Bristol, UK

Abstract. Temporal monitoring of computer network data for statistical anomalies provides a means for detecting malicious intruders. The high volumes of traffic typically flowing through these networks can make detecting important changes in structure extremely challenging. In this article, agile algorithms which readily scale to large networks are provided, assuming conditionally independent node and edge-based statistical models. As a first stage, changes in the data streams arising from edges (pairs of hosts) in the network are detected. A second stage analysis combines any anomalous edges to identify more general anomalous substructures in the network. The method is demonstrated on the entire internal computer network of Los Alamos National Laboratory, comprising approximately 50,000 hosts, using a data set which contains a real, sophisticated cyber attack. This attack is quickly identified from amongst the huge volume of data being processed.

1 Introduction

Detection of intruders within enterprise computer networks is a challenging and important problem in cyber security. Perimeter security systems are meant to prevent intruders from gaining access, but these systems are notoriously permeable and it is inevitable that some intruders will succeed in penetrating the perimeter. Once an attacker has gained access to a host inside the perimeter, they will typically traverse [7] from host to host, either to collect valuable data to exfiltrate, to establish a persistent presence on the network, or to escalate access; this traversal can be achieved, for example, through pass-the-hash techniques [3]. It is this type of anomalous traversal pattern within the network, as the intruder attempts to compromise multiple hosts, that this work aims to identify.

There are two broad approaches to intrusion detection: the first searches for known patterns or structures observed in previous attacks, commonly known as signature-based detection; the second, anomaly-based detection, screens for deviations from a model of the normal state of the system. The anomaly-based approach has a clear advantage in that new types of attacks that have not previously been observed can still potentially be identified; [9] and [6] provide an overview of work in this area related to network intrusion.

H. Blockeel et al. (Eds.): IDA 2014, LNCS 8819, pp. 321–332, 2014.
© Springer International Publishing Switzerland 2014

Many of the existing methods for detecting anomalous substructures in a computer network examine heuristics calculated over the whole graph, [8]. In [10], scan statistics are used for analysing the email communications of users. In [11], graph-structured hypothesis testing is used to find anomalous structures in a network based on anomaly scores for nodes. The approach proposed here is to model the time series of connections between each pair of nodes (hosts). For tractable inference, conditionally independent discrete time models are used to model connection frequencies (*cf.* [2, 5, 7]). Several challenges exist when taking an anomaly-based approach to intrusion detection: To enable deployment on large computer networks in real time, methods must be computationally fast and scalable. Additionally, interactions between humans and technology are complex and it is imperative that models are able to capture a wide range of features such as diurnal, periodic, and bursty behaviour.

Following anomalous edge detection, a second stage analysis is performed to identify network traversal behaviour. Temporally connected components of anomalous edges provide anomalous subgraphs which could indicate an intruder moving laterally through the network. This aggregation of anomalous edges over a sliding time window is similar to the methodology of [7], which looks for anomalous edges that form a path within the network to detect traversal of an intruder. A path is defined as a sequence of edges where the destination node of one edge is the source node for the next edge. In the present work, rather than looking only for path-like traversal through the network initiating from a single infected node, the proposed method aims to detect more abstract structures.

The analysis of a computer network data set from Los Alamos National Laboratory (LANL) forms the basis of this article; these data are described in Sect. 2. A hierarchical Bayesian model for first the node and then the edge behaviour is proposed in Sect. 3. A scheme for anomaly detection in which subgraphs are formed from potentially anomalous edges is detailed in Sect. 4. Finally in Sect. 5, the proposed method is demonstrated on the LANL data set, and a real cyber attack is successfully identified.

2   Data

Los Alamos National Laboratory (LANL) owns a large computer network, holding extremely sensitive data. The network faces regular cyber attacks, some of which penetrate the network boundary, and so fast anomaly detection methods are imperative. The data set analysed in this article consists of NetFlow records [12] of the connections made between individual computers (hosts). Each record marks a connection event from a source IP address to a destination IP address. Following the approach of [7], the data are aggregated into discrete time series of counts of the connections made between pairs of computers over consecutive ten minute intervals, one discrete time series for each pair.

The data comprise an initial four weeks of NetFlow records when there were no known compromised hosts on the network; these will be referred to as the training data. Further, there are NetFlow data for ten days during which a

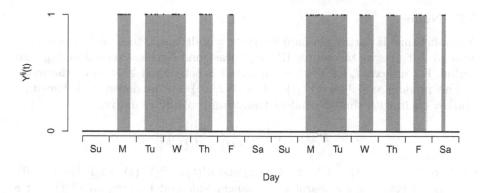

Fig. 1. Activity status of an edge in the LANL computer network over a fortnight. The areas shaded are the intervals where the source node was active.

sophisticated, persistent threat attack occurred, referred to as the test data. Forensic analysts believe that the attack started early on in the test data, and six hosts were known to be infected. Note that in such cases there is no absolute ground truth, and further hosts may have been infected but never discovered. Within the training and test data there are a total of 57,459 unique hosts (nodes) observed and on average there are approximately 10,000 active nodes and 40,000 observed edge events in each ten minute interval. Although any source computer can connect to nearly any other computer in the network, analysis is restricted to communications between nodes that were observed in both the training and the test data as the types of anomalies sought require the intruder to make connections between computers in order to traverse the network.

A plot of edge activity for a typical host pair is shown in Fig. 1 and clearly demonstrates the bursty nature of the data. Many edges have very sparse activity, with 80% of the edges active less than 10% of the time. Another feature of the data are volatile connection counts; it is common for edges to be quiet for long periods of time, followed by bursts of very high connection counts.

3   Hierarchical Markov Model

A piecewise Markov chain model is used to capture the seasonal behaviour of the source node. Note that seasonality could in practice be modelled at an edge level rather than a node level and all of the models described below easily extend to this case. However, if plausible it is preferable to borrow strength across edges from the same source node by learning seasonality at the node level; this is particularly true when many of the edges have very sparse data. Furthermore, there may be some computer networks where it would be appropriate to assume that seasonality is shared across all nodes in the network, or perhaps across pools of hosts conducting similar activity.

3.1 Node

Assuming time is partitioned into intervals, a node is said to be *active* in a given interval if it acts as the source IP for at least one NetFlow record during that period. For interval t, let $Y^i(t) = 1$ if node i is active and $Y^i(t) = 0$ otherwise.

The random variables $\{Y^i(t) : t = 1, 2, \ldots\}$ are modelled as a two-state Markov chain with time-dependent transition probability matrix

$$P^i(t) = \begin{pmatrix} 1 - \psi^i(t) & \psi^i(t) \\ 1 - \phi^i(t) & \phi^i(t) \end{pmatrix},$$

where $\phi^i(t) = \mathbb{P}(Y^i(t) = 1 | Y^i(t-1) = 1)$ and $\psi^i(t) = \mathbb{P}(Y^i(t) = 1 | Y^i(t-1) = 0)$.

Seasonal variability is learnt as a changepoint model acting on $P^i(t)$ over a finite seasonal period of length S. A vector of ℓ ordered seasonal changepoints $s_{1:\ell}$ take values in $\{1, 2, \ldots, S - 1\}$. Notationally let $s_0 = 0$ and $s_{\ell+1} = S$.

Within each seasonal segment the changepoint model assumes a fixed but unknown parameter pair (ψ_j, ϕ_j) for $j = 0, \ldots, \ell$, implying a piecewise-homogeneous Markov chain overall. Note that the dependency on node i is suppressed to simplify notation.

The presence or absence of changepoints at each position are assumed to be independent Bernoulli(ν) trials. Conjugate priors for the transition probabilities, $\phi_j, \psi_j \stackrel{iid}{\sim} \text{Beta}(\alpha, \beta)$, allow the unknown transition probability matrices for the node to be integrated out.

For simplicity denote $Y_T = (Y^i(0), \ldots, Y^i(T))$. Reversible jump Markov chain Monte Carlo sampling [1] of the unknown number of unknown seasonal changepoints is used to make inference from the posterior density at time T,

$$\pi(s_{1:\ell}, \ell | Y_T) \propto \nu^\ell (1 - \nu)^{S-\ell-1} \prod_{j=0}^{\ell} \prod_{i=0}^{1} \frac{\text{B}(\alpha + n_{i1}^j, \beta + n_{i0}^j)}{\text{B}(\alpha, \beta)}$$

where $\text{B}(\alpha, \beta)$ is the Beta function and (n_{i0}^j, n_{i1}^j) are respectively the number of transitions from state i to states 0 and 1 observed after time T in the jth seasonal period defined by $\{s_j, \ldots, s_{j+1} - 1\}$.

The maximum a posteriori (MAP) sample obtained will be used as an estimate for the seasonal changepoints of each node. The seasonal changepoints can initially be learnt on a batch of training data, and then updated periodically.

Within each seasonal period between the MAP changepoints, the Markov chain probabilities have a known Beta posterior distribution at time T,

$$[\psi_j | Y_T] \equiv \text{Beta}(\alpha + n_{01}^j, \beta + n_{00}^j), \quad [\phi_j | Y_T] \equiv \text{Beta}(\alpha + n_{11}^j, \beta + n_{10}^j). \quad (1)$$

When performing inference sequentially, these distributions can be updated over time as each new data point is observed.

3.2 Edge

Activity Status. Conditional on the source node being active in an interval, the activity status for an edge is assumed to be independent of the time of day.

Even when restricting attention to intervals when the source node is active, connections along an edge are still bursty, as illustrated in Fig. 1. To capture the bursty behaviour, an additional two-state Markov chain for the activity status of the edge is applied to those time points where the source node is active.

Let $N^{ij}(t)$ be the random variable denoting the number of connections from node i to node j at the tth time point and let $Y^{ij}(t) \in \{0,1\}$ be the indicator variable for whether the edge is active at t: $Y^{ij}(t) = 1 \iff N^{ij}(t) > 0$. A two-state Markov chain model for the activity status of each edge has transition probability matrix

$$P^{ij} = \begin{pmatrix} 1 - \psi^{ij} & \psi^{ij} \\ 1 - \phi^{ij} & \phi^{ij} \end{pmatrix}.$$

Conjugate beta distribution priors, $\phi^{ij}, \psi^{ij} \overset{\text{iid}}{\sim} \text{Beta}(\bar{\alpha}, \bar{\beta})$, allow the unknown transition probability parameters to be integrated out. The Beta posterior distributions for the unknown parameters ψ^{ij} and ϕ^{ij} at time T are then

$$\left[\psi^{ij}|Y_T^{ij}\right] \equiv \text{Beta}(\bar{\alpha} + n_{01}^{ij}, \bar{\beta} + n_{00}^{ij}), \quad \left[\phi^{ij}|Y_T^{ij}\right] \equiv \text{Beta}(\bar{\alpha} + n_{11}^{ij}, \bar{\beta} + n_{10}^{ij}). \quad (2)$$

As with the node model, these posterior distributions act as a prior when preforming inference sequentially and can be updated with each new data point.

Negative Binomial Distribution for Counts. For those periods in which an edge (i,j) is active, the communication counts $\{N^{ij}(t)\}$ will be considered to be independent realisations from a fixed probability model. [5] proposes using a negative binomial distribution to capture the over-dispersion apparent in NetFlow connection counts. If $Y^{ij}(t) = 1$, then it is assumed

$$N^{ij}(t) - 1 \sim \text{NB}(r^{ij}, \theta^{ij}). \quad (3)$$

If $n^{ij} = \sum_{u=0}^{t} N^{ij}(u)$ and $a^{ij} = \sum_{u=0}^{t} Y^{ij}(u)$, then the conjugate prior $\theta^{ij} \sim \text{Beta}(\tilde{\alpha}, \tilde{\beta})$ implies a conditional posterior distribution for θ^{ij} given r^{ij},

$$\left[\theta^{ij}|n^{ij}, a^{ij}, r^{ij}\right] \equiv \text{Beta}(\tilde{\alpha} + n^{ij} - a^{ij}, \tilde{\beta} + r^{ij}a^{ij}). \quad (4)$$

An exponential prior is chosen for r^{ij}. Full Bayesian inference would require marginalising over r^{ij} using numerical integration or Monte Carlo simulation. For efficiency, here r^{ij} is determined by maximising the joint posterior of r^{ij} and θ^{ij} using the counts obtained from training data, via a root finding algorithm.

4   Monitoring

For a past window of length w, at time t let $G_t = (V_t, E_t)$ be the graph consisting of all the communicating nodes, V_t, and directed edges, E_t, active during the time window $\{t - w + 1, \ldots, t\}$. For each edge $(i,j) \in E_t$, a predictive p-value, p_t^{ij}, is obtained from the counts observed over $\{t - w + 1, \ldots, t\}$, signifying how far the edge has deviated from its usual behaviour in the last w intervals; this is explained in Sect. 4.1. An anomaly subgraph of the network is then formed from all edges that have a p-value below a threshold and further inference is conducted on the anomaly subgraph; this is discussed in Sect. 4.2. This latter procedure can be considered as a second stage of analysis, similar to that advocated in [2].

The window length w can be chosen to suit the concerns of the analyst, but realistically should be small relative to the history of the graph. The use of sliding windows to monitor network graphs over time is detailed in [7].

4.1 Predictive Distributions

At each time interval we would like to test whether the observed communication counts along edges $\{N^{ij}(t)\}$ are typical draws from their respective probability distributions, or if some of those relationships may have changed. Defining the random variable $N^{ij}(t, w) = \sum_{u=0}^{w-1} N^{ij}(t - u)$, surprise can be measured by the predictive distribution p-value $p_t^{ij} = \mathbb{P}\left(N^{ij}(t, w) \geq k_{t,w}\right)$, where $k_{t,w}$ is the sum of the observed communication counts from i to j in the previous w intervals. Note that interest is focused on one-sided, upper tail p-values: when intruders move around the network they should increase communication counts between edges.

Let $Y^i(t, w) = \sum_{u=0}^{w-1} Y^i(t - u) \in \{0, \ldots, w\}$ be the number of intervals in which source node i was active in the window looking back from time t. The marginal probability mass function $\mathbb{P}(Y^i(t, w) = y^i)$ can be obtained by summing over the Markov chain probabilities of all possible activity status w-tuple combinations. For example if $Y^i(t - 3) = 0$ then

$$\mathbb{P}(\bar{Y}^i(t, 3) = 2) = (1 - \psi_j^i)\psi_j^i\phi_j^i + {\psi_j^i}^2(1 - \phi_j^i) + \psi_j^i\phi_j^i(1 - \phi_j^i),$$

given t is in the jth seasonal period.

The unknown transition probabilities are integrated out using the posterior distributions (2) given the data observed so far. Note that when integrating out $\phi^i(t)$ and $\psi^i(t)$, the expression will depend on which seasonal interval the window of time falls in; furthermore, there will be times when the window $\{t - w, +1 \ldots, t\}$ may cross over one or more seasonal intervals.

Suppose node i is active for y^i intervals. Let $Y^{ij}(t, w) \in \{0, \cdots, y^i\}$ be the number of intervals in which edge (i, j) is active. The marginal probability mass function $[Y^{ij}(t, w)|Y^i(t, w) = y^i]$ is also obtained by summing the corresponding Markov chain probabilities for that edge and integrating out the parameters. For $k_{t,w} = 0$, $p_t^{ij} = 1$. For $k_{t,w} > 0$ the p-value is calculated as

$$p_t^{ij} = \sum_{y^i=1}^{w} \sum_{y^{ij}=1}^{y^i} \left[\mathbb{P}(N^{ij}(t, w) \geq k_{t,w}|Y^{ij}(t, w) = y^{ij}) \right. \tag{5}$$
$$\left. \mathbb{P}(Y^{ij}(t, w) = y^{ij}|Y^i(t, w) = y^i)\mathbb{P}(Y^i(t, w) = y^i) \right].$$

If $k_{t,w} \leq y^{ij}$ then trivially $\mathbb{P}(N^{ij}(t, w) \geq k_{t,w}|Y^{ij}(t, w) = y^{ij}) = 1$. Otherwise let $N_-^{ij}(t, w) = N^{ij}(t, w) - Y^{ij}(t, w)$ and $k'_{t,w} = k_{t,w} - y^{ij}$, where $k'_{t,w} > 0$. Then as $N_-^{ij}(t, w)$ is the sum of y^{ij} independent negative binomial variables, (3),

$$\left[N_-^{ij}(t,w)|Y^{ij}(t,w) = y^{ij}, r^{ij}, \theta^{ij}\right] \equiv \text{NB}(r^{ij}y^{ij}, \theta^{ij}).$$

The distribution of the unknown parameter θ^{ij} is given by (4). This can be integrated out, implying a beta negative binomial distribution for $N_-^{ij}(t,w)$,

$$\left[N_-^{ij}(t,w)|Y^{ij}(t,w) = y^{ij}, r^{ij}\right] \equiv \text{BNB}(y^{ij}r^{ij}, \tilde{\alpha} + n^{ij} - a^{ij}, \tilde{\beta} + r^{ij}a^{ij}).$$

The predictive p-values therefore have closed form solutions, which is critical for fast inference on large scale networks.

4.2 Anomaly Graphs

Classifying anomalies for single edge p-values that fall below a threshold could result in many false alarms. Setting a threshold extremely low to limit the number of false alarms obtained could result in true anomalous events being missed. As intruders traverse the network they are introducing anomalous signal on multiple, connected edges. The idea of an anomaly graph is to combine p-values to identify security-relevant anomalous substructures in the network.

For a p-value threshold $T_p \in (0,1)$, an anomaly subgraph of the network, $S_t = (V_t^S, E_t^S)$, is formed from edges that have a p-value below the threshold,

$$E_t^S = \{(i,j) \in E_t | p_t^{ij} < T_p\},$$
$$V_t^S = \{i \in V_t | \exists j \neq i \in V_t \text{ s.t. } (i,j) \in E_t^S \text{ or } (j,i) \in E_t^S\}.$$

In practice, the threshold T_p can be chosen so that the number of nodes in the anomaly graph $\{|V_t^S|\}$ does not exceed a desired level.

A graph is said to be connected if every vertex is reachable from every other vertex, [4]. A weakly connected component of a graph is a maximally connected subgraph with the property that if all the directed edges were replaced with undirected edges, the resulting subgraph would be connected. Each of the weakly connected components of S_t can be considered as a potentially anomalous attack indicating lateral movement of an attacker in a network. Let $A_{k,t} = (V_{k,t}, E_{k,t})$ denote the kth weakly connected component. Under the null hypothesis of normal behaviour the set of p-values $p_{A_{k,t}} = \{p_t^{ij} | (i,j) \in E_{k,t}\}$ obtained from the edges of a weakly connected component $A_{k,t}$ are independent and approximately uniformly distributed on $(0, T_p)$. Note that they are only approximately uniformly distributed due to the discreteness of the counts.

The p-values can be combined to give an overall anomaly score for each connected component. Fisher's method is commonly used to combine p-values obtained from independent tests into a single test statistic. Using this method, for each connected component a test statistic is obtained,

$$X_{k,t}^2 = -2 \sum_{(i,j) \in E_{k,t}} \log \frac{p_t^{ij}}{T_p} \sim \chi_{2|E_{k,t}|}^2$$

The upper tail probability of $X_{k,t}^2$ yields a p-value $p_{k,t}^F$ for that component.

A popular alternative method is Stouffer's Z-score method. Stouffer's method has the flexibility to allow the p-values to be weighted. In a computer network there are nodes that are inherently more active and hence would naturally appear in the anomaly graph more frequently. At the other extreme there are nodes that are rarely active, and hence edges in the anomaly graph originating from these nodes are more rare. The p-values do not take into account the level of overall connectivity of a node and so weighting them when combining the p-values would incorporate the surprise associated with each edge appearing in the anomaly graph. Given weights for each edge in the anomaly graph, w_t^{ij}, define

$$Z_{k,t} = \frac{\sum_{(i,j) \in E_{k,t}} w_t^{ij} Z_t^{ij}}{\sqrt{\sum_{(i,j) \in E_{k,t}} w_t^{ij^2}}} \sim N(0,1),$$

then p-values for the components can be obtained from the upper tail probabilities of $Z_{k,t}$. Ultimately how to set the weights would be dependent on the network and could be guided by an analyst with detailed knowledge of the network layout. A reasonable weighting mechanism could be

$$w_t^{ij} = \left(1 - \frac{\sum_{t'=0}^{t} \mathbb{I}((i,k) \in S_{t'}, k \neq i)}{t+2}\right) \left(1 - \frac{\sum_{t'=0}^{t} \mathbb{I}((k,j) \in S_{t'}, k \neq j)}{t+2}\right),$$

corresponding to a linearly increasing interaction in the weights according to how often the source and destination nodes appear in the anomaly graph.

Finally, the p-values given by (5) for edges originating from the same source node depend on each other through $Y^i(t, w)$. So when combining the p-values, an alternative solution is to condition on the activity status of the source node so that, $p_t^{ij} = \mathbb{P}\left(N^{ij}(t, w) \geq k | Y^i(t, w) = y^i\right)$. Using Stouffer's Z-score method to combine the p-values and weighting the edges according to how often the source node appears in the anomaly graph would retain some level of surprise from the source node. For the LANL data most of the surprise from the p-values comes from the counts along the edges, so conditioning on the node activity level when combining the p-values has little effect.

5   Results

For analysis of the LANL data, the counts are aggregated into ten minute intervals following [7]. LANL have a work scheme whereby employees can take every other Friday off; the effect on NetFlow patterns is apparent in Fig. 2, which shows the activity status of an example node from the network with respect to a fortnightly seasonal period. The seasonal changepoint prior parameter ν is set at 0.009, corresponding a priori to an average of 18 changepoints per seasonal fortnight period. The parameters α and β for the conjugate Beta priors on the node activity transition probabilities are set to a default value of 1, as very small values can lead to overfitting of changepoints. Similarly, $\bar{\alpha}$ and $\bar{\beta}$ for the Beta priors on the transition probabilities for each edge are set at 1. The parameters

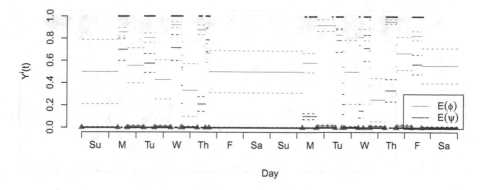

Fig. 2. Activity status of a node from the LANL computer network data with the MAP seasonal changepoints indicated by red triangles. The dotted lines are the posterior expectations (1): $\mathbb{E}_\pi(\phi) \pm \mathrm{s.d.}_\pi(\phi)$ in green and $\mathbb{E}_\pi(\psi) \pm \mathrm{s.d.}_\pi(\psi)$ in blue.

of the model were initially learnt from the four weeks of training data and then updated recursively throughout the test period. Fig. 2 shows the MAP seasonal changepoints and posterior expectation of the transition probabilities for one IP.

Following [7], the window size used to monitor the network is chosen to be 30 minutes, which corresponds to $w = 3$. Analysts at LANL suggest this to be an appropriate choice for a window size, as it corresponds to an approximate time required to see intruders moving around the network.

Many edges in the LANL network exhibit occasional extreme counts as part of normal behaviour. Standard exponential family models will not capture these heavy tails well; further work related to modelling the edges better is discussed in Sect. 6. As a result, the p-values obtained in this analysis are unstable in the tails of the distribution and this would distort the Fisher score when combining p-values in the anomaly graph, since they will be far from uniformly distributed under normal behaviour. Hence for the analysis, the empirical cumulative distribution function obtained from all p-values obtained less than the threshold T_p up until the current time are used to recalibrate the p-values. This provides stabilised p-values which can be combined using Fisher's or Stouffer's method to score each component in the anomaly graph as described in Sect. 4.2.

The threshold, T_p, for the anomaly subgraph was chosen so that the median of the number of nodes in the anomaly graphs in the training period was ten. The median was considered rather than the average as the empirical distribution of the number of nodes in the anomaly graphs has very heavy tails.

Figure 3 shows the Fisher combined p-values less than 0.05 from the components of the anomaly graphs over the ten day test period. The points in this plot which are close together in time are often the same event being repeatedly detected in sequential time windows; in that the majority of the actors in the connected components are the same. For example, the points in Fig. 3 joined together by black lines all constitute one event, detected over several windows.

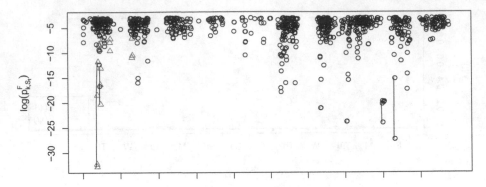

Fig. 3. Anomaly scores for components of the anomaly graphs over the 10 day test period using Fisher's method. The triangles indicate components that contain known infected nodes.

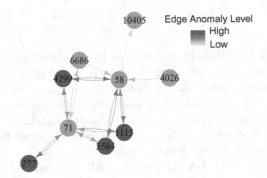

Fig. 4. Heat map of the edges of the anomalous component that was first detected. The red nodes are the known infected hosts.

The most anomalous component detected on the first day of the test data contained four of the known infected hosts. Figure 4 shows the anomaly graph of this event at the window in which it was first detected, just one hour after the initial infection was known to have occurred. Analysis of the attack suggests that the four infected computers (labelled 277, 1115, 3584 and 4299) did not become active until an hour after the time of the initial infection. The same event is then detected in two subsequent time windows thereafter, and again half an hour later. Some of the four infected nodes plus two additional anomalous nodes (130 and 10580) appear in much larger connected components in the later detections. The strength of the detection of this anomaly is sufficiently high that it would be possible to set a threshold to detect this real attack with no false alarms, an exceptional result given the sophisticated nature of the attack.

The central nodes (71 and 58) in Fig. 4 are core servers in the LANL network and are thus connected to a large portion of the network. The four anomalous hosts were connecting to these core servers simultaneously for data reconnaissance purposes. Combining anomalous activity along edges to look for coordinated anomalous traversal in the network allows the anomaly to be detected even though it is centred around busy hubs of the network. When thresholding is based on the single p-values of edges there are many false alarms. For example, a single edge p-value threshold that would detect just the lowest edge p-value in Fig. 4 would raise 57 false alarms.

6   Conclusion

For anomaly detection in computer networks, a system was developed for the operational detection of intruders on internal computer networks and shown to give excellent performance. Discrete time hierarchical Bayesian models were used to model the seasonal and bursty behaviour of the nodes and edges. A second stage of analysis combined predictive p-values obtained from the edge models to detect locally anomalous components in the network indicative of an intruder traversing the network. The system was demonstrated on Los Alamos National Laboratory's enterprise network and a sophisticated attack was detected very soon after the traversal was known to have begun.

The discrete count model for each edge was approximated as a negative binomial distribution. However, highly variable behaviour is observed along some edges across the network. For example, the servers or central machines in the network are usually constantly active and the distribution of counts observed along edges connecting to these central machines is very different from that observed along edges that are more user driven. A more complete modelling effort will provide a future extension to the method, where different models are applied to different categories of edges. Organisational knowledge, such as lists of hosts dedicated to specific functions, should be invaluable.

References

1. Green, P.J.: Reversible jump Markov chain Monte Carlo computation and Bayesian model determination. Biometrika 82, 711–732 (1995)
2. Heard, N.A., Weston, D.J., Platanioti, K., Hand, D.J.: Bayesian anomaly detection methods for social networks. Annals of Applied Statistics 4(2), 645–662 (2010)
3. Hummel, C.: Why crack when you can pass the hash. SANS 21 (2009)
4. Kolaczyk, E.D.: Statistical Analysis of Network Data: Methods and Models. Springer, New York (2000)
5. Lambert, D., Liu, C.: Adaptive thresholds: Monitoring streams of network counts. Journal of the American Statistical Association 101(473), 78–88 (2006)
6. Lazarevic, A., Ozgur, A., Ertoz, L., Srivastava, J., Kumar, V.: A comparative study of anomaly detection schemes in network intrusion detection. In: Proceedings of the Third SIAM International Conference on Data Mining, pp. 25–36 (2003)
7. Neil, J., Storlie, C., Hash, C., Brugh, A., Fisk, M.: Scan statistics for the online detection of locally anomalous subgraphs. Technometrics 55(4), 403–414 (2013)
8. Noble, C.C., Cook, D.J.: Graph-based anomaly detection. In: Proceedings of the Ninth ACM SIGKDD International Conference on Knowledge Discovery and Data Mining, pp. 631–636. ACM (2003)

9. Patcha, A., Park, J.: An overview of anomaly detection techniques: Existing solutions and latest technological trends. Computer Networks 51(12), 3448–3470 (2007)
10. Priebe, C.E., Conroy, J.M., Marchette, D.J.: Scan statistics on Enron graphs. Computational and Mathematical Organization Theory 11(3), 229–247 (2005)
11. Sexton, J., Storlie, C., Neil, J., Kent, A.: Intruder detection based on graph structured hypothesis testing. In: 2013 6th International Symposium on Resilient Control Systems (ISRCS), pp. 86–91. IEEE (2013)
12. Sperotto, A., Schaffrath, G., Sadre, R., Morariu, C., Pras, A., Stiller, B.: An Overview of IP flow-based intrusion detection. IEEE Communications Surveys Tutorials 12(3), 343–356 (2010)

Indirect Estimation of Shortest Path Distributions with Small-World Experiments

Antti Ukkonen

Helsinki Institute for Information Technology HIIT
Aalto University, Finland
antti.ukkonen@aalto.fi

Abstract. The distribution of shortest path lenghts is a useful characterisation of the connectivity in a network. The small-world experiment
is a classical way to study the shortest path distribution in real-world
social networks that cannot be directly observed. However, the data observed in these experiments are distorted by two factors: attrition and
routing (in)efficiency. This leads to inaccuracies in the estimates of shortest path lenghts. In this paper we propose a model to analyse small-world
experiments that corrects for both of the aforementioned sources of bias.
Under suitable circumstances the model gives accurate estimates of the
true underlying shortest path distribution without directly observing the
network. It can also quantify the routing efficiency of the underlying population. We study the model by using simulations, and apply it to real
data from previous small-world experiments.

1 Introduction

Consider the real-life social network where two people are connected if they
mutually know each other "on a first name basis".

What is the distribution of shortest path lengths in this network?

In the late 1960s, decades before the social networking services of today, Jeffrey
Travers and Stanley Milgram conducted an experiment to study this question
[15]. Participants recruited from Omaha, Nebraska, were asked to forward messages to a target person living in Boston, Massachusetts, via chains of social
acquaintances. Given only basic demographic information of the target, such as
name, hometown and occupation, the starting persons were instructed to pass
the message to a friend (their neighbor in the underlying social network) whom
they considered as likely to forward the message further so that it eventually
reaches the target. The participants were also instructed to notify the experi-
menters whenever they forwarded the message.

Out of the 296 messages that were initially sent, 64 reached the target person, and took only a relatively small number of hops [15]. The main hypothesis
resulting from this experiment was that we live in a "small-world", meaning
that a lot of people are connected to each other by *short* paths in the underlying social network. Because of this we refer to such experiments as *small-world
experiments*. Similar experiments have been repeated a number of times [12,5].

H. Blockeel et al. (Eds.): IDA 2014, LNCS 8819, pp. 333–344, 2014.
© Springer International Publishing Switzerland 2014

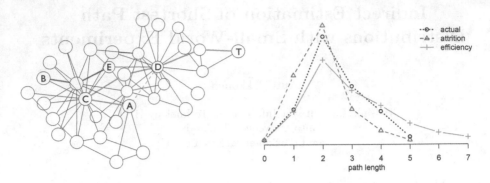

Fig. 1. *Left:* The Karate-club graph with two paths (green and red, solid lines) that are examples of observations in a small-world experiment on this network. *Right:* The shortest paht distribution of the Karate-club graph (black, dotted), together with the effects of attrition (red, dashed) and routing efficiency (green, solid) to the length distribution of observed completed paths.

The observations in such a small-world experiment are a *set of paths* that each indicate the trajectory of one message through the social network. Some of these paths are *completed*, that is, the message reached the target individual, while others are discarded on the way. The length distribution of the completed paths can be seen as a characterization of the *social connectivity* of the population. Indeed, the popular notion of "six-degrees-of-separation" is based on the median of the length distribution of the completed paths that were observed by Travers and Milgram [15].

However, the lengths of completed paths are in practice distorted by two factors: *attrition* and *routing efficiency* of the participants. Attrition refers to messages being discarded for one reason or another during the experiment, causing some paths to terminate before reaching the target. By routing efficiency we refer to the ability of the participants to pass the message to an acquaintance who is in fact closer to the target when distance is measured in terms of the true shortest path distance.

Figure 1 illustrates examples of attrition and routing efficiency with the well known Karate-club network [17]. There are two starting persons, nodes A and B, with T as the target person, as well as two paths that originate at A and B. The path that started from B was terminated early due to attrition: node E failed to forward the message further. In real small-world experiments a message is discarded at every step with an average probabiliy that varies between 0.25 and 0.7, depending on the study [15,5]. On the other hand, the path that starts from node A does reach the target T in five steps, but it does this via node C. Observe that the true shortest path distance from C to T is four steps, while the node A is in fact only three steps away from T (via node D, dashed edge). Node A made thus a "mistake" in forwarding the message to C instead of D. This will also happen in practice. Indeed, even when all participants belong to the same organization, the messages are forwarded to an acquaintance who is truly closer to the target only in a fraction of the cases [10].

Attrition and routing efficiency *will distort the lengths of paths that we observe in a small-world experiment*. In particular, attrition makes long paths less likely to appear[1], even if they might exist in the underlying social network. As a consequence the length distribution of observed completed paths is *shifted towards short paths*, giving an "optimistic" view of the social connectivity. Routing (in)efficiency, on the other hand, implies that even in the absence of attrition we would not observe the true shortest paths. Finding these requires access to the entire network topology; information that individuals most likely do not have. Instead, we observe so called *algorithmic shortest paths* (see also [11,13]), i.e., those that the participants are able to discover using only information about their immediate neighbors and the target person. These, however, are longer than the shortest paths. The lengths of observed completed paths are thus *shifted towards longer paths*, making the network seem less connected.

Indeed, attrition and routing efficiency have contradictory effects on the observed path lengths, as illustrated on the right side of Fig. 1. Attrition makes the observed paths shorter, while routing (in)efficiency makes them longer when compared to the actual shortest path distribution. In practice we observe paths that have been affected by both. Motivated by this discrepancy between the observations and the shortest path distribution, we address the following question:

Can we recover the true shortest path distribution given the observed paths from a small-world experiment?

This is an interesting task for a number of reasons. First, it is another way to address the original question of Travers and Milgram [15]. Using only the completed paths gives one view of the social connectivity, by estimating the actual shortest paths distribution results in a less biased outcome. Second, the question is related to recent work on reconstructing networks and properties thereof from observed traces of activity in the network [8,9,14,4]. The small-world experiment can be seen as another type of such activity. Moreover, a better understanding of the process that underlies the small-world experiment may lead to improvements in other propagation processes over networks, such as spreading of epidemics or opinions.

Our Contributions: We propose a new model to analyse the observations of a small-world experiment that accounts for the bias caused by attrition and routing efficiency. The main difference to previous approaches [10,7] is the use a well-defined probabilistic model that can estimate the true shortest path distribution. Our technical contribution is that of devising an intuitive parametrization of the process that underlies a small-world experiment, as well as a means to express the likelihood of the observed paths in terms of the process parameters. By fitting the proposed model to data from previous small-world experiments [15,5] we compute *estimates for the shortest path distribution* of the underlying social

---

[1] If a path terminates with probability r at every step, a path of length l appears with probability $(1 - r)^l$. This decreases rapidly as l increases. See also [7].

network, as well as *quantify the routing efficiency* of the population that participated in the experiment. To the best of our knowledge this has not been done before.

2   A Model for Small-World Experiments

The input to our model consists of the set D of both *completed* and *failed* paths that are observed in a small-world experiment. A path is completed if it reaches the target, and failed if it terminates due to attrition. Of every path in D, we know thus its *outcome*, and its *length*. These are the only two characteristics of a path that our model is based upon.

We first discuss the parameters of the *message-forwarding process* that underlies a small-world experiment. As with any generative model, we do not claim that this process accurately represents reality, it is merely a useful and tractable representation of it. Then, we show how the process parameters *induce the parameters of a multinomial distribution* over different types of paths. The likelihood of D is determined by this multinomial distribution in a standard manner.

Model Parameters and Dynamics

We assume the structure of the underlying social network to be hidden. However, at every step each of the messages must be at some *shortest path distance* from the target person. These distances are unknown, but we can model how they evolve as the messages are forwarded. When a node forwards the message, the shortest path distance to the target can *decrease by one, stay the same, or increase by one*. Note that the distance can *not* increase or decrease by any other amount in a single step.

We assume that each node chooses the recipient so that the distance decreases with probability q_-, increases with probability q_+, and remains the same with probability $1 - q_- - q_+$. We also enforce the constraint $q_+ < q_-$ (the participants are assumed to be in some sense "benevolent"), as well as $q_- + q_+ \leq 1$. The probabilities q_- and q_+ are the first and second parameters of our model, and they capture routing (in)efficiency. Moreover, a node might not forward the message in the first place. We assume that a message can be discarded at every step with the constant probability r, the attrition rate, which is the third parameter of our model.

The parameters q_-, q_+, and r capture how the shortest path distance from the current holder of the message to the target person evolves, but the process must start from somewhere. The distances from the starting nodes to the target node are unknown. We assume that *the initial distances from the starting nodes to the target* follow some predefined distribution $\tau(\cdot)$. In particular, we take these to be Weibull distributed with parameters k and λ, because it has been argued [3] that this produces a good fit for shortest path distributions that are observed in different types of random networks. The k and λ parameters of τ are the fourth and fifth parameters of our model. All five parameters are summarised in Table 1.

Table 1. Summary of model parameters

$q_-,\ q_+$	probabilities for message to approach and move away from target
r	probability of message to be discarded
$k,\ \lambda$	shape and scale parameters of the initial distance distribution τ

In summary, we model the small-world experiment using the following two-stage process. Initially the messages are distributed at distances from the target that is given by $\tau(\cdot \mid k, \lambda)$. Upon receiving the message a node 1) decides if it is going to discard the message, and if not, 2) the node forwards the message to a neighbor. An independent instance of this process launches from every starting node, and continues until the message reaches the target or is discarded.

Path Probabilities

Recall that of every path in the input D we know its outcome o and length l. A path is of *type* (o, l) if it has outcome o and length l, where o is either $\top$ (a completed path) or $\bot$ (a failed path). For example, the red and green paths in Fig. 1 are of types $(\bot, 2)$ and $(\top, 5)$, respectively.

Next we derive the probability to observe a path of type (o, l), assuming that the forwarding process adheres to the parameters discussed above. To this end we express the message-forwarding process as a discrete-time Markov chain. A graphical representation of this chain together with the transition probabilities is shown in Fig. 2. The states of the chain are all possible distances to the target up to some maximal distance m, denoted $0, 1, \ldots, m$, a special attrition state, denoted A, and a special terminal state, denoted E. At every step the chain is in one of the states. The message reaches the target when the chain enters the state 0. Likewise, the message is discarded when the chain enters the state A. The state E is needed for technical reasons.

Let Q denote the transition probability matrix associated with the Markov chain of Fig. 2 so that Q_{xy} is the probability of taking a transition from state x to state y. From the basic theory of Markov chains we know that given any initial probability distribution π_0 over the states, the distribution after running

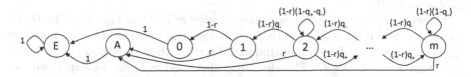

Fig. 2. The Markov chain that represents the message-forwarding process. The message is discarded with probability r independently of its distance to the target. The distance decreases with probability $(1 - r)q_-$, increases with $(1 - r)q_+$, and remains the same with probability $(1 - r)(1 - q_- - q_+)$. A neighbor of the target node can discard the message or pass it to the target. In the state m the distance can no longer increase, so we introduce a self-loop with probability $(1 - r)(1 - q_-)$.

the chain for l steps is equal to $\pi_l = \pi_0^\mathsf{T}(\mathbf{Q})^l$. Recall that $\tau(\cdot \mid k, \lambda)$ is our guess of the distribution of distances from the starters to the target. We initialize π_0 so that $\pi_0(\mathsf{A}) = \pi_0(\mathsf{E}) = \pi_0(0) = 0$, and for the states $1, \ldots, \mathsf{m}$ we set $\pi_0(\mathsf{x}) = \tau(\mathsf{x} \mid k, \lambda)$. The probabilities to observe a path of type (o, l) given the model parameters is now

$$\Pr((o,l) \mid q_-, q_+, r, k, \lambda) = \begin{cases} \pi_l(0), \text{ if } o = \top, \text{(completed path)}, \\[2mm] \pi_l(\mathsf{A}), \text{ if } o = \bot, \text{(failed path)}, \end{cases} \tag{1}$$

where $\pi_l(0)$ and $\pi_l(\mathsf{A})$ are the probability masses at states 0 and A after running the chain for l steps, respectively. The above equation is derived simply by noticing that the states 0 and A correspond to endpoints of completed and failed chains, and the probability of a path of length l to land in either of these is obtained by running the chain for l steps. (The special terminal state E is needed to guarantee that π_l indeed has the desired value at states 0 and A.)

Finally, to compute the probabilities in practice, the maximum distance m must be set to a large enough value. In practice we did not observe m to have a strong effect. In the experiments we use $m = 30$.

Likelihood of the Input D

As in [7], we view a small-world experiment as a simple sampling procedure, where $|D|$ paths are *drawn independently[2] from a categorical distribution*, where every category corresponds to a path type, and the probabilities of individual categories are induced by the model parameters (Table 1) as discussed above. Let $\theta = (q_-, q_+, r, k, \lambda)$ denote a vector with all parameters, and define the probability of a type $p_\theta(o, l) = \Pr((o, l) \mid \theta)$ as in Eq. 1.

In theory the number of distinct types is infinite, because our model does not impose an upper limit on the length of a path. However, we only consider types up to some length, because the probability of long paths decreases rapidly. Given θ, we choose $l_{\max}$ to be the smallest integer so that the inequality $\sum_{l=1}^{l_{\max}} (p_\theta(\top, l) + p_\theta(\bot, l)) \geq 1 - \epsilon$ holds for some small ϵ (in the experiments we let $\epsilon = 10^{-8}$). That is, we assume that the paths in D are an i.i.d. sample from a discrete distribution with categories $C(\theta) = \{(\top, 1), (\bot, 1), \ldots, (\top, l_{\max}), (\bot, l_{\max})\}$. Note that depending on θ, $l_{\max}$ can be larger than the longest path we observe in D.

Furthermore, Let $c_D(o, l)$ denote the *number of paths* of type (o, l) in D. Of course $c(o, l) = 0$ for any type (o, l) that does not occur in D. The numbers $c_D(o, l)$ are a sample from a multinomial distribution with parameters $|D|$ and $p_\theta(o, l)$ for every $(o, l) \in C$. Therefore, the likelihood of D can be expressed as

$$\Pr(D \mid \theta) = \frac{|D|!}{\prod_{(o,l) \in C(\theta)} c_D(o,l)!} \prod_{(o,l) \in C(\theta)} p_\theta(o,l)^{c_D(o,l)}. \tag{2}$$

---

[2] This may not fully hold in real small-world experiments, as lengths and outcome of paths having e.g. the same source or target may be correlated [5]. However, we consider this to be a reasonable simplification to make the model tractable.

The likelihood depends on the model parameters θ thus through the probabilities $p_\theta(o, l)$. To compute the likelihood, we first find $p_\theta(o, l)$ for every $(o, l) \in C(\theta)$ using the Markov chain of Fig. 2, and then apply Equation 2.

Parameter Estimation

For parameter estimation we can use any available optimisation technique. After trying out several alternatives, including different numerical optimisation algorithms as well as MCMC techniques, the Nelder-Mead method was chosen[3]. It does not require partial derivatives, finds a local optimum of the likelihood function, and is reasonably efficient for our purposes. It also allows to use fixed values for some of the paramters, and solve the model only for a subset of them to obtain *conditional estimates*. For instance, we can fix q_- and q_+ and solve the model only for r, k and λ.

It is worth pointing out that the model can be non-identifiable. (Meaning its true parameters are hard to find even given an infinite amount of data.) Some of the parameters have opposite effects. For instance, in terms of the observed paths in D, it might not matter much if a) the initial distances to the target are long and the routing efficiency is high, or b) the initial distances are short but the routing efficiency is low. This means that a simultaneous increase in both e.g. the median of $\tau(k, \lambda)$ as well as the routing efficiency parameter q_- can result in only a very small change in $\Pr(D \mid \theta)$.

3 Experiments

Estimation Accuracy

As there in general is no ground truth available in a small-world experiment, it is important that the estimates obtained from the model are at least somewhat stable. The estimates are affected by size of the input data, as well as indentifiability issues of the model itself. We start by quantifying these effects by using paths that are obtained from simulations of the message forwarding process with known parameters.

As expected, estimating all model parameters simultaneously is a hard problem. The top half of Table 2 shows both the true parameter values as well as the median of their estimates from 250 inputs of 5000 paths each. (The estimate of r is always very accurate, and is omitted from the table.) We can see that while the median is often fairly close to the true value, the quantiles indicate a high variance in the estimates, especially for q_-.

However, by fixing some parameters it is possible to improve the quality of the resulting estimates. The bottom half of Table 2 shows conditional estimates for q_- and q_+ *given* k and λ, and vice versa, the estimates for k and λ given q_- and q_+. Now the estimates are very accurate in all of the cases. This suggests that the model is most useful when we have some prior knowledge either about the routing efficiency, or the shortest path distribution. For example, [10] suggests to

---

[3] We use the implementation provided by the optim function of GNU R, http://www.r-project.org.

Table 2. Estimating parameters from simulated data (5000 paths)

true values	simultaneous estimates (with 5% and 95% quantiles)			
q_- q_+ r k λ	q_-	q_+	k	λ
0.5 0.0 0.25 4 5	0.60 (0.44,0.82)	0.03 (0.00,0.10)	3.71 (3.11,4.46)	5.53 (4.56,6.76)
0.5 0.1 0.25 4 5	0.47 (0.29,0.69)	0.07 (0.00,0.21)	4.10 (3.53,5.08)	4.86 (3.83,6.01)
0.3 0.0 0.25 4 5	0.43 (0.30,0.90)	0.05 (0.00,0.13)	3.45 (2.34,4.10)	6.13 (4.87,11.93)
0.3 0.1 0.25 4 5	0.33 (0.15,0.70)	0.12 (0.00,0.31)	3.86 (2.79,5.50)	5.27 (3.60,8.51)
0.5 0.0 0.50 4 5	0.65 (0.35,1.00)	0.04 (0.00,0.17)	3.65 (2.79,5.05)	5.80 (3.92,8.38)
0.3 0.0 0.50 4 5	0.39 (0.18,0.88)	0.05 (0.00,0.17)	3.67 (2.44,5.41)	5.64 (3.71,11.95)

true values	conditional estimates (with 5% and 95% quantiles)			
q_- q_+ r k λ	$q_- \mid k,\lambda$	$q_+ \mid k,\lambda$	$k \mid q_-q_+$	$\lambda \mid q_-,q_+$
0.5 0.0 0.25 4 5	0.50 (0.49,0.53)	0.00 (0.00,0.03)	4.01 (3.83,4.21)	4.99 (4.88,5.10)
0.5 0.1 0.25 4 5	0.50 (0.46,0.54)	0.10 (0.05,0.15)	4.01 (3.79,4.24)	4.98 (4.87,5.14)
0.3 0.0 0.25 4 5	0.31 (0.29,0.33)	0.00 (0.00,0.05)	4.02 (3.75,4.24)	5.00 (4.84,5.18)
0.3 0.1 0.25 4 5	0.30 (0.27,0.33)	0.10 (0.04,0.16)	4.00 (3.75,4.30)	4.99 (4.80,5.22)
0.5 0.0 0.50 4 5	0.51 (0.46,0.59)	0.00 (0.00,0.18)	4.00 (3.64,4.42)	5.00 (4.71,5.31)
0.3 0.0 0.50 4 5	0.31 (0.27,0.37)	0.00 (0.00,0.22)	3.99 (3.55,4.58)	5.01 (4.61,5.53)

use $q_- = 0.3$, and [1] provides the exact shortest path distribution of the entire Facebook social network, which could be used in place of $\tau(\cdot \mid k, \lambda)$.

The next question is how sensitive are the estimates of k and λ to errors in our assumptions of q_- and q_+? That is, suppose the true q_-, denoted q_-^{True}, is 0.5, but we solve the model with fixed a $q_- = 0.35$ for example. Let $\hat{\mu}$ denote the median of the estimated shortest path distribution, while μ^{True} is the median of the true distribution. (This is a more intuitive quantity than k and λ when interpreting the shortest path distribution.) The left panel in Figure 3 shows $\Delta(\mu) = \hat{\mu} - \mu^{\text{True}}$ as a function of $\Delta(q_-) = q_- - q_-^{\text{True}}$ when k and λ are estimated given different q_- from 1000 paths generated by the model. (The plot shows distributions over 100 runs for every q_-. Input was generated with $q_- = 0.5$, $q_+ = 0.05$, $r = 0.25$, $k = 4$, $\lambda = 5$.) We observe that in this range of q_- an under- or overestimate of 0.15 will make the conditional estimate of $\hat{\mu}$ about one step too low/high.

The variance of the estimates will also depend on the size of the input D. The right panel in Figure 3 shows effect of $|D|$ on the variance of $\hat{\mu}$ when k and λ are estimated with q_- and q_+ fixed to their correct values. (The paths were generated using the same parameter values as above.) We find that the conditional estimate of $\hat{\mu}$ is both *unbiased* and *consistent*. In practice a high accuracy requires > 1000 paths, but even only 300 paths seem to give reasonable results.

Estimating Shortest Path Distributions

We continue by estimating the shortest path distribution[4] in random networks. The top row of Fig. 4 shows conditional estimates for an Erdős-Rényi [6] as well

---

[4] Notice that here the resulting shortest path distributions reflect the initial distances to a single target from the starters. To obtain the all-pairs shortest path distribution, all paths should have independently sampled starters and targets.

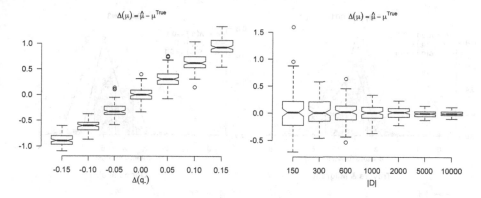

Fig. 3. *Left:* Effect of the error when fixing q_- to the conditional estimate of $\hat{\mu}$. *Right:* Effect of $|D|$ to the conditional estimate of $\hat{\mu}$. (Boxes show 1st and 3rd quartiles.).

as a Barabási-Albert [2] graph. In both cases we considered three values for q_-: the correct one, and ones that under- and overestimate the true value by 0.1. We find that in both cases the $\Delta(q_-) = 0$ estimate (solid line) captures the qualitative properties of the true distribution (bars). And like above, an under- or overestimate in q_- leads to a slight under- or overestimate in the path lenghts.

Finally, we apply the model to observed data from previous small-world experiments. Travers & Milgram (TM) provide the numbers of completed and failed paths that we need for our model (in Table 1 of [15]), while for Dodds et al. (DMW) we obtain the numbers through visual inspection of Fig. 1 in [5]. We fit both the full model, as well as conditional estimates given fixed q_- and q_+. Resulting shortest path distributions are shown in the bottom row of Fig. 4.

While the full estimation is known to be unstable, it is interesting to see that it produces reasonable estimates. For TM the estimates for q_-, q_+, and r are 0.92, 0.08 and 0.21, respectively, while for DMW we obtain $q_- = 0.52$, $q_+ = 0.00$, $r = 0.71$. The attrition rate estimates are very close to the ones reported in literature [15,5], while the routing efficiency parameters tell an interesting story. It appears that in the TM experiment the participants almost always chose the "correct" recipient, while in the DMW experiment they did this only half of the time. This is not inconceivable, as the TM study was carried out by regular mail, while DMW used email as the means of communication. Subjects in the TM study might indeed have been more careful when choosing the recipient as participating took more effort also otherwise.

The conditional estimates suggest that the shape of the estimated shortest path distributions is more or less the same in both experiments if we assume an identical q_-. The conditional estimate for $q_- = 0.5$ in TM is qualitatively somewhat similar to the full DMW estimate. This applies also to the conditional estimate for $q_- = 0.92$ in DMW and the full TM estimate.

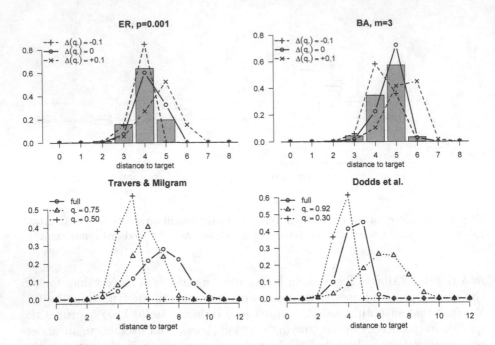

Fig. 4. *Top row:* True shortest path distribution (bars) together with three estimates for different values of $\Delta(q_-)$ computed from 1000 paths in an Erdos-Renyi (left) and a Barabasi-Albert (right) graph. *Bottom row:* Estimated shortest path distributions given data from the Travers & Milgram [15] as well as the Dodds et. al. [5] experiments.

4   Discussion

Our model has common aspects with the method devised in [10] as they also use a similar Markov chain to infer frequencies of observed completed paths given the q_- parameter. But there are some important differences. We allow the message also to move away from the target, and our model fully separates attrition from routing efficiency. Moreover, we propose to compute maximum-likelihood parameter estimates from observed paths. Finally, compensating for attrition has received attention in previous literature as well [5,7], but using very a different technique (importance sampling).

Small-world experiments show also that humans can find short paths in a decentralized manner. It is not obvious why this happens. There are two factors that play a role in the process: *structure* of the underlying social network and the *strategy* (or algorithm) used to forward the messages. In real small-world experiments it has been observed that participants tend to pass the message to an aqcuaintance who has some common attributes with the target. Especially geographical location and occupation have been reported as important criteria [5]. For such a "greedy" routing strategy to find short paths, the network must have certain structural properties that reflect the similarity of the nodes in terms of social attributes [11,16,13]. However, our model is independent of both routing strategy and network structure, meaning that we do not have to make assumptions about either.

Estimating the shortest path distribution from a set of very short (biased) random walks over an unobserved network is a hard problem, and the lack of a ground truth makes the results difficult to evaluate. We claim, however, that the proposed model can be a useful tool when analysing small-world experiments. Extending the model to deal with other types of data, such as information cascades [8,9,14,4] is an interesting open question, and so is improving the stability of the unconditional parameter estimates.

Acknowledgements. I would like to thank Aristides Gionis for his comments to a very early draft of this paper, and Hannes Heikinheimo for the valuable suggestions that helped to substantially improve the final version.

References

1. Backstrom, L., Boldi, P., Rosa, M., Ugander, J., Vigna, S.: Four degrees of separation. In: WebSci., pp. 33–42 (2012)
2. Barabási, A.-L., Albert, R.: Emergence of scaling in random networks. Science 286(5439), 509–512 (1999)
3. Bauckhage, C., Kersting, K., Rastegarpanah, B.: The Weibull as a model of shortest path distributions in random networks. In: MLG (2013)
4. Bonchi, F., De Francisci Morales, G., Gionis, A., Ukkonen, A.: Activity preserving graph simplification. Data Min. Knowl. Discov. 27(3), 321–343 (2013)
5. Dodds, P.S., Muhamad, R., Watts, D.J.: An experimental study of search in global social networks. Science 301(5634), 827–829 (2003)
6. Erdős, P., Rényi, A.: On random graphs. Publicationes Mathematicae Debrecen 6, 290–297 (1959)
7. Goel, S., Muhamad, R., Watts, D.J.: Social search in "small-world" experiments. In: WWW, pp. 701–710 (2009)
8. Gomez-Rodriguez, M., Leskovec, J., Krause, A.: Inferring networks of diffusion and influence. Transactions on Knowledge Discovery from Data 5(4), 21 (2012)
9. Gomez-Rodriguez, M., Schölkopf, B.: Submodular inference of diffusion networks from multiple trees. In: ICML (2012)
10. Killworth, P.D., McCarty, C., Bernard, H.R., House, M.: The accuracy of small world chains in social networks. Social Networks 28(1), 85–96 (2006)
11. Kleinberg, J.M.: Navigation in a small world. Nature 406(6798), 845 (2000)
12. Korte, C., Milgram, S.: Acquaintance links between white and negro populations: Application of the small world method. Journal of Personality and Social Psychology 15(2), 101–108 (1970)
13. Liben-Nowell, D., Novak, J., Kumar, R., Raghavan, P., Tomkins, A.: Geographic routing in social networks. Proceedings of the National Academy of Sciences 102(33), 11623–11628 (2005)
14. Mathioudakis, M., Bonchi, F., Castillo, C., Gionis, A., Ukkonen, A.: Sparsification of influence networks. In: KDD, pp. 529–537 (2011)
15. Travers, J., Milgram, S.: An experimental study of the small world problem. Sociometry 32(4), 425–443 (1969)

16. Watts, D.J., Dodds, P.S., Newman, M.E.J.: Identity and search in social networks. Science 296(5571), 1302–1305 (2002)
17. Zachary, W.: An information flow model for conflict and fission in small groups. Journal of Anthropological Research 33(4), 452–473 (1977)

Parametric Nonlinear Regression Models
for Dike Monitoring Systems

Harm de Vries[1,2,3], George Azzopardi[2,3], André Koelewijn[4], and Arno Knobbe[1]

[1] Leiden Institute of Advanced Computer Science,
Leiden University, Leiden, The Netherlands
[2] TNO, Groningen, The Netherlands
[3] Johann Bernoulli Institute for Mathematics and Computer Science,
University of Groningen, Groningen, The Netherlands
[4] Deltares, Delft, The Netherlands

Abstract. Dike monitoring is crucial for protection against flooding disasters, an especially important topic in low countries, such as the Netherlands where many regions are below sea level. Recently, there has been growing interest in extending traditional dike monitoring by means of a sensor network. This paper presents a case study of a set of pore pressure sensors installed in a sea dike in Boston (UK), and which are continuously affected by water levels, the foremost influencing environmental factor. We estimate one-to-one relationships between a water height sensor and individual pore pressure sensors by parametric nonlinear regression models that are based on domain knowledge. We demonstrate the effectiveness of the proposed method by the high goodness of fits we obtain on real test data. Furthermore, we show how the proposed models can be used for the detection of anomalies.

Keywords: Structural health monitoring, dike monitoring, nonlinear regression, anomaly detection.

1 Introduction

Dikes are artificial walls that protect an often densely populated hinterland against flooding disasters. Especially the Netherlands, with large areas below sea level, has a rich history of dike failures that resulted in drowning deaths and devastation of infrastructure. Although dike technology has improved over the years, only 44% of the 2875 kilometers of main Dutch dikes met the government's dike regulations in 2006 [1]. Traditional dike monitoring involves visual inspection by a dike expert at regular time intervals. Dike patrolling is, however, a time consuming and costly process that does not always reveal weak spots of a dike. In this light, the IJkdijk foundation[1] has been established in 2007 with the ambition to enhance dike monitoring by sensor systems. The largely successful program initiated an EU-funded project, called UrbanFlood[2], that also implements sensor systems in dikes, but intends to construct so-called Early Warning

---

[1] Official website: http://www.ijkdijk.nl
[2] Official website: http://www.urbanflood.eu

H. Blockeel et al. (Eds.): IDA 2014, LNCS 8819, pp. 345–355, 2014.

Systems (EWS) for floodings [2]. As a consequence of both projects, more and more dikes across Europe are being equipped with sensor systems, and therefore there is an urgent need for algorithms that are capable of detecting damage to the dike as early as possible.

Geophysical models [3,4] can be used to assess the stability of a dike by simulation of the underlying physical processes. Such models are computationally intensive and thus not appropriate for (near) real-time dike monitoring. To overcome such problems, data driven techniques were considered to detect indicators for instability of a dike. In [2], the authors proposed neural clouds in order to detect outliers in the sensor values. The main drawback of this approach is that sensor values are highly influenced by environmental conditions, and thus outliers often correspond to rare environmental conditions rather than changes in the internal structure of the dike. The same research group also proposed an anomaly detection technique that uses one-step-ahead prediction of (non-linear) auto regressive models [5]. Although such methods achieve high model fits, they are not appropriate to detect gradual changes in a response of a sensor.

In this paper, we conjecture that in order to detect internal changes in a dike, we first have to model the *normal* relationship between some environmental conditions and dike sensors. We present a case study of a set of pore pressure sensors that are installed in a sea dike in Boston (UK), and which are continuously affected by water levels, the foremost influencing environmental factor. Although the pore pressure signals vary significantly among the set of sensors, we hypothesize that essentially two physical processes play a role. We estimate the one-to-one relationships by parametric nonlinear regression models that aim to reflect the underlying physical phenomena. In contrast to black box modeling techniques, such as Transfer Function Modeling and Neural Networks, the proposed models are intuitive, interpretable and provide more insight into the dynamics of the dike. Moreover, we demonstrate that the models can be effectively used for anomaly detection.

2 Background

Fig. 1 shows a schematic overview of the concerned dike in Boston (UK) that includes the placement of seven sensors that measure the pore pressure at time intervals of 15 minutes. Although not shown in Fig. 1, there is a sensor that measures the water level nearby the dike with a sample interval of 15 minutes. In Fig. 2b, we show the water levels of the month of October 2011. It is characteristic of the dike in Boston that water levels follow half-daily tides with extreme differences (up to seven meters) between high and low tides. Note that the amplitude of the half-daily tides also varies with an approximately two-weekly period due to the lunar cycle. It is also worth mentioning that the sensor cannot detect water levels below -1.6 meters, which is reflected in the data by the flat lower envelope of the water level signal. The estuary near the dike falls dry at that point, although the actual sea levels are a little bit lower.

In Fig. 2a, we illustrate all seven pore pressure signals recorded in October 2011. The relationship between a pressure signal and a water level signal is

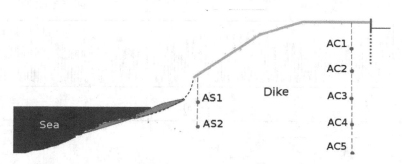

Fig. 1. Schematic overview of the sensor setup in the Boston (UK) dike

influenced by the location of the concerned pressure sensor. For instance, the AC1 sensor, which is placed at the top of the dike, does not respond to the water level. On the other hand, the AC4 pore pressure, which is located at a very deep level, seems to follow the same tidal fluctuations as the water levels.

The available data set consists of one year of sensor values, and ranges from October 2011 till October 2012. The data set has a lot of missing values. In particular, approximately 10% of the water levels and 20% of the pore pressure values are missing. Moreover, measurements of different sensors are not synchronized. As a preprocessing step, we therefore linearly interpolate the water level signal in order to align it with the pore pressure signal in question. In this way, we also fill small gaps in the water level signal of at most 2 samples (i.e. 30 minutes). The models we propose use some history of water levels to model the current pore pressure. We exclude the sample from the training set if either the pore pressure value is missing, or there is a gap of at least two water level measurements in the history. In Fig. 2, a sequence of missing values is visible as a straight line.

3 Model Estimation

In general, we expect that two physical phenomena play a role in the response of a pore pressure sensor:

Short-term effect. An almost immediate response to the water levels due to water pressure at the front of the dike. Therefore, the regular rise and fall of the water levels cause peaks in the pore pressure signals.

Long-term effect. A much slower effect that accounts for the degree of saturation of the dike. A dike that is exposed to high water levels absorbs water, which increases the degree of saturation of the dike, which in turn increases the pore pressure.

Fig. 2b clearly shows that the long-term effect does not play a role in all sensors. For example, the sensors AC4 and AC5, that reside deep in the dike, do not

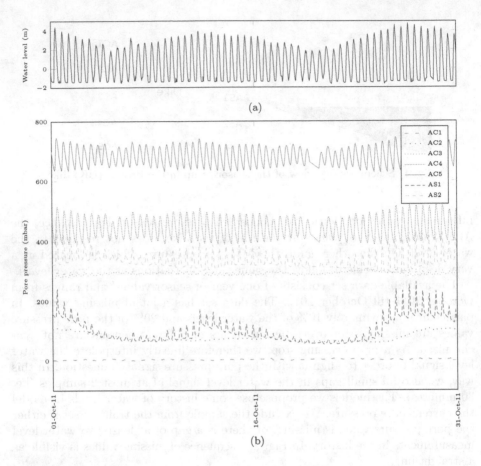

Fig. 2. The raw signals of one month of data of (a) the water level sensor and (b) the seven pore pressure sensors installed in the dike in Boston (UK). The straight lines indicate missing values.

have significant changes in the baseline. The saturation degree at that location in the dike is not heavily influenced by the dike's exposure to water levels. In the following, we refer to these sensors as *short-term effect sensors*, and we propose a model for them in Section 3.1.

The other sensors follow a mixture of both effects. A typical example is the AS1 sensor, of which an example month of data is shown in Fig. 2b. There are significant changes in the baseline of the signal that seem to follow the two-weekly cycle of the water levels, but there are still half-daily peaks that are superimposed on this baseline. We model these *mixed effect sensors* in Section 3.2.

3.1   Short-Term Effect Sensors

In the following, we select a subset of the complete data set as training data. Formally, we consider N pore pressure values $y[1], \ldots, y[N]$ (equally spaced in time) and water level measurements $x[1], \ldots, x[N]$ that are aligned in time. We model the current pore pressure value $y[n]$ as a function of recent history of water levels $x[n - M + 1], \ldots, x[n]$. Here, $M > 0$ represents the number of historical water level measurements. It should be chosen large enough in order to reliably predict the pore pressure. We choose $M = 100$, which translates to roughly one day of water levels. The training set contains $P = N - M + 1$ examples, and consists of a set of water level input vectors $X = \{\boldsymbol{x}_i = x[i+M-1], \ldots, x[i] \mid i = 1, \ldots, P\}$ and pore pressure output $\mathbf{y} = \{y_i = y[i + M - 1] \mid i = 1, \ldots, P\}$.

We propose to model the short-term effect sensor by the following parametric nonlinear regression model[3]:

$$f(\boldsymbol{\theta}; \boldsymbol{x}_i) = b + a \sum_{m=0}^{M-1} \exp(-\lambda \, m) \, \boldsymbol{x}_i[m + 1] \ \text{ with } \ \boldsymbol{\theta} = \begin{bmatrix} b & a & \lambda \end{bmatrix} \qquad (1)$$

where λ controls the rate of decay, and a and b are affine transformation parameters. Our model corresponds to the solution of a first-order constant coefficient differential equation[4]:

$$y'(t) = -\lambda y(t) + ax(t) \ \text{ with initial condition } \ y(0) = b \qquad (2)$$

By rewriting the right hand side to $\lambda\,(cx(t) - y(t))$ with $c = \frac{a}{\lambda}$, our model assumptions become clear. First, the water level linearly relates (by factor c) to the pressure on the front of the dike. Second, the rate of change of the pore pressure is proportional to the difference between the current pressure on the front of the dike and the pore pressure of the sensor.

We estimate the parameters of the model by minimizing the sum of squared residuals:

$$S(\boldsymbol{\theta}) = \sum_{i=1}^{P} r_i^2(\boldsymbol{\theta}) \ \text{ where } \ r_i(\boldsymbol{\theta}) = y_i - f(\mathbf{x}_i; \boldsymbol{\theta}). \qquad (3)$$

which is identical to the Maximum Likelihood Estimator (MLE) under white Gaussian error terms. We optimize the cost function by a Gauss-Newton solver, which is appropriate to optimize a least-squares problem [6].

We fit the model on 12 days of AC4 sensor values. We obtain the parameters $b_{mle} = 430.13$, $a_{mle} = 3.17$ and $\lambda_{mle} = 0.1142$, and show the predicted values by

---

[3] For initial rest (i.e. $b = 0$) the proposed model is a Linear Time Invariant (LTI) system with an exponential decaying impulse response function. Linear constant coefficient differential equations can be represented by causal LTI systems if and only if they satisfy the initial rest condition. We refer the interested reader to http://ocw.mit.edu/resources/res-6-007-signals-and-systems-spring-2011/lecture-notes/MITRES_6_007S11_lec06.pdf.

[4] See http://web.mit.edu/alecp/www/useful/18.03/Supplementary-CG.pdf for more details.

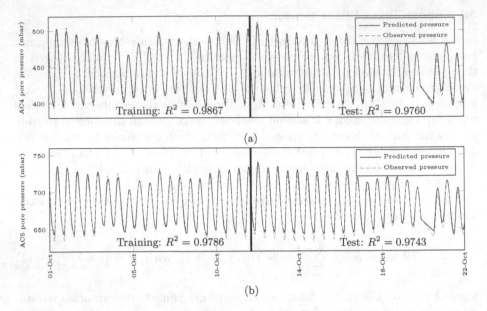

Fig. 3. A comparison between the predicted and observed values for (a) the AC4 and (b) the AC5 sensors. The black bar separates the training and test sets.

our model in Fig. 3a. Note that the model almost perfectly follows the observed pore pressure. We quantify the goodness of fit by:

$$R^2 = 1 - \frac{\sum_{i=1}^{P}(y_i - f(\boldsymbol{\theta}; \mathbf{x}_i))^2}{\sum_{i=1}^{P}(y_i - \bar{y})^2} \quad \text{with mean} \quad \bar{y} = \sum_{i=1}^{P} \frac{y_i}{P}, \quad (4)$$

which, roughly speaking, measures how successful the model is in explaining the variation of the data. For the above example, we find $R^2 = 0.9867$, which indicates that the estimated model fits very well. We consider as test data the 12 days that follow the training data. For this period we also obtain a very high value of $R^2 = 0.9760$, which demonstrates that the proposed regression model is not prone to over-fitting.

To give an impression of the duration of the exponential decay, note that $\lambda = 0.1142$ corresponds to a mean lifetime of $\tau = 1/0.1142 = 8.756$ measurements, which in our case amounts to slightly over 2 hours. The corresponding half-life is $\tau_{1/2} = 8.756/\ln(2) = 6.070$ which corresponds to roughly 1.5 hours.

In a similar way, we estimate the parameters for the model on 12 days of AC5 pore pressure values, and obtain $b_{mle} = 671.78$, $a_{mle} = 3.09$ and $\lambda_{mle} = 0.1458$. We show the predicted values obtained by our model in Fig. 3b, and point out that the predicted values are quite close to the observed pore pressure measurements. This is confirmed by the goodness of fit, $R^2 = 0.9786$. For the next 12 days, we obtain $R^2 = 0.9743$, which demonstrates its effectiveness on unseen data.

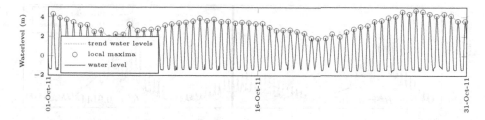

Fig. 4. Extraction of the top envelope of the water levels. The circles indicate the local maxima which are connected by straight lines using linear interpolation.

3.2 Mixed-Effect Sensors

We extend the models that we proposed above to sensors that are also affected by the degree of saturation of the dike. Intuitively speaking, a dike is only saturated if it is exposed to high water levels for a longer period. That is to say, we assume that the process varies slowly and it is not influenced by half-daily tides, but it is related to the two-weekly cycle of the water levels.

We extract the top envelope of the water levels in order to capture the general trend (the underlying fortnightly cycle). To this end, we extract the local maxima by taking into account the fact that local maxima are expected to be separated by 12 hours. Fig. 4 shows the extracted local maxima for the water levels in October 2011. This extraction decreases the resolution of the signal considerably. We use linear interpolation to fill in the gaps between the extracted local maxima, such that we obtain the same resolution as the given water level signal. We denote by $\hat{x}[1], \ldots, \hat{x}[N]$ the estimated general trend of the water levels, and plot it in Fig. 4.

We model the long-term effect with a first-order constant coefficient differential equation that is similar to the short-term model defined in Eq. 2, but here we use the general trend of the water levels $\hat{\mathbf{x}}$ as input signal. The underlying assumption is that the rate of change in saturation degree is proportional to the difference between the current trend in water level (i.e. how much water the dike is currently exposed to) and the current saturation degree. In other words, a dike that is exposed to high water levels absorbs water much faster whenever it is not saturated. We propose to model the mixed-effect sensors by superimposing the short-term model on the long-term one:

$$f(\boldsymbol{\theta}; \mathbf{x}, \hat{\mathbf{x}}) = b + a_l \sum_{m_l=0}^{M_l-1} \exp(-\lambda_l m_l)\,\hat{\mathbf{x}}[m_l + 1]^{p_l}$$
$$+ a_s \sum_{m_s=0}^{M_s-1} \exp(-\lambda_s m_s)\,\mathbf{x}[m_s + 1]^{p_s}, \qquad (5)$$

where $\boldsymbol{\theta} = \begin{bmatrix} b\ a_l\ \lambda_l\ p_l\ a_s\ \lambda_s\ p_s \end{bmatrix}^\top$, and subscripts s and l indicates the variables for short- and long-term effect, respectively. Note that we have included new

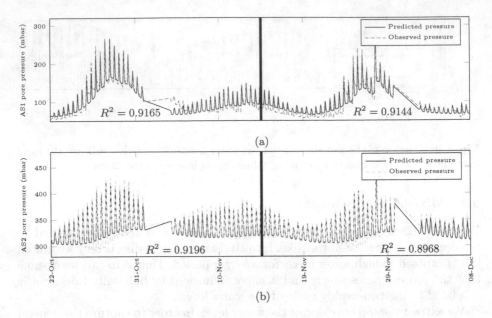

Fig. 5. A comparison between predicted and observed pore pressure values for (a) the AS1 sensor and (b) the AS2 sensor. The black bar separates the training and test sets.

parameters p_l and p_s that raise the water levels $\hat{\mathbf{x}}$ and $\mathbf{x}$ to the powers of p_l and p_s, respectively. This means that the relation between water level (general trend of water level) and the pressure (saturation degree) on the front of the dike is assumed to be $c(x[n])^p$. Exploration of the sensor signals revealed that the response of some sensors is much higher to water levels above a particular threshold. There are several explanations for such an effect, which include the vertical height of the sensor that is below or above a particular water level, a change in slope of the dike front or a change in material covering the dike around that height. From the actual setup of the dike, as shown in Fig. 1, it is not immediately clear which of these the underlying reason might be.

In general, we expect that the long-term effect is based on a longer history than the short-term effect; i.e. $M_l \gg M_s$. Here, we set $M_l = 2000$ and $M_s = 100$, which correspond to roughly 21 and 1 day(s), respectively. The mixed-effect model is more richly parameterized than the short-effect model, and therefore we also need more data to reliably estimate the parameters. For the mixed-effect sensor, the training set is of size $P = N - M_l + 1$, and we decided to use approximately 21 days of training data; i.e. $P = 2000$. The training set consists of the general trend of water level input vectors $\hat{X} = \{\hat{\mathbf{x}}_i = x[i + M_l - 1], \ldots, x[i] \mid i = 1, \ldots, P\}$, water levels input vectors $X = \{\mathbf{x}_i = x[i + M_l - 1], \ldots, x[i + M_l - M_s] \mid i = 1, \ldots, P\}$, and pore pressure output $\mathbf{y} = \{y_i = y[i + M_l - 1] \mid i = 1, \ldots, P\}$.

Table 1. The estimated parameters for the mixed-term sensors, as well as the goodness of fit R^2 for training and unseen data

Sensor	b	a_l	λ_l	p_l	a_s	λ_s	p_s	R^2	R^2_{unseen}
AS1	57.75	$4.32e{-}11$	0.0089	10.59	$2.92e{-}6$	0.83	7.47	0.9165	0.9144
AS2	283.31	$6.18e{-}11$	$7.15e{-}4$	9.41	$2.2e{-}3$	1.24	4.60	0.9196	0.8968

We estimate the set of parameters $\boldsymbol{\theta}$ by MLE, which minimizes the sum of squared residuals as defined in Eq. 3, but this time the residuals $r_i(\boldsymbol{\theta}) = y_i - f(\boldsymbol{\theta}; \mathbf{x}_i, \hat{\mathbf{x}}_i)$ are defined with respect to the new model $f(\boldsymbol{\theta}; \mathbf{x}, \hat{\mathbf{x}})$. Table 1 presents the estimated parameters for the AS1 and AS2 sensors, as well as the goodness of fit for training and test data. The values determined by the proposed models for training and test sets are shown in Fig. 5. The AS1 sensor model only partly captures the long-term effect. In particular, there is a significant difference around November 3-4. The short-term effect is also not modeled very accurately, but this might be a consequence of the imperfect long-term effect model. Our observation is confirmed by the goodness of fit $R^2 = 0.9165$, which is slightly worse than the goodness of fit for short-term effect sensors. For unseen data we obtain $R^2 = 0.9144$, which indicates that the model captures at least some of the underlying dynamics of the AS1 sensor.

The AS2 sensor is dominated by the short-term effect, and only has a minor contribution from the long-term one. Fig. 5b shows that the estimated model captures the short-term effect, but fails to learn the long-term one. Nevertheless, the goodness of fit for both training and test set is in the order of $R^2 = 0.91$.

We have excluded the results of the AC2 and AC3 sensors, since they are in line with the AS2 sensor; the proposed model is not robust enough to fully capture the long-term effect. We believe that this is mainly due to other environmental factors (e.g. outside temperature, humidity, and air pressure) that were not considered (because they were not available) in the proposed model.

4   Anomaly Detection

In the previous section, we showed that we can reliably estimate a model for the short-term effect sensors. As an example application, we show that such models can be effectively used to detect changes in the response of the sensor — the so-called anomalies.

We employ the following semi-supervised strategy to detect anomalies in the AC4 sensor. We first estimate the parameter set $\boldsymbol{\theta}$ of the model on data of October 2011 (that is explicitly labeled as normal). We then use the estimated model $f(\boldsymbol{\theta}; \mathbf{x})$ to predict future pore pressure values $\tilde{y}[n]$ for the next months. A pointwise anomaly score is calculated by measuring the Euclidean distance $\sqrt{(y[n] - \tilde{y}[n])^2}$ between the predicted and observed pore pressure. We plot the anomaly score of the AC4 sensor till August 2012 in Fig. 6a, and mark two anomalies in this plot by colored rectangles. The red rectangle indicates a rather

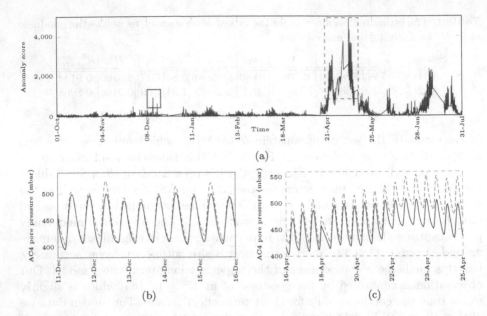

Fig. 6. (a) Anomaly score for the AC4 sensor that is computed as the pointwise Euclidean distance between the predicted and observed pore pressure measurements. (b-c) A zoomed-in visualization of the predicted (purple dashed) and observed (black solid) pressure measurements for the respective red solid and green dashed marked regions in (a).

small anomaly that corresponds to a small offset between the predicted and observed pressure around December 12, Fig. 6b. On the other hand, the green rectangle indicates a more serious anomaly since the anomaly score is high over a long period of time. Indeed, Fig. 6c illustrates a significant discrepancy between predicted and observed pore pressure from April 16 till April 25. Note that the anomaly score returns to almost zero around half of May, and thus the detected anomaly is not a structural change in the response of the sensor. April 2012 was characterized by extreme rainfall[5], and we speculate that the anomaly is caused by outflow from the locks just upstream of the monitoring site.

5   Discussion and Conclusions

In this paper, we proposed parametric nonlinear regression models that describe the relationship between a water height sensor and individual pore pressure sensors. The models that we propose are highly effective (in the order of goodness of fit $R^2 = 0.97$) for pore pressure sensors (AC4 and AC5) that exhibit short-term physical phenomenon. Moreover, we demonstrated that the proposed model can be effectively used for the detection of anomalies.

---

[5] See http://www.metoffice.gov.uk/climate/uk/summaries/2012/april

While we also achieved reasonably high goodness of fit for what we refer to as long-term effect sensors (AS1 and AS2), we believe that the proposed model can be further enriched by incorporating information about other environmental factors, such as rainfall, humidity and outside temperature. Although the presented models are tailored to sensors installed in a sea dike, we think that, due to the general nature of the applied techniques, they are applicable to other sensing and monitoring systems.

Acknowledgement. This work has been funded by TNO. We thank Jeroen Broekhuijsen for making the data available and for valuable discussions.

References

1. Van Baars, S., Van Kempen, I.: The causes and mechanisms of historical dike failures in the Netherlands (2009)
2. Pyayt, A., Mokhov, I., Lang, B., Krzhizhanovskaya, V., Meijer, R.: Machine learning methods for environmental monitoring and flood protection. World Academy of Science, Engineering and Technology (78), 118–124 (2011)
3. Melnikova, N., Shirshov, G., Krzhizhanovskaya, V.: Virtual dike: multiscale simulation of dike stability. Procedia Computer Science 4(0), 791–800 (2011); Proceedings of the International Conference on Computational Science (2011)
4. Tua, P., van Gelder, P., Vrijlinga, J., Thub, T.: Reliability-based analysis of river dikes during flood waves
5. Pyayt, A., Kozionov, A., Kusherbaeva, V., Mokhov, I., Krzhizhanovskaya, V., Broekhuijsen, B., Meijer, R., Sloot, P.: Signal analysis and anomaly detection for flood early warning systems (2014)
6. Nocedal, J., Wright, S.: Numerical optimization, vol. 2. Springer, New York (1999)

Exploiting Novel Properties of Space-Filling Curves for Data Analysis

David J. Weston

Department of Computer Science and Information Systems, Birkbeck College,
University of London, London, United Kingdom
dweston@dcs.bbk.ac.uk

Abstract. Using space-filling curves to order multidimensional data has been found to be useful in a variety of application domains. This paper examines the space-filling curve induced ordering of multidimensional data that has been transformed using shape preserving transformations. It is demonstrated that, although the orderings are not invariant under these transformations, the probability of an ordering is dependent on the geometrical configuration of the multidimensional data. This novel property extends the potential applicability of space-filling curves and is demonstrated by constructing novel features for shape matching.

Keywords: space-filling curves, peano curves, shape preserving transformations.

1 Introduction

Space-filling curves can be used to map multidimensional data into one dimension that preserves to some extent the neighbourhood. In other words points that are close, in the Euclidean sense, in the multidimensional space are likely to be close along the space-filling curve. This property has been found to be useful in many application domains, ranging from parallelisation to image processing [1].

This paper examines the ordering of point sets mapped to a space-filling curve that have been transformed using shape preserving transformations. It is shown that the probability of an ordering is related to the geometry of the points in the higher dimensional space. Crucial to the analysis is the definition of *betweenness* and the ability to measure a corresponding *in-between* probability. The motivation for this paper is to demonstrate that the spatial configuration of multivariate data can be usefully encoded with these *in-between* probabilities with a view to develop novel data analysis algorithms. To this end a practical example based on shape matching is described which uses features derived from in-between probabilities.

The remainder of this paper is structured as follows. The following section space-filling curves are described in more detail and relevant literature is reviewed. In Section 3 betweenness and the in-between probability are defined. Section 4 presents experiments to demonstrate the geometric underpinnings of

H. Blockeel et al. (Eds.): IDA 2014, LNCS 8819, pp. 356–367, 2014.

the in-between probability. Section 5 concludes with a discussion regarding applying the approach to other data analytic tasks. Note some figures and definitions have been reproduced from [20].

2   Background and Related Work

This section briefly describes the construction of space-filling curves and discusses related work.

2.1   Space-Filling Curves

A space-filling curve is a continuous mapping of the unit interval $[0,1]$ onto a higher dimensional Euclidean space, where the image of the unit interval consists of every point within a compact region. For two dimensional space this means the image has non-zero area and the mapping is typically defined to fill the unit square and in three dimensions the image fills the unit cube, etc.

For simplicity only mappings onto two dimensional space are considered, but it is worth noting that the ideas in this paper generalise to higher dimensional space.

Space-filling curves are typically defined recursively where the unit square is subdivided into equal sized sub-tiles and ordered. The first three iterations of the recursion for the Siérpinski curve are shown in Figure 1. The lines joining the centres of the ordered sub-tiles are collectively referred to as the polygon approximation to the space-filling curve. The Siérpinski curve is the limit of this polygon approximation curve as the size of the sub-tiles tends to zero.

Not all recursively defined orderings have a curve as the limit, one example is raster order shown in Figure 2 (in the limit this mapping is space-filling but not a curve, see e.g. [13] for an detailed explanation of this issue). In the computing literature these orderings are often referred to as discrete space-filling curves due to the fact that the polygon approximation curve visits all the sub-tiles. In order to allow for the use of discrete space-filling curves, the multidimensional data will be represented in a (sufficiently finely) discretized space.

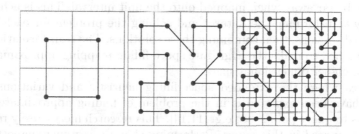

Fig. 1. First three iterations for Siérpinski curve construction

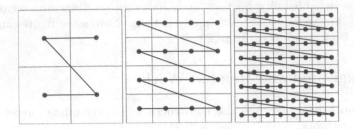

Fig. 2. Raster scan order

Typically space-filling curves are used to map data to the unit interval, hence it is the *inverse* of the space-filling curve mapping that is required, source code for calculating the inverse of various space-filling curves in two and higher dimensions can be found in e.g. [1,15,16].

2.2   Related Work

Combinatorial problems in multidimensional Euclidean space can be approached using the *general space-filling curve heuristic* [4]. This heuristic involves using a space-filling curve to map data onto the unit interval and then solve the one dimensional version of the problem, which is often much easier. A notable example is the planar Travelling Salesman Problem [15] in which, given the locations for a set of cities, the problem is to find the shortest tour. A tour begins and ends at the same city and visits all the other cities only once. In two or more dimensions this is a well known $\mathcal{NP}$ problem, however in one dimension this problem has polynomial computational complexity. Indeed in one dimension the shortest tour can be constructed by simply sorting the city locations into ascending (or descending) order. It is the neighbourhood preserving properties of the space-filling curve mapping that ensure that the optimal one dimensional tour, once it is projected back to the original dimension, produces a reasonable sub-optimal solution.

An extension this heuristic is called the *Extended Space-filling Heuristic* [14] and is designed to address the problem that points close in the higher dimensional space may be far away when mapped onto the unit interval. This is achieved by repeatedly transforming the dataset and solving the problem for each of these transformed versions, then combining these solutions. The transformation of the data is designed to make the aggregate space-filling mapping approximate more closely the higher dimensional space.

One area where the extended space-filling heuristic and variations of this heuristic have been explored is in the problem of finding approximate nearest neighbours to query points, see e.g. [11,14]. This research most closely resembles the work proposed in this paper. Performing shape preserving shape transformations to the data (and the query point) will obviously not affect the nearest neighbour when measured in the original high dimensional space however it will effect the point order. The motivation for transforming the data is to increase the probability that the 'true' nearest neighbour is close to query point along the unit interval. In contrast this paper proposes *measuring* these probabilities, since they carry information about the spatial configuration of the dataset.

Shape Matching. In Section 4 a shape matching task is used to further demonstrate that spatial information of a point set can be captured using probabilities based on space-filling curve induced point orderings. In this section the use of space-filling curves to map shapes to one-dimension is discussed.

There are not many instances in the literature where space-filling orderings are used to represent shapes and in most cases shape normalization is performed before the shape is mapped to one-dimension. This is done to reduce as far as possible the effect of the change in point ordering due to affine transforms, see e.g. [6,9,19]. In [17,18] the space of all possible rotations and translations is searched (interestingly using another space-filling curve) to find a match.

Matching using one-dimensional representations of shapes which used cross-correlation was proposed in [8]. Class specific regions of the representation, known as *key feature points*, can be extracted by overlaying one-dimensional representation from shapes of the same class. Intervals that have lower variance are considered to be informative for identifying the class. A portion of the one-dimensional representation with the lowest variance is extracted to produce a representation of reduced length and high similarity across the class. An extension to the key feature point [7] denoted *rotational key feature points* involves concatenating representations from rotated instances of the same shape and identifying key feature points.

3 Betweenness and the In-between Probability

This section first presents a demonstration for the in-between probability using the Siérpinski curve before presenting a more formal definition.

Consider 3 points a, b, c. The point b, is *in-between* a and c, if it is on the shortest path on the curve between a and c. The darkened part of the curve in Figure 3 shows examples of shortest paths on polygon approximation to the Siérpinski curve.

The probability b is in-between a and c is simply the proportion of shape preserving affine transforms that map b to the region between the transformed locations of a and c. For example, in Figure 4 each image shows a shape preserving transformation of a right triangle. This figure shows that the configuration of the in-between region varies depending on the locations of a, c. Only in the first and last image is b in-between and a, c.

3.1 In-between Probability

This section presents the in-between probability more formally and for clarity only the two dimensional case is considered.

Fig. 3. Examples of shortest paths (shaded) along a polygon approximation to the Siérpinski curve between two points

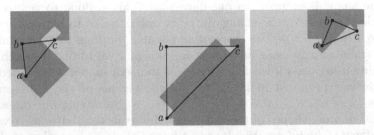

Fig. 4. Affine transformed right triangle with region in-between a, c darkened

Let (a, b, c) be a 3-tuple of unique points in the unit square, e.g. the vertices of a triangle shown in Figure 4. Let the shape preserving transformations be scale, translation, rotation and reflection (and composites of these transformations).

There are two minor technical considerations. First for simplicity the space-filling curves used in this paper are defined over the unit square, hence no point should be transformed outside the unit square, otherwise its location along the curve cannot be measured. The set of allowable transformations for a tuple (a, b, c), i.e. those that map all three points into the unit square, is denoted $\mathcal{S}_{\{a,b,c\}}$.

For $s \in \mathcal{S}_{\{a,b,c\}}$, let $a' = s(a)$ this is the location point a after the shape preserving transformation s is applied. The second minor technical consideration relates to the use of discrete space-filling curves. These mappings require the unit square to be discretized, hence all transformed points are rounded to their nearest tile centre.

Figure 3 shows that the Siérpinski curve wraps around to meet itself, whereas raster order does not (Figure 2). In order to capture this difference two types of betweenness, *circular* and *linear*, are defined.

The *linear* in-between probability for tuple (a, b, c) and space-filling curve f is defined as, $p(X_l = i; (a, b, c), f, \mathcal{S}_{\{a,b,c\}})$
where $i \in 0, 1$ and X_l is a random variable defined as,

$$
X_l = \begin{cases} 1 & \begin{array}{l} \text{if } f^{-1}(a') < f^{-1}(b') < f^{-1}(c') \\ \text{or } f^{-1}(c') < f^{-1}(b') < f^{-1}(a'), \end{array} \\ \\ 0 & \text{otherwise.} \end{cases}
\tag{1}
$$

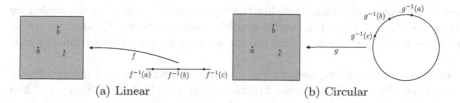

(a) Linear   (b) Circular

Fig. 5. The in-between mapping

In words, for a particular space-filling curve mapping f,
$p(X_l = 1; (a, b, c), f, S_{\{a,b,c\}})$ is the probability the pre-image of b is in-between
the pre-images of a and c under valid shape preserving transformations. Recall
that space-filling curves are defined to map points from the unit interval onto
the higher dimensional space, hence the inverse space-filling curve mapping is
required, see Figure 5(a).

Using similar notation, the *circular* in-between probability is defined as,
$p(X_c = i; (a, b, c), g, S_{\{a,b,c\}})$, where $i \in 0, 1$, g is a space-filling curve mapping
and X_c is a random variable which is defined as,

$$
X_c =
\begin{cases}
1 & \begin{aligned}&\text{if } g^{-1}(b') \text{ is on the shortest path connecting,}\\&\text{but not including, } g^{-1}(a') \text{ and } g^{-1}(c')\end{aligned} \\
& \\
0 & \text{otherwise.}
\end{cases}
\tag{2}
$$

See Figure 5(b) for a graphical representation of *circular* betweenness.

4   Spatial Configurations and In-between Probabilities

The previous section defined the in-between probability, in this section the relationship between spatial configurations of points and their corresponding in-
between probabilities is investigated experimentally. First by empirically estimating the in-between probability distribution for triangles in the plane then
by investigating how well the spatial configuration of large sets of points can
be usefully captured using betweenness probabilities in the practical setting of
shape matching.

For all the following experiments the set of shape preserving transformations
is sampled as follows:

The unit square is subdivided into 2048×2048 tiles and all transformed locations are rounded to the nearest tile centre. This level of granularity was chosen
to allow shapes described in Section 4.2 to be scaled up to an order of magnitude. First, with probability $\frac{1}{2}$ the shape is reflected through the x-axis. Then,
the shape's centre of gravity is translated to a location that has been sampled
uniformly at random from the unit square. The shape is then rotated uniformly
about its centre of gravity. A scale is sampled uniformly in the range 1 to a maximum scale S, where S is chosen such that a shape scaled to any value greater

than S will not fit completely within the unit square. A shape is not scaled by a value less that 1 since this would amplify aliasing effects. Finally the transformation is rejected if the points do not all map to positions within the unit square.

For linear betweenness, assume $x_1, \dots, x_\eta$ are identically and independently drawn from the probability mass function $p(X_l = i; (a, b, c), f, \mathcal{S}_{\{a,b,c\}})$. Then the maximum likelihood estimate is simply,

$$\hat{p}(X = i; (a, b, c), \mathcal{S}_{\{a,b,c\}}) = \frac{1}{\eta} \sum_{t=1}^{\eta} \mathbf{1}(x_t = i),$$

where $\mathbf{1}(\cdot)$ is the indicator function and the number of samples, η, is set to 20,000. A similar formula can be obtained for circular betweenness.

4.1   Estimating the in-between probabilities for triangles

In this section the in-between probability for different triangular configurations of points is investigated empirically, more precisely the relationship between the shape of a set of 3 points (a, b, c) and the circular in-between probability $p(X_c = i; (a, b, c), f, \mathcal{S}_{\{a,b,c\}})$, where f denotes a Siérpinski curve mapping.

A simple way to represent shape of triangles in two dimensions is to use Bookstein shape coordinates. In these coordinates the location of points a, c are fixed to the locations $a = (-\frac{1}{2}, 0)$ and $c = (\frac{1}{2}, 0)$, the location of b is the free parameter. Note, since reflections are one of the shape preserving transformations, the location b can be restricted to the positive half plane to get the full distribution. To obtain a larger set of triangular shapes the domain b is -3 to 3. This coordinate system is shrunk by a factor of $\frac{1}{3}$ and translated in order to fit into the unit square.

Figure 6 shows the circular in-between probability mass function for the Siérpinski curve, shown in both a surface plot and a contour plot. Each location in the plot corresponds to b a vertex of the triangle which has as a base

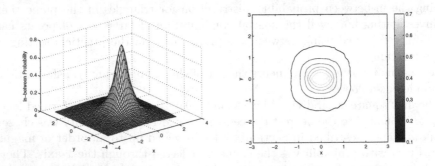

Fig. 6. The Siérpinski curve circular in-between probability mass function in both a surface plot and a contour plot. The location of points a and c are $(-\frac{1}{2}, 0)$ and $(\frac{1}{2}, 0)$ respectively.

the segment joining $(-\frac{1}{2}, 0)$ to $(\frac{1}{2}, 0)$. The symmetry about the x-axis is due to introducing reflection invariance. It can be seen there is a clear dependency between the probability and shape of the triangle (a, b, c). The maximum occurs at the midpoint between a and c. The contour plot demonstrates that, in general, a particular value for the in-between probability does not correspond to a particular shape of triangle. The locus of shapes with the same in-between probability starts approximately elliptical and becomes progressively rounder the further away b is from the line segment joining a to c.

4.2 Shape Matching

The objective of the following experiments is twofold. First to demonstrate the joint in-between distribution for data comprising of more than three points is also related to its spatial configuration. Second to directly compare novel shape descriptors based on this joint in-between distribution with state-of-the-art shape descriptors.

Fig. 7. Example images from the MPEG-7 Core Experiment CE-Shape-1 Part B dataset

The MPEG-7 Core Experiment CE-Shape-1 Part B dataset is a widely used benchmark dataset for image retrieval that contains 1400 shapes, [10]. There are 70 classes of shape, each with 20 instances, examples of shapes from this dataset can be seen in Figure 7. Performance for this benchmark dataset is measured using the *bulls-eye* score, which is calculated as follows.

For each target shape retrieve the 40 most similar shapes, count the number of shapes that are from the same class as the target. The maximum score for one target shape is 20 and the overall maximum score is 28,000. The bulls-eye score is typically shown as the percentage of the maximum score.

The approach that has the current highest bulls-eye score, which is 97.4%, is described in [3]. This approach uses two different shape descriptors; *shape contexts* (SC) and *inner distance shape contexts* (IDSC). The main purpose of [3] is to introduce an algorithm called *co-transduction* which efficiently combines shape dissimilarities derived from these two descriptors. Combining approaches is beyond the scope of this paper, however motivated by the success of the shape descriptors used, the following experiments include results for SC and IDSC for comparison. The reader is referred to [5] and [12] for detailed descriptions for SC and IDSC respectively.

In the following experiments, shape matching is achieved using the approach described in [12]. Briefly, each shape is represented by $n = 100$ points extracted at regular intervals from the boundary and for each point a descriptor is measured.

Shape matching proceeds in the following fashion, let shape S_1 consist of the points $p_1, \ldots, p_n$ and the shape S_2 the points $q_1, \ldots, q_n$. A dissimilarity matrix $c_{i,j}$ is generated where each entry is a measure of the difference between the descriptors for point p_i and for point q_j. The level of dissimilarity between shapes S_1 and S_2 involves finding an optimal mapping between the point sets from S_1 and S_2 which is solved using dynamic programming.

Novel Descriptors. The concern in this section is how to construct a dissimilarity matrix using betweenness probabilities. Taking any two points, p_i and p_j from S_1, it is possible in principle, to build a distribution over the number of the remaining points that lie in-between them along the space-filling curve. This distribution contains information about the spatial locations of the remaining points relative to p_i and p_j. However this would be unwieldy to measure and store, instead two simple descriptors are proposed.

The *mean* descriptor. Let $f_\mu(p_i, p_j)$ be the expected number of points in-between p_i and p_j. Then the descriptor for point p_i is the set $\{f_\mu(p_i, p_1), \ldots, f_\mu(p_i, p_n)\}$.

The 10% descriptor, $f_{10\%}(p_i, p_j)$ is the probability that 10% of the total number of points or fewer are in-between p_i and p_j. The descriptor for point p_i is the set $\{f_{10\%}(p_i, p_1), \ldots, f_{10\%}(p_i, p_n)\}$.

There are, of course, plenty of alternative features that could have been constructed. The advantage of the two described above is their very obvious relationship with the underlying in-between probabilities. Furthermore in both cases the descriptor assigns a one dimensional vector to each point much like SC and IDSC.

To measure the dissimilarity, $c_{i,j}$, between p_i from shape S_1 and q_j from shape S_2, the descriptor sets of p_i and q_j are sorted into order and the absolute difference between the entries is taken, i.e.

$$c_{i,j} = \sum_{k=1}^{n} |f(p_i, p_{\pi_{p_i}(k)}) - f q_j, q_{\pi_{q_j}(k)})|,$$

where π_{p_i} and π_{q_j} denote the values in the descriptor sets of p_i and q_j sorted into ascending order respectively.

The dissimilarity matrix c is all the information needed to use the matching process described above.

Results. To allow for a direct comparison between IDSC, SC and the proposed shape descriptors, shape matching for both IDSC and SC is performed such that it is invariant to rotation and reflection.

For each descriptor, the space-filling curve mapping that yielded the highest bulls-eye score is shown in Table 1. For the 10% descriptor this was the Siérpinski curve and for the *mean* descriptor this was raster order. In both these cases the performance was not at the same level as SC and IDSC. Note that the bulls-eye score for IDSC is slightly higher than that reported by [12], it is also interesting to note that both SC and IDSC have very similar performance.

Table 1. Bulls-eye scores

Method	Siérpinski-10%	Raster-mean	SC	IDSC
Score	77.72%	78.80%	85.22%	85.81%

Table 2. Bulls-eye scores using additional clustering step

Method	Siérpinski-10%	Raster-mean	SC	IDSC
Score	86.14%	87.15%	90.93%	91.17 %

For this particular retrieval task, plugging in an additional clustering phase has been shown to greatly improve performance [2]. Table 2 show the results that includes a clustering step referred to as Graph Transduction [2]. All the approaches have been dramatically improved and with our novel descriptors obtaining the greatest boost. The results shown in Table 2 clearly demonstrate that our descriptors are capable of encoding in a meaningful way the spatial configuration of a point set.

Finally it should be noted that space-filling approaches have been applied to this image retrieval task, namely the key feature point and the rotational key feature point, which were described in Section 3. These approaches have have bulls-eye scores of 85.3% and 99.3% respectively. However these results cannot easily be compared to the results shown above and indeed the majority of methods applied to this MPEG-7 shape retrieval task since both the key feature point and the rotational key feature point require the use of additional information about shape classes.

5   Conclusion

It should be remarked that although the examples described in this paper have been in two dimensions the methodology extends naturally to higher dimension. In order to perform analysis of n-dimensional data all is needed is an n-dimensional space filling curve and the ability to affine transform points in n-dimensional space.

This paper has shown that the in-between probability is related to the spatial configuration of a dataset. This has been demonstrated by investigating the in-between probability of triangles in the plane and by using features derived from the in-between probability to successfully perform an image retrieval task. Although these features did not achieve state-of-the-art performance, the very fact that these features captured sufficient information about the configuration to perform the task suggests that in-between probabilities are likely to be useful in other data analytic tasks.

For example the median of a point set could be defined as the data point which is most likely to be in-between all other pairs of points in the dataset.

Taking this concept further, the degree to which a point is in-between all point pairs can be used identify outliers.

Indeed any data analysis processes that requires a concept of neighbourhood in the Euclidean sense, such as those that use Voronoi graphs, are all candidates for our approach to be deployed.

References

1. Bader, M.: Space-Filling Curves: An Introduction with Applications in Scientific Computing, vol. 9. Springer (2012)
2. Bai, X., Yang, X., Latecki, L.J., Liu, W., Tu, Z.: Learning context-sensitive shape similarity by graph transduction. IEEE Transactions on Pattern Analysis and Machine Intelligence 32(5), 861–874 (2010)
3. Bai, X., Wang, B., Yao, C., Liu, W., Tu, Z.: Co-transduction for shape retrieval. IEEE Transactions on Image Processing 21(5), 2747–2757 (2012)
4. Bartholdi III, J., Platzman, L.: Heuristics based on spacefilling curves for combinatorial problems in euclidean space. Management Science 34(3), 291–305 (1988)
5. Belongie, S., Malik, J., Puzicha, J.: Shape matching and object recognition using shape contexts. IEEE Transactions on Pattern Analysis and Machine Intelligence 24(4), 509–522 (2002)
6. Ebrahim, Y., Ahmed, M., Abdelsalam, W., Chau, S.: Shape representation and description using the Hilbert curve. Pattern Recognition Letters 30(4), 348 (2009)
7. Ebrahim, Y., Ahmed, M., Chau, S., Abdelsalam, W.: Significantly improving scan-based shape representations using rotational key feature points. In: Campilho, A., Kamel, M. (eds.) ICIAR 2010. LNCS, vol. 6111, pp. 284–293. Springer, Heidelberg (2010)
8. Ebrahim, Y., Ahmed, M., Chau,S., Abdelsalam, W.: A view-based 3D object shape representation. In: 4th International Conference on Image Analysis and Recognition, ICIAR 2007, Montreal, Canada, August 22-24, pp. 411–422. Springer-Verlag New York Inc. (2007)
9. El-Kwae, E.A., Kabuka, M.R.: Binary object representation and recognition using the Hilbert morphological skeleton transform. Pattern Recognition 33(10), 1621–1636 (2000)
10. Latecki, L.J., Lakamper, R., Eckhardt, T.: Shape descriptors for non-rigid shapes with a single closed contour. In: Proceedings. IEEE Conference on Computer Vision and Pattern Recognition 2000, vol. 1, pp. 424–429 (2000)
11. Liao, S., Lopez, M., Leutenegger, S.: High dimensional similarity search with space filling curves. In: Proceedings of the International Conference on Data Engineering. pp. 615–622 (2001)
12. Ling, H., Jacobs, D.W.: Shape classification using the inner-distance. IEEE Transactions on Pattern Analysis and Machine Intelligence 29(2), 286–299 (2007)
13. Peitgen, H., Jürgens, H., Saupe, D.: Chaos and Fractals: New Frontiers of Science. Springer (2004)
14. Perez-Cortes, J., Vidal, E.: The extended general spacefilling curves heuristic. In: 1998. Proceedings. Fourteenth International Conference on Pattern Recognition, August 16-20, vol. 1, pp. 515–517 (16)
15. Platzman, L., Bartholdi, J.: Spacefilling curves and the planar travelling salesman problem. Journal of the Association for Computing Machinery 36(4), 719–737 (1989)

16. Sagan, H.: Space-Filling Curves. Springer (1994)
17. Tian, L., Chen, L., Kamata, S.: Fingerprint matching using dual Hilbert scans. In: SITIS 2007: Proceedings of the 2007 Third International IEEE Conference on Signal-Image Technologies and Internet-Based System, pp. 593–600. IEEE Computer Society, Washington, DC (2007)
18. Tian, L., Kamata, S.: A two-stage point pattern matching algorithm using ellipse fitting and dual Hilbert scans. IEICE Transactions on Information and Systems E91-D(10), 2477–2484 (2008)
19. Tian, L., Kamata, S.I., Tsuneyoshi, K., Tang, H.: A fast and accurate algorithm for matching images using Hilbert scanning distance with threshold elimination function. IEICE Transactions on Information and Systems 89(1), 290–297 (2006)
20. Weston, D.: Shape Matching using Space-Filling Curves. Ph.D. thesis, Imperial College, London (July 2011)

RealKrimp — Finding Hyperintervals
that Compress with MDL for Real-Valued Data

Jouke Witteveen[1], Wouter Duivesteijn[2], Arno Knobbe[3], and Peter Grünwald[4]

[1] ILLC, University of Amsterdam, The Netherlands
[2] Fakultät für Informatik, LS VIII, TU Dortmund, Germany
[3] LIACS, Leiden University, The Netherlands
[4] CWI and Leiden University, The Netherlands

Abstract. The MDL Principle (induction by compression) is applied
with meticulous effort in the KRIMP algorithm for the problem of itemset mining, where one seeks exceptionally frequent patterns in a binary
dataset. As is the case with many algorithms in data mining, KRIMP is
not designed to cope with real-valued data, and it is not able to handle such data natively. Inspired by KRIMP's success at using the MDL
Principle in itemset mining, we develop REALKRIMP: an MDL-based
KRIMP-inspired mining scheme that seeks exceptionally high-density patterns in a real-valued dataset. We review how to extend the underlying
Kraft inequality, which relates probabilities to codelengths, to real-valued
data. Based on this extension we introduce the REALKRIMP algorithm:
an efficient method to find hyperintervals that compress the real-valued
dataset, without the need for pre-algorithm data discretization.

Keywords: Minimum Description Length, Information Theory, Real-
Valued Data, REALKRIMP.

1 Introduction

When data result from measurements made in the real world, they quite often are taken from a continuous, real-valued domain. This holds, for example,
for meteorological measurements like temperature, precipitation, atmospheric
pressure, etcetera. Similarly, the sensors in a smartphone (GPS, accelerome-
ter, barometer, magnetometer, gyroscope, light sensor, etcetera) monitor data
streams from a domain that is, for all practical purposes, real-valued. Most data
mining algorithms, however, specialize in data from a discrete domain (binary,
nominal), and can only handle real-valued data by discretization. Native support
for real-valued data would be an asset to such algorithms.

An example of a popular discrete algorithm is KRIMP [1], an algorithm that
finds local patterns in the data. Specifically, KRIMP seeks *frequent itemsets*:
attributes that co-occur unusually often in the dataset. KRIMP employs a mining
scheme to heuristically find itemsets that *compress* the data well, gauged by a
decoding function based on the Minimum Description Length Principle [2,3].

In an effort to extend the applicability of KRIMP to continuous data, we introduce REALKRIMP: a KRIMP-inspired mining scheme with a strong foundation in

H. Blockeel et al. (Eds.): IDA 2014, LNCS 8819, pp. 368–379, 2014.

information theory, that finds interesting hyperintervals in real-valued datasets. This interestingness is expressed by an MDL-based model for compression in real-valued data. The resulting REALKRIMP algorithm can be seen as a KRIMP-inspired model for frequent patterns in continuous data, where the role of the frequent itemsets is played by hyperintervals in the continuous domains, that show an exceptionally high density.

2   Related Work

The Minimum Description Length (MDL) principle [2,3] can be seen as the more practical cousin of Kolmogorov complexity [4]. The main insight is that patterns in a dataset can be used to compress that dataset, and that this idea can be used to infer which patterns are particularly relevant in a dataset by gauging how well they compress: the authors of [1] summarize it by the slogan *Induction by Compression*. Many data mining problems can be practically solved by compression. Examples of this principle have been given for classification, clustering, distance function design (all in [5]), feature selection [6], finding temporally surprising itemsets [7], and defining a parameter-free distance measure on time series [8]. Clearly, the versatility of compression as a data mining tool has been recognized by the community. All the work done so far within the data mining community, however, has in common that the structure being compressed stems from a domain that is either finite [5,6,7] or at most countably infinite [5,8]. This is in sharp contrast with the use of MDL in statistics and machine learning, which has included continuous applications such as density estimation and regression from the very beginning [2]. The present paper provides a continuous-data MDL application in data mining.

In the data mining subtask of finding a small subset of dataset-describing patterns, arguably the most famous contribution is KRIMP [1], as described in the introduction of this paper. An alternative and closely related approach to data summarizing is tiling [9]. Tiling seeks a group of, potentially overlapping, itemsets that together cover all the ones in a binary dataset. Similar as it may be to KRIMP, tiling does not concern itself with model complexity or MDL. While KRIMP approaches the binary dataset in an asymmetric fashion, only regarding the items that are present (the ones), two methods inspired by KRIMP fill the void by approaching the dataset in a symmetric fashion. Pack [10] combines decision trees with a refined version of MDL, and typically selects more itemsets than KRIMP. Conversely, LESS [11] sacrifices performance in terms of the involved compression ratio, in order to end up with a set of low-entropy patterns that is typically an order of magnitude smaller than the set found with KRIMP.

3   Relating Codes and Probabilities

An important piece of mathematical background for the application of MDL in data mining, which is relevant for both KRIMP and REALKRIMP, is the *Kraft Inequality*, relating code lengths and probabilities (cf. [12]). In the following sections, we inspect the inequality in its familiar form, where it is only applicable to at most countable spaces, and then show the derivation of a suitable code length function for Euclidean (hence uncountable) spaces.

Consider a *sample space*, i.e., a set of possible outcomes Ω. Let $\mathcal{E}$ be a partition of Ω that is finite or countably infinite. We think of $\mathcal{E}$ as the level of granularity at which data are observed. If Ω is countable, then in most MDL applications, $\mathcal{E} = \Omega$; but if $\Omega = \mathbb{R}^d$, then we will always receive data only up to a given precision determined by the data generating and processing system at hand; then $\mathcal{E}$ is some coarsening of $\mathbb{R}^d$; in practice, the modeler or miner may not know the details (such as the precision) of this coarsening but as we will see, the MDL principle can still be applied without such knowledge. A *probability mass function* p on a countable set $\mathcal{E}$ is simply a function $p : \mathcal{E} \to [0, 1]$ so that $\sum_{y \in \mathcal{E}} p(y) = 1$. We call p *defective* if, instead, $\sum_{y \in \mathcal{E}} p(y) < 1$.

Let $\mathcal{A}$ denote a finite alphabet, and let $\mathcal{C}$ denote a finite or countably infinite *prefix-free subset* of $\bigcup_{i \geq 0} \mathcal{A}^i$, i.e., a subset of the strings over $\mathcal{A}$ such that there exist no two elements z, z' of $\mathcal{C}$ such that z is a strict prefix of z'. A *description method* [3] for $\mathcal{E}$ with *code word set* $\mathcal{C}$ is defined implicitly by its *decoding function*, a surjection $D : \mathcal{C} \twoheadrightarrow \mathcal{E}$. While we allow that some $y \in \mathcal{E}$ can be encoded in more than one way, we do require unique decodability, so that the inverse function D from $\mathcal{C}$ to $\mathcal{E}$ does exist. Note that these requirements are standard in all applications of MDL [3]. We call a description method a (prefix) *code* if D is 1-to-1; a natural way to turn a given description method D into a code is to encode each $y \in \mathcal{E}$ by the *shortest* z with $D(z) = y$. The *length function* corresponding to code C, $\ell_C : \mathcal{E} \to \mathbb{Z}_{\geq 0}$, assigns to an outcome in $\mathcal{E}$ the length of its encoding under C.

Theorem 1 (Kraft). *For every code C over an alphabet $\mathcal{A}$, a (possibly defective) probability mass function p on $\mathcal{E}$ exists that makes short encoded lengths and high probabilities of outcomes correspond as follows:*

$$\text{for all } y \in \mathcal{E}: \ -\log_{|\mathcal{A}|} p(y) = \ell_C(y). \tag{1}$$

Proofs of this result exist [12] for the case when $\mathcal{C}$ is finite or countably infinite. One can also prove the converse of Theorem 1: for every probability mass function p on $\mathcal{E}$, there is a code C such that (1) holds. This allows a bi-directional *identification* of code length functions and probability mass functions [3]. Thus, as in most papers on MDL and Shannon information theory, we simply *define* codelength functions in terms of probability mass functions: every probability mass function p on $\mathcal{E}$ defines a code with for all $s \in \mathcal{C}$, lengths given by

$$\ell(s) = -\log p(s), \tag{2}$$

This is also the manner in which the relation between code lengths and probabilities is introduced in the KRIMP paper [1, Theorem 1].

What if outcomes are continuous? We start with the basic case with a sample space equal to $\mathbb{R}$. No code allows encoding data points $x \in \mathbb{R}$ with infinite precision, so one proceeds by encoding a discretization of x. We define the *uniform discretization* $\mathcal{E}_k$ of $\mathbb{R}$ at level k as the partition $\{ [n/2^k, (n+1)/2^k) \mid n \in \mathbb{Z} \}$ of $\mathbb{R}$. Every possible outcome x is a member of exactly one element of $\mathcal{E}_k$, denoted $[x]_k$.

Given an arbitrary distribution P on (a connected subset of) $\mathbb{R}$, identified by its density p, and a data point $x_0 \in \mathbb{R}$, the probability of $[x_0]_k$ is given by

$$P([x_0]_k) = \int_{x \in [x_0]_k} p(x)\mathrm{d}x \approx p(x_0)2^{-k} \tag{3}$$

As follows directly from the definition of (Riemann) integration, provided p is continuous, the approximation (3) gets better as $k \to \infty$. This makes it meaningful to define a length-like function, the *lengthiness*, on $\mathbb{R}$ by:

$$\ell^k(x) := -\log p(x) + k. \tag{4}$$

As k gets larger, (4) becomes a better approximation to the actual codelength $-\log P([x]_k)$ achieved at precision k. The lengthiness ℓ^k would only become a proper length function, i.e., one that satisfies Theorem 1, in the limit as k approaches infinity, where it would assign infinite length to all elements of X. However, crucially, the lengthiness does not alter its behavior with varying k, other than that it is shifted upwards or downwards. Hence, to *compare* elements of X, non-limit values for the lengthiness can be used as a length proxy.

To extend the idea above to encode data vectors $x = (a_1, \ldots, a_n)$ in $\mathbb{R}^n$ for some $n > 1$, we define $\mathcal{E}$ as a set of hyperrectangles of side width 2^{-k} and define $[x]_k$ to be the single hyperrectangle in $\mathcal{E}$ containing x. Approximating the integral over $[x]_k$ as in (3), we then get a lengthiness of

$$\ell^k(x) := -\log p(x) + n \cdot k. \tag{5}$$

We should note that one can formalize this discretization process in detail for general noncountable (rather than just Euclidean) measurable spaces and general types of discretization. (rather than just uniform; in practice our data may be discretizable in a different manner). This requires substantially more work but leads to exactly the same conclusions as to how to apply MDL to continuous-valued data; for details see [13].

4 Two-Part MDL Code for Hyperintervals

Given a set of candidate hypotheses $\mathcal{H}$ and data ω, the MDL Principle for hypothesis selection tells us to select, as best explanation for the data, the $H \in \mathcal{H}$ minimizing the two-stage description length

$$\ell_1(H) + \ell_2(\omega \mid H). \tag{6}$$

The first term, ℓ_1, is the codelength function corresponding to some code C_1 for encoding hypothesis H. For each $H \in \mathcal{H}$, the second term, $\ell_2(\cdot \mid H)$, is the length function for a code $C_{2,H}$ for encoding the data 'with the help of H', i.e., a code such that, the better H fits ω, the smaller the codelength $\ell_2(\omega \mid H)$.

To find interesting patterns in an *uncountable* dataset ω, consisting of N *records* of the form $x = (a_1, \ldots, a_n)$, where each *attribute* a_i is taken from a real-valued domain, the data can be discretized at level k, turning (6) into

$$\ell_1(H) - \log p(\omega \mid H) + N \cdot n \cdot k, \tag{7}$$

where we approximate the actual codelength function by the 'lengthiness' (5) and the factor N appears because we discretize N data points.

In the original KRIMP paper, the resulting patterns are itemsets in finite-dimensional binary-valued space. In REALKRIMP, the patterns are bounded hyperrectangles, with edges parallel to coordinate axes. Unlike KRIMP, REAL-KRIMP does not demand any point to be covered by the hyperrectangle. Effectively we strive to find relevant endpoints of intervals of attributes. Hence, we refer to such hyperrectangles as *hyperintervals*:

Definition 1 (Hyperinterval). *Let $\bar{\mathbb{R}} = [-\infty, +\infty]$ represent the extended real numbers. Given a set of $2n$ extended reals $h_1^L, h_1^U, h_2^L, h_2^U, \ldots, h_n^L, h_n^U$ in $\bar{\mathbb{R}}$, the hyperinterval $H \subseteq \mathbb{R}^n$ encoded by the $2n$-tuple $(h_1^L, h_1^U, h_2^L, h_2^U \ldots, h_n^L, h_n^U)$ is the subset $H = [h_1^L, h_1^U] \times \ldots \times [h_n^L, h_n^U]$ of $\mathbb{R}^n$ in which the ith dimension is restricted to $[h_i^L, h_i^U]$.*

Just as KRIMP strives to find itemsets that have a relatively high support, REAL-KRIMP should strive to find hyperintervals with a relatively high record density. We want to attain better compression, for increasing difference between density within a hyperinterval and density outside of the hyperinterval (signposted by the records in ω).

Description Length $\ell(\omega)$ of Data without Hypothesis Let M denote the volume of the smallest hyperinterval covering the entire dataset. Without prior information on the dataset, we do not want to discriminate between records a priori, so we assign the same length of $-\log 1/M$ to each record, using the code corresponding to a uniform distribution. That makes the complexity of the dataset equal to

$$N \cdot -\log 1/M = N \log M \tag{8}$$

where we can ignore the discretization constant $N \cdot n \cdot k$ since it needs to be added to both (8) and to (7), which are to be compared.

Description Length $\ell(\omega \mid H)$ of Data given Hypothesis Suppose that we are given a hyperinterval H lying within the interval of volume M; we denote the number of records it covers by N_{in} and its volume by M_{in}. Additionally, we write $N_{\text{out}} = N - N_{\text{in}}$ and $M_{\text{out}} = M - M_{\text{in}}$. Since $N_{\text{in}}, N_{\text{out}}, M_{\text{in}}$, and M_{out} are determined by H and here we assume H as given, we can base our code on these quantities. Each record x is now naturally equipped with the following length:

$$\ell(x) := \begin{cases} -\log 1/M_{\text{in}} = \log M_{\text{in}} & \text{for } x \in H \\ -\log 1/M_{\text{out}} = \log M_{\text{out}} & \text{for } x \notin H. \end{cases} \tag{9}$$

Note that we can code records x with these lengths only once we know, for each record x, whether $x \in H$ or not. Hence, to describe ω given H, we need to describe a binary vector $(b_1, \ldots, b_N)$ of length N where $b_i = 1$ if the ith record is in H. The standard way of doing this is first to describe N_{in} using a uniform code on $\{0, 1, \ldots, N\}$, which takes $\log(N+1)$ bits irrespective of the value of N_{in}, adding another constant (independent of H) to the codelength that is irrelevant for comparisons and hence may be dropped; we then code $(b_1, \ldots, b_N)$ by giving its index in lexicographical order in the set of all N-length bit vectors with N_{in} 1s, which takes $\log \binom{N}{N_{\text{in}}}$ bits, which is itself equal, up to yet another constant term, to $N \cdot \text{Entropy}(N_{\text{in}}/N) = N \log N - N_{\text{in}} \log N_{\text{in}} - N_{\text{out}} \log N_{\text{out}}$ [3]. Combining with Equation (9), the complexity of the dataset is equal to:

$$N_{\text{in}} \log M_{\text{in}}/N_{\text{in}} + N_{\text{out}} \log M_{\text{out}}/N_{\text{out}} + N \log N \qquad (10)$$

Description Length of Hypotheses I We gauge the complexity of specifying the model itself through specifying its boundaries. Consider an attribute a_i with minimal value L and maximal value U in ω. We may define a probability density function $\bar{p}$ on the maximal value of a_i within H as $\frac{x-L}{1/2(U-L)^2}$, a choice justified below. Given this maximal value u, we take a uniform probability density function on the minimal value of a_i within H, which has constant probability density $\frac{1}{u-L}$. Any combination of boundary values for a_i within H now has probability $\frac{x-L}{1/2(U-L)^2} \cdot \frac{1}{x-L} = \frac{2}{(U-L)^2}$, which is independent of the values themselves, thus justifying our choice for $\bar{p}$. Following this procedure for all attributes, the likelihood of every hyperinterval becomes $2^n/M^2$, which corresponds to the 'length':

$$-\log \bar{p}(H) = -\log 2^n/M^2 = 2 \log M - n \log 2 \qquad (11)$$

While we could safely ignore the discretization constant when deriving the raw complexity of the dataset (8) and the complexity of the dataset given a model (10), we cannot do so for the codelength of the hyperinterval. This is because shortly, we will also look at hyperintervals that are defined only on a dimension $n' < n$. Hence we add a discretization constant twice (for the points describing the minimal and maximal values) to (11) to make it a proper 'lengthiness' function. To determine k, note that we should be more demanding towards the detail in the model as the number of records increases, so ideally (11) should increase with N when the discretization constant is taken into account. We take $-\log M/N^n$ for the discretization constant $n \cdot k$, turning (11) into:

$$2 \log M - n \log 2 - 2 \log \frac{M}{N^n} = 2n \log N - n \log 2 \qquad (12)$$

This choice is quite natural, since it has an additional interpretation as the codelength arising from a rather different way of encoding H, namely by specifying, for each dimension $1 \le i \le n$, two records: one giving the lower boundary for attribute a_i in interval H, and one given the upper boundary.

Description Length of Hypotheses II: Unbounding Irrelevant Dimensions. When determining whether something exceptional is going on in a particular subset of the dataset, typically, only a sharply reduced subset of $n' \ll n$ attributes is relevant. We proceed to generalize the derived complexities and lengths, allowing REALKRIMP to assess whether a dimension should be bounded or unbounded.

To gauge the informativeness of bounding a dimension, we turn to Equation (12). With our choice of discretization constant, the length for the model specification was derived as $2n \log N - n \log 2$. This was based on n dimensions, so the length per dimension, which we denote as Δ, is given by

$$\Delta = (2n \log N - n \log 2)/n = \log N^2/2$$

The quantity Δ represents the information contained in the specification of a single dimension of the hyperinterval. For a specification of $n' \leq n$ dimensions, the complexity of the model, as originally given in (12), becomes:

$$n'\Delta = n' \log \frac{N^2}{2}. \tag{13}$$

When encoding data based on such an n'-dimensional hyperinterval H with $n' < n$, the form of (10) remains intact, but we need to specify what N_{in} and M_{in} mean when some of the dimensions remain unbounded. We consider any hyperinterval to span the full range of any unbounded dimension. Hence, N_{in} is the number of records that are covered on the specified dimensions; coverage on the unbounded dimensions is implied. Also, M_{in} is calculated from a hyperinterval that, in the unbounded dimensions, spans the full range available in the dataset.

When encoding a hyperinterval H, we must encode which dimensions will be specified/unbounded. We do that by taking a uniform prior over the 2^n available models in the class. Hence, we obtain a constant complexity for each choice of specified/unbounded dimensions, making model comparison solely dependent on the lengths defined in those models. Therefore, a dimension is considered relevant, when specification of hyperinterval boundaries in that dimension delivers a reduction in description length bigger than Δ.

The hyperintervals that compress the dataset are those for which (8), the complexity of the database, is larger than the sum of (10), the complexity of the data given the hyperinterval, and (13), the complexity of specifying the hyperinterval. When this inequality holds, enough information is present in the hyperinterval to justify the cost of its specification, and we have found an underlying concept in the dataset. Subtracting $N \log N$ from both sides and rearranging terms, we find that we are interested in hyperintervals for which

$$N \log \frac{M}{N} - n' \log \frac{N^2}{2} \quad > \quad N_{in} \log \frac{M_{in}}{N_{in}} + N_{out} \log \frac{M_{out}}{N_{out}} \tag{14}$$

Here, everything that does not depend on the choice of hyperinterval is gathered on the left-hand side, leaving everything that does on the right-hand side.

Algorithm 1. The RealKrimp Algorithm

Input: A real-valued dataset ω
Output: Hyperintervals that compress ω well
1: Sample the dataset.
2: Compute all (Euclidean) distances between records in the sample.
3: Pick two neighboring (in distance) rows in the sample.
4: Extend a hyperinterval H covering these rows based on other rows in the sample, in a compression-increasing direction (measuring compression on the entire dataset).
5: Calculate the coverage of each dimension in H.
6: In order of decreasing coverage, determine if compression improves when letting a dimension go unbounded.
7: Report the resulting hyperinterval if it is interesting according to Equation (14).
8: Until no more interesting hyperintervals can be found, restart from step 3 with two rows not covered by any of the previously reported hyperintervals.

5   The RealKrimp Algorithm

One of the most prominent problems in theory mining in general, is the *pattern explosion* problem: if we set the interestingness constraints tight, then we find only a few well-known patterns, but if we set the constraints looser, we are quickly overwhelmed with an amount of interesting patterns that is unsurveyable for any data miner. With the real-valued MDL criterion, we can also find many interesting hyperintervals very easily. When untreated, the pattern explosion problem hinders a practical application of a pattern mining method. Naturally, the KRIMP paper [1] discusses the problem. KRIMP strives to find a set of itemsets compressing the dataset, and obviously the candidate space is enormous. The chosen approach to the explosion, is to forego finding the *best* set of patterns — heuristically finding a pattern set that compresses *well* is good enough.

REALKRIMP aims for a similar goal in an uncountable space, which amplifies the pattern explosion problem. For every hyperinterval that we could find, every boundary can have infinitely many values leading to the hyperinterval covering exactly the same records, with an arbitrarily small change in the volume of the hyperinterval. To deal with this serious problem in the applicability of real-valued MDL, in this section we introduce the REALKRIMP algorithm: a mining scheme that confines its attention to those interesting hyperintervals that locally maximize the inequality of (14); no better compression is obtained by a hyperinterval that is either an extension or a restriction of the considered hyperinterval. The REALKRIMP algorithm is given in Algorithm 1. Our implementation of REALKRIMP is written in Python 3, and available for the general public at http://github.com/joukewitteveen/hint.

The first seven lines of the algorithm detail how a single hyperinterval can be found. Since we are interested in finding a set of well-compressing hyperintervals, the algorithm subsequently loops back to step 3 in an attempt to heuristically find additional compressing hyperintervals. The endurance with which the algorithm proceeds to attempt this is governed by a user-set parameter. Additionally, to gloss over small complexity bumps in the hyperinterval space, a perseverance

-s 1000/-s 500

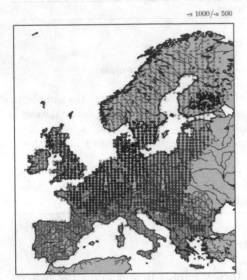 

(a) The whole dataset (blue), a hyperin- (b) Köppen classification of Europe [16]
terval (red), and a sub-hyperinterval (dark
red)

Fig. 1. Spatial distribution over Europe of hyperintervals found by REALKRIMP, juxtaposed with the Köppen classification of Europe

parameter can be set that allows the algorithm to escape local optima in step 4. Lastly, by varying the sample size employed in step 1, substantial influence can be exercised over the total runtime of the algorithm. To discuss all these details in full and properly incorporate them into the pseudocode would substantially bloat the discussion in this section, at the expense of either the theory in previous sections or the experimental results in the next section. Instead, we refer the reader who is interested in details on all individual steps to [13].

6   Experiment

We experiment on the *Mammals* dataset [14], which combines information from three domains: ① the location of grid cells covering Europe (latitude, longitude); ② the climate within these grid cells (monthly temperatures, precipitation, annual trends as captured by the BIOCLIM scheme [15]); ③ the presence or absence of species of mammals in the grid cells. The data from these three domains were pre-processed into one coherent flat-table dataset by Heikinheimo et al. [14]. This version of the dataset is the one we also use in our experiments. We feed the 19 BIOCLIM features from the second domain and all features from the third domain to the REALKRIMP algorithm, to see its performance on a mixture

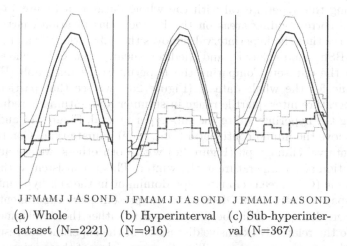

JFMAMJJASOND JFMAMJJASOND JFMAMJJASOND
(a) Whole   (b) Hyperinterval (c) Sub-hyperinter-
dataset (N=2221) (N=916)   val (N=367)

Fig. 2. Climate data for the hyperintervals of Figure 1a. The solid lines represent the mean monthly temperature quartiles; the dashed lines represent the precipitation quartiles. Temperatures range from -5 to $20°C$; precipitation ranges from 0 to 175mm.

of numeric and binary features. We withhold the location information and the other 48 climate features from the algorithm for evaluation purposes.

For the sake of the distance computation in line 2 of the REALKRIMP algorithm, we need to define a way to handle the binary features from the third domain. Given two sets of present species, we assign a distance determined by the species in the symmetric difference between those sets. For each species in this symmetric difference, we add an amount to the distance equal to the binary entropy of that species in the dataset; if a species in the symmetric difference occurs k times in the dataset, it adds $-\frac{k}{N} \log \frac{k}{N} - \frac{N-k}{N} \log \frac{N-k}{N}$ to the distance. We consider this distance computation a parameter of the algorithm; treating binary attributes by computing its entropy is a convenient domain-agnostic way, but given particular domain information one might prefer other solutions.

A REALKRIMP run resulted in many interesting hyperintervals, including the one depicted by the red dots in Figure 1a. Inspecting the boundaries of all 120 features that define the hyperinterval is infeasible; instead we make some observations that stand out. One bioclimatic variable is left unbounded: the mean diurnal range. The species that are necessarily present in the hyperinterval are the *Vulpes vulpes* (Red Fox), the *Capreolus capreolus* (European roe deer), and the *Lepus europaeus* (European hare). Applying the REALKRIMP algorithm recursively, we find the sub-hyperinterval depicted by the dark red dots in Figure 1a. We inspect the relation between the whole dataset, the hyperinterval, and the sub-hyperinterval, by aggregating information from the 48 climate variables withheld from REALKRIMP to draw up the *climographs* of Figure 2. These climographs can be used to illustrate the differences between groups in the Köppen climate classification [16].

Comparing the hyperinterval with the whole dataset in Figure 1a and considering the corresponding areas on the Köppen classification chart of Figure 1b, we observe that the hyperinterval removes the subarctic (Dfc and Dfd, teal), semi-arid (BSh, sand-colored), and Mediterranean (Csa, Csb, yellow) climate types from the dataset. Comparing the hyperinterval climograph (Figure 2b) with the one for the whole dataset (Figure 2a), we see that particularly the mean temperature inter-quartile range in summer and autumn is reduced.

Reducing the hyperinterval to the sub-hyperinterval, Figures 1a and 1b show that we remove the humid continental (Dfb, blue) climate type. Comparing the sub-hyperinterval climograph (Figure 2c) with both others, we see an increase in precipitation and temperature in the winter. This is consistent with the temperate oceanic (Cfb, green) climate type dominant in the sub-hyperinterval.

Due to space constraints we decided to not present more experimental results in this paper; doing so would be detrimental to either the development of the theory or to the relatively extensive discussion of the experimental results we do discuss in the paper, and we are willing to pay neither of these prices. However, more experiments were performed, and can be accessed elsewhere. More experimental results on artificial data can be found in the technical report [13]. Part of these experiments were performed on benevolent artificial data, whose underlying structure comes in an ideal form to be represented by hyperintervals, and part were performed on antagonistic artificial data, whose underlying structure is particularly problematic for REALKRIMP. These artificial experiments illustrate what can be expected from REALKRIMP when presented with a variety of patterns to discover. More experimental results on the Mammals dataset can be inspected online, at http://www.math.leidenuniv.nl/~jwitteve/worldclim/. The main page displays the map corresponding to found hyperintervals. Clicking on such a hyperinterval will display the results of a REALKRIMP run mining for sub-hyperintervals.

7 Conclusions

We introduce REALKRIMP: an algorithm that finds well-compressing hyperintervals in a real-valued dataset, based on the Minimum Description Length Principle. The hyperintervals are bounded hyperrectangles, with edges parallel to coordinate axes, and the interesting ones are those with a relatively high density of records in the dataset. In order to allow REALKRIMP to search for compressing hyperintervals, the formal relation between codes and probabilities on Euclidean spaces is expressed by the *lengthiness*, a codelength-like function. We then discuss the MDL Principle for hypothesis selection and its application within REALKRIMP, and describe a two-part MDL code for hyperintervals. The REALKRIMP algorithm employs this code to heuristically mine for well-compressing hyperintervals. Hence, REALKRIMP can be seen as a real-valued cousin of the well-known KRIMP algorithm.

On the Mammals data, REALKRIMP finds hyperintervals defined on BIOCLIM and zoogeographical attributes. Evaluation of the hyperintervals on withheld attributes shows that the found regions are spatially coherent, that they correspond to climate types on the Köppen classification chart, and that they display meteorological behavior that is to be expected with these climate types. These observations provide evidence that REALKRIMP finds hyperintervals representing real-life phenomena on real-life data from a real-valued domain.

Acknowledgments. This research is supported in part by the Deutsche Forschungsgemeinschaft (DFG) within the Collaborative Research Center SFB 876 "Providing Information by Resource-Constrained Analysis", project C1. Tony Mitchell-Jones and the Societas Europaea Mammalogica kindly provided the European mammals data [17].

References

1. Vreeken, J., van Leeuwen, M., Siebes, A.: KRIMP: Mining Itemsets that Compress. Data Mining and Knowledge Discovery 23, 169–214 (2011)
2. Rissanen, J.: Modeling by Shortest Data Descriptions. Automatica 14(1), 465–471 (1978)
3. Grünwald, P.D.: The Minimum Description Length Principle. MIT Press, Cambridge (2007)
4. Li, M., Vitányi, P.: An Introduction to Kolmogorov Complexity and its Applications. Springer, New York (1993)
5. Faloutsos, C., Megalooikonomou, V.: On Data Mining, Compression and Kolmogorov Complexity. Data Mining and Knowledge Discovery 15(1), 3–20 (2007)
6. Pfahringer, B.: Compression-Based Feature Subset Selection. In: Proc. IJCAI Workshop on Data Engineering for Inductive Learning, pp. 109–119 (1995)
7. Chakrabarti, S., Sarawagi, S., Dom, B.: Mining Surprising Patterns Using Temporal Description Length. In: Proc. VLDB, pp. 606–617 (1998)
8. Keogh, E., Lonardi, S., Ratanamahatana, C.A.: Towards Parameter-Free Data Mining. In: Proc. KDD, pp. 206–215 (2004)
9. Geerts, F., Goethals, B., Mielikäinen, T.: Tiling Databases. In: Proc. DS, pp. 278–289 (2004)
10. Tatti, N., Vreeken, J.: Finding Good Itemsets by Packing Data. In: Proc. ICDM, pp. 588–597 (2008)
11. Heikinheimo, H., Vreeken, J., Siebes, A., Mannila, H.: Low-Entropy Set Selection. In: Proc. SDM, pp. 569–579 (2009)
12. Cover, T.M., Thomas, J.A.: Elements of Information Theory. Wiley (2006)
13. Witteveen, J.: Mining Hyperintervals – Getting to Grips With Real-Valued Data, Bachelor's thesis, Leiden University (2012)
14. Heikinheimo, H., Fortelius, M., Eronen, J., Manilla, H.: Biogeography of European land mammals shows environmentally distinct and spatially coherent clusters. Journal of Biogeography 34(6), 1053–1064 (2007)
15. Nix, H.A.: BIOCLIM — a Bioclimatic Analysis and Prediction System, research report, CSIRO Division of Water and Land Resources, pp. 59–60 (1986)
16. Peel, M.C., Finlayson, B.L., McMahon, T.A.: Updated World Map of the Köppen-Geiger Climate Classification. Hydrology and Earth System Sciences 11, 1633–1644 (2007)
17. Mitchell-Jones, T., et al.: The Atlas of European Mammals, Poyser natural history (1999)

Real-Time Adaptive Residual Calculation for Detecting Trend Deviations in Systems with Natural Variability

Steven P.D. Woudenberg[1], Linda C. van der Gaag[1],
Ad Feelders[1], and Armin R.W. Elbers[2]

[1] Department of Information and Computing Sciences, Utrecht University, The Netherlands
[2] Department of Epidemiology, Central Veterinary Institute, Wageningen UR, The Netherlands

Abstract. Real-time detection of potential problems from animal production data is challenging, since these data do not just include chance fluctuations but reflect natural variability as well. This variability makes future observations from a specific instance of the production process hard to predict, even though a general trend may be known. Given the importance of well-established residuals for reliable detection of trend deviations, we present a new method for real-time residual calculation which aims at reducing the effects of natural variability and hence results in residuals reflecting chance fluctuations mostly. The basic idea is to exploit prior knowledge about the general expected data trend and to adapt this trend to the instance of the production process at hand as real data becomes available. We study the behavioural performance of our method by means of artificially generated and real-world data, and compare it against Bayesian linear regression.

1   Introduction

In many fields of production, data are collected in a real-time fashion. Such data often result from monitoring a specific production process for the purpose of early identification of potential problems. In our applications field of animal production, examples include monitoring the daily egg production at a poultry farm, monitoring weight gain of fattening pigs in a pig production unit, and monitoring the daily milk yield on a dairy farm. In these examples, a sudden drop in production could be due to a temporary problem from an external factor, yet may also point to more serious underlying health issues. The main motivation for our research is the early detection of disease in animal production, for which reliable real-time detection of abnormal values is prerequisite.

Various methods are in use for identifying unexpected values from real-time collected data in general. These methods often build on the calculation of residuals, that is, the differences between expected and actually observed data values. Straightforward methods for residual calculation suffice if the process being monitored should maintain a more or less constant level, such as when monitoring the temperature of a cold storage plant; to account for chance fluctuations from the required level, only some degree of variance should be allowed for the observed data values. The identification of unexpected values in real-time data becomes more challenging if these data are expected to exhibit some non-constant yet fixed trend. By explicitly modelling the expected data trend, residuals are readily calculated; if the trend is captured sufficiently accurately in fact, the residuals again tend to describe random fluctuation. To the calculated residuals,

H. Blockeel et al. (Eds.): IDA 2014, LNCS 8819, pp. 380–392, 2014.

well-known methods for deviation detection, such as the Shewhart control chart [7], the CUSUM method [8] and ARIMA models [2], can be applied.

Real-time detection of unexpected values from animal production data is significantly more challenging than finding deviations in a data sequence which is expected to exhibit a fixed trend, because production levels reflect not just chance fluctuations but the influence of natural variability among animals and between animal groups as well. This variability makes future production levels hard to predict, even though a general overall trend may be known. While the daily egg production of a flock of laying hens is known to follow a specific overall trend for example, the exact day at which the production of a particular flock will peak can differ by more than ten weeks from other flocks. Yet, the calculation of residuals reflecting chance fluctuations mostly again is crucial for reliably identifying trend deviations in a sequence of observed production levels.

Based upon the above considerations, Mertens and his colleagues were the first to develop a tailored method for detecting trend deviations from daily egg production data [5] and from egg weight data [6] in poultry. Their method assumes a general overall trend which is modelled as a mathematical function. The parameters of this function are estimated from the real data points observed during a start-up period of the process being monitored; unexpected values during this start-up period are not detected as yet. The results are quite promising, but their method is not easily generalised to other problems, as the start-up period required strongly depends on the expected data trend. While for monitoring daily egg production the necessary start-up period proved to be some three weeks, the period required for other applications may be impracticably long.

In this paper, we further elaborate on the idea of supplementing a deviation detection method with prior knowledge about the overall expected data trend for systems with natural variability. We focus more specifically on the calculation of residuals resembling the chance fluctuations involved only; these residuals are then used as input for readily available deviation detection methods. The basic idea of our method is to exploit a general expected trend, which is again modelled by a mathematical function, and use it for predicting future data points before calculating residuals. As real data become available, the general trend is adapted gradually to the process instance being monitored. Compared to the method by Mertens et al., our method does not require a start-up period and is therefore applicable to a wider range of more involved data trends.

We would like to note that Bayesian regression [1] provides an alternative approach to adapting a general trend to observed data. While for problems which are linear in their parameters a Bayesian approach would be quite efficient, for nonlinear regression problems such an approach would require posterior simulation algorithms which are computationally quite demanding [3]. For many problems in our applications field of animal production in fact, employing Bayesian regression would prove impractical.

The paper is organised as follows. In Section 2 we outline our method and focus on the choice of parameters involved. We report results from experiments on artificially generated data in Section 3, and briefly compare these against Bayesian regression. In Section 4 our approach is applied to real-world production data from poultry farms. The paper ends in Section 5 with our conclusions and directions for further research.

2   Real-Time Adaptive Prediction for Residual Calculation

Our method for adaptive prediction of future production levels has been designed to improve residual calculation for real-time monitored production data reflecting natural variability. The idea is to exploit a general expected data trend and use it for predicting data points against which residuals are to be established. This expected trend is supplied to our method as prior knowledge about the production process in general. As real data points become available over time, the trend used for future predictions is gradually adapted to the process instance being monitored, thereby modelling posterior knowledge. With our method we aim at adequately capturing the instance's real data trend, which will result in residuals resembling random fluctuations without reflecting natural variability; we note that such residuals are essential for arriving at high quality deviation detection [9]. In this section, we outline our adaptive prediction method and discuss some of its details; Fig. 1 provides a schematic overview of the method.

2.1   Overview of the Method for Residual Calculation

In our applications field, production data vary considerably, both through chance fluctuation and through natural variability. Yet, the trends exhibited by these data are relatively robust, that is, while the function class of the data trend is the same for all instances of the production process, the values of the parameters involved differ among instances. Our method for residual calculation builds upon this observation and exploits the general trend to provide control over the prediction of future data points. More specifically, the method starts with a mathematical function $f(t)$ describing the general trend, adapts this function as new data points y_t, $t = 1, \ldots, T$, become available, and uses the adapted function for the prediction of production levels at time $t + 1$.

The function f to be used is chosen from a pre-specified class of functions and is coined the *initial prediction function* (step 1).

In the field of animal production, research has resulted in function classes describing the overall data trends from various production processes; such classes have been detailed for example, for the egg production of laying hens [4], for the daily milk yield of dairy cows [11] and for the weight of fattening pigs [10]. When a function class is not readily known from the literature, historical data and expert knowledge can be instrumental in choosing an appropriate class for the expected data trend for a production process in general. Given the selected function class, an initial prediction function is constructed by choosing parameter values so as to obtain the best possible estimate of the general overall trend. This function thus captures the expected data trend for a new process instance about which no further knowledge is available a priori. When historical data is available from similar previous instances, the average parameter values over these instances may be used; also, when further knowledge is available about the new instance, this knowledge can be included upon constructing the initial prediction function.

Our method adopts least-squares regression as an approach to analysing trends in the production data. To supply the regression with knowledge about the expected trend, mock data points are drawn from the initial prediction function (step 2).

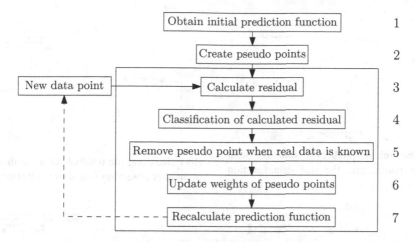

Fig. 1. A schematic overview of our method for real-time adaptive residual calculation

These so-called *pseudo points* are chosen over the full time span of the process instance being monitored; in Section 2.2 we will return to the number and positioning of these points.

At its start, no further knowledge about a process instance is available than the general expected trend, and the initial prediction function capturing this trend is used for prediction. As monitoring progresses, real data points become available which provide evidence about the process instance at hand, and the prediction function is adapted to incorporate this posterior knowledge. More specifically, for a newly observed data point y_t at time t, its residual is calculated against the predicted data point $f(t)$ from the current prediction function f (step 3); the basic idea is illustrated for the fifth observed data point in Fig. 2(b). Based upon the established residual, the point y_t is classified as either an expected or an unexpected production level (step 4); any deviation detection method can be employed to this end. If a pseudo point had been specified for time t, this mock point is discarded and excluded from further processing (step 5): since the pseudo points were created to represent prior estimates for yet to be observed production levels, they are superseded by real data. As evidence of the process instance is building up, the prediction uncertainty for future data points decreases and the control provided by the remaining pseudo points is weakened accordingly. To decrease the weights of these points in the regression over time, a *devaluation function* is employed (step 6).

The final step of our method is to calculate the updated prediction function by applying weighted least-squares regression, for the given function class, to the available real data points and remaining pseudo points (step 7). Fig. 2(c) and (d) illustrate the adaptive behaviour of the prediction function at different time points in the process being monitored. We note that all real data points are used for the regression, whether classified as expected or not: since in the beginning of the production process the deviation detection method is more prone to yield false positives, excluding incorrectly classified expected data points can be more harmful than including sporadic unexpected ones.

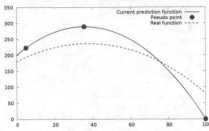

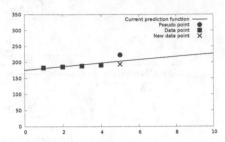

(a) The initial prediction function with pseudo points, and the real data function (steps 1, 2)

(b) Calculating the residual for the fifth data point after observing four data points (step 3)

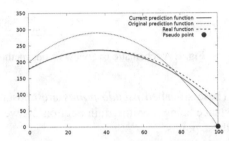

(c) The new prediction function constructed from five observed data points and two remaining pseudo points (step 7)

(d) The new prediction function constructed from fifty observed data points and a single remaining pseudo point (step 7)

Fig. 2. A schematic illustration of real-time adaptive residual calculation

2.2   On pseudo Points

The number and positioning of the pseudo points to be used are strongly dependent of the function class at hand. For the regression procedure to arrive at satisfactory prediction functions for example, pseudo points should be positioned over the entire expected run of the process being monitored. Fig. 3 illustrates, on the left, the possible effect of positioning pseudo points at the first time points of the process only: even though the correct function class is enforced by the regression, with few observed data points the resulting function may deviate significantly from the initial prediction function. To arrive at a satisfactory function moreover, the regression procedure requires pseudo points near each inflection point of the prediction function. The role of the pseudo points is not just to enforce the shape of the initial prediction function however, but also to provide control over the general form of all subsequent prediction functions. The first pseudo point therefore, should not be placed at the very onset of the process being monitored as it will immediately be superseded by an observed point and prematurely lose its control; especially if the new observation is unexpected, could discarding the early pseudo point have a strong effect on future prediction functions.

As more and more real data points are becoming available, the importance of the remaining pseudo points decreases, although they retain some role in enforcing the general shape of the prediction function. For this purpose, the pseudo points are assigned weights, between 0 and 1, for the regression, which are adapted over time by a

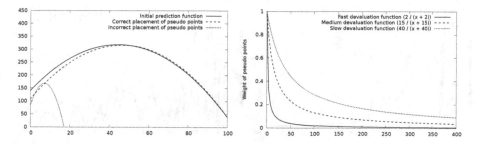

Fig. 3. The initial prediction function, with functions constructed from correctly positioned pseudo points and from pseudo points placed only at early time points, respectively *(left)*. Three devaluations functions for the values of the pseudo points *(right)*.

devaluation function. As they originate from the same source of information, all pseudo points are assigned the same weight; the weights of all observed data points are fixed at 1.0. The devaluation function is designed so as to achieve an initial strong decline of the weights of the pseudo points to capture the idea of real data points taking over the overall data trend, which is followed by a more gradual decline to guarantee that the pseudo points still retain some control over the function form. A first-quadrant hyperbola branch as illustrated in Fig. 3 on the right, captures the general idea and hence constitutes a suitable devaluation function. We note that the more common exponential decay function is less suited to our purposes, as it either models an initial rapid decline followed by the function approximating zero, or an overall gradual decline.

Where the initial prediction function captures prior knowledge about the data trend expected from the process being monitored, do the number, positioning and devaluation of the pseudo points describe the importance of this knowledge for the deviation detection. Prior knowledge about a process with little variability for example, is best modelled by a combination of many pseudo points and a gradually decreasing devaluation function, as this combination will result in weak adaptation of the prediction function to observed data points; for a process involving more variability, either fewer pseudo points, a higher devaluation rate or both would result in increased adaptivity.

3 Experiments with Artificially Generated Data

The adaptive behaviour of our method was investigated experimentally using artificial data. Since our method aims at calculating appropriate residuals, the experiments focused on the differences between predicted data points and observed points. For the experiments, we chose function classes which are linear in their parameters, to allow ready comparison of our method with Bayesian linear regression as an alternative approach to real-time adaptation of an expected trend. We show that the results from our method are comparable to those obtained with Bayesian linear regression. In Section 4 we then show that our method can also be applied with more involved function classes, for which non-linear Bayesian regression would become inhibitively demanding [3].

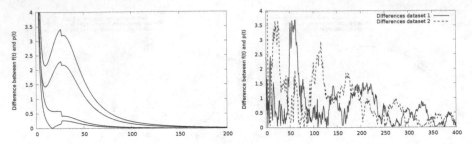

Fig. 4. Differences between the prediction function and the real data function without chance fluctuations *(left)* and with fluctuations generated with $\sigma_\varepsilon = 7$ *(right)*

3.1 Set-up of the Experiments

From each function class under study, 30 functions were generated, to be taken as the real data functions in 30 runs of our method. Each function f_i was generated by drawing, for each parameter θ, a value from a Gaussian distribution $\mathcal{N}(\mu_\theta, \sigma_\theta)$ mimicking natural variability. Although we studied our method's behaviour for various function classes, space limitations allow detailed discussion of the experiments with a single class only. The reported experiments are loosely based on the linear trend of the feed intake of laying hens. The parameter values for natural variability were drawn from independent Gaussian distributions with $\mu_{slope} = 1.24$, $\sigma_{slope} = 0.1$, $\mu_{intercept} = 21$ and $\sigma_{intercept} = 20$. Chance fluctuations were modelled by random noise: from each function f_i, a sequence of data points $g_i(t) = f_i(t) + \varepsilon, t = 1, \ldots, 400$, were generated, with ε drawn from the Gaussian noise distribution $\mathcal{N}(0, \sigma_\varepsilon)$. For each run i of our method, the initial prediction function p_i was established as the function with averaged parameter values over all other functions: for each parameter θ, the value $\bar{\theta}_i$ for p_i was calculated as $\bar{\theta}_i = \sum_{j \neq i} \theta_j / 29$, where θ_j denotes the value of θ in f_j.

3.2 Experimental Results

Our method starts with the selection of pseudo points from the initial prediction function to convey its prior knowledge to the regression procedure. As the first pseudo point should not be placed too close to the function's onset, we placed it at time point 25. As the prediction function is linear, a single additional pseudo point is required, which we placed at time point 200. For the first 200 observations therefore, this point provides prior knowledge; without any inflection points expected, 200 values were considered to be more than sufficient for replacing all prior knowledge. With the two pseudo points forcing the linear form of the expected trend, the devaluation function can model a sharp decrease of the weights of the pseudo points and was chosen to be $d(t) = \frac{4.0}{t+4.0}$.

Experiments without Chance Fluctuations. For four of the 30 runs of our method without added chance fluctuations, Fig. 4 on the left plots the differences between the predicted data points and the true data points $f_i(t)$ under study. The runs with the largest and the smallest differences are shown; the results are shown for the first 200 time points only as from then onwards the differences further converge to zero. All runs revealed a sharp initial decrease of the calculated differences, caused by rapid adaptation of the prediction function to the data. The slope of the decrease is dependent of the position and weight of the first pseudo point. We recall that the positioning and weight of the pseudo points serve to describe the importance of the prior knowledge conveyed by the initial prediction function. For a process likely to yield aberrant values in its start-up phase for example, this prior knowledge is called upon to exert rather strong control over the construction of the prediction functions. The first pseudo point had then best be placed at a somewhat later time and its weight had best decrease more gradually. Alternatively, additional pseudo points can be placed in the start-up phase of the process.

The runs further revealed a marked local maximum in the differences between the predicted data points and the true data. This maximum has its origin in the control exerted by the remaining pseudo point after some true data points have become available from the first phase of the (artificially simulated) process. The actual difference attained depends on the position of this pseudo point and the slope of the initial prediction function. By definition, least-squares regression will be inclined to construct a prediction function which does not deviate too much from the pseudo point and will allow some deviations from the yet limited number of true data points. The larger the slope of the initial prediction function, the more a later pseudo point will contribute to the sum of squared differences minimised by the regression procedure. As more and more true data points are becoming available and the weight of the remaining pseudo point decreases, the regression will allow larger deviations from this point. The differences between the predicted data points and the true data will then gradually decrease.

We would like to note that, after the first true data point had been processed, the largest difference found with our method was smaller than 3.5. This difference was found with the data function $1.35 \cdot t + 48.7$. Compared to the smallest difference 28.2 between this function and the initial prediction function $1.20 \cdot t + 20.6$, a maximum difference smaller than 3.5 with the adapted prediction function suggests that our method is well able to handle at least some degree of natural variability.

Experiments with Chance Fluctuations. The second set of runs with our method served for studying the effects of chance fluctuations on the calculated residuals. For this purpose, the regression was supplied with data points from the functions g_i instead of from f_i, with $\sigma_\varepsilon = 7$. Since we are interested in the ability of our method to yield residuals resembling chance fluctuations mostly, we compare the data points predicted by the functions p_i with the data points drawn from the functions f_i from which the original data sets were created. If the established differences approach zero, we can then conclude that our method is able to capture the real data trend sufficiently accurately so as to result in appropriate residuals which no longer reflect the influence of natural variability. We note that upon real-world application of our method, the differences between the predicted points $p_i(t)$ and the actually observed data points $g_i(t)$ are considered, as the data points $f_i(t)$ without the chance fluctuations are not known in reality. Fig. 4 on the right shows the differences between p_i and f_i for two of the thirty runs, including the one with the largest differences. When compared to Fig. 4 on the left, a largely similar behaviour of the calculated differences is seen. Even though the chance fluctuations affect the calculated prediction functions, only a limited effect on the differences between the prediction function and the trend of the true data function is found.

In our further experiments, including the classes of parabolic and exponential functions, essentially the same behavioural characteristics as reviewed above were found. Similar characteristics were also found with function classes for which linear regression could not be applied and for which the Gauss-Newton method was used for fitting prediction functions. Since more complex function classes offer more freedom for the prediction functions to differ from a true data trend, these findings suggest robustness of our method. Our experiments consistently showed in fact, that the residuals calculated by our method for real-time adaptive prediction captured little influence of the natural variability involved and represented the random fluctuations quite well. We recall that such residuals are essential for achieving a high quality of deviation detection.

3.3 Bayesian Linear Regression

Bayesian linear regression is a well-known approach to updating linear regression functions, and in essence constitutes a possible alternative to our method. This Bayesian approach takes an initial regression function which is updated after each newly observed data vector by applying an updating rule (see for example Bishop [1]) to its parameters:

$$\mathbf{m}_N = \mathbf{S}_N \left(\mathbf{S}_0^{-1} \mathbf{m}_0 + \beta \mathbf{X}^T \mathbf{y} \right)$$

with $\mathbf{S}_N^{-1} = \mathbf{S}_0^{-1} + \beta \mathbf{X}^T \mathbf{X}$. The vector $\mathbf{m}_0$ contains the prior values for the parameters of the initial regression function, and $\mathbf{m}_N$ is the vector of posterior values after processing a new observation vector $\mathbf{y}$. Using the covariance matrix $\mathbf{S}_0$ of $\mathbf{m}_0$, the new observation vector, the tuning parameter β, and the matrix $\mathbf{X}$ containing the data on the (possibly transformed) input variables, the function parameters are updated. The vector $\mathbf{m}_N$ of updated parameter values and its associated covariance matrix $\mathbf{S}_N$ now serve as priors for the updating rule when a subsequent data vector is observed. This Bayesian scheme of updating regression functions is computationally more attractive than repeatedly performing least squares regression after every new observation. A more elaborate introduction to Bayesian linear regression is found in [1].

The tuning parameter β of the Bayesian updating scheme describes the importance of the prior information in relation to newly observed data. For a regression function to adapt slowly to new observations for example, a small value should be chosen for the tuning parameter. As a general guideline, $\beta = \frac{1}{\sigma^2}$ is suggested, where σ is the expected standard deviation of the observations, for example obtained from datasets or from domain experts. In essence, the tuning parameter β of the Bayesian scheme serves the same purpose of moderating the importance of prior information as the combination of pseudo points and weight devaluation scheme of our approach.

In the experiments described above, we applied Bayesian linear regression to the generated data as well. Values for the parameters of the initial prediction function and for the tuning parameter β were based on the available prior knowledge. In all experiments, the results from Bayesian linear regression were quite similar to those obtained with our method. We would like to note however, that our method allowed more flexibility for tuning adaptive behaviour than Bayesian linear regression: where the behaviour of our approach could be fine-tuned by the number, positioning and weight devaluation of the pseudo points, the Bayesian approach offered just a single adjustable parameter.

4 Monitoring a Flock of Laying Hens in Poultry Farms

We recall that the motivation for our research was the challenge of detecting abnormal values in the production data of a flock of laying hens, for the purpose of early detection of health problems. Typical production parameters for laying hens are the egg production, mortality rate, and feed and water intake. Due to space limitations we focus here on feed intake and egg production; additional application results are reported in [12].

For applying our method of adaptive residual calculation to the poultry domain, we had available complete production data from ten healthy flocks; in addition we had partial data from three further flocks, two of which were infected with Low Pathogenic Avian Influenza and one was without health problems. The available partial data ranged from the time at which the hens were 32 weeks of age, to the moment they were culled some 60 days later. For validation purposes, we elicited, from poultry experts, classifications as expected or unexpected for all data points from the three partial datasets.

Based on the available data and the knowledge elicited from our experts, we found that the trend of the feed intake of a flock was fairly well described by a linear function, although more complex functions had been proposed [4]:

$$f(t) = \frac{a}{1 + b \cdot e^{-a \cdot c \cdot t}} + d \cdot t + e \cdot t^2$$

We decided to use the linear function for ease of study, but are aware that more accurate results can be expected from using the more complex function class. For estimating the two parameter values for the initial prediction function, the data from the ten healthy flocks were used; as only partial datasets were available for the remaining three flocks, all starting at week 32, we defined an alternative intercept for the initial prediction function to account for the limited data range. Prior knowledge about the function was expressed by two pseudo points, one of which was positioned at day 5, near the first true data point, and the other one was placed at day 65, near the last observed data point. With the small number of true data points available and just two pseudo points, the prior knowledge should more gradually devalue in the regression than in the experiments from Section 3; the devaluation function was chosen as $d(t) = \frac{8.0}{t+8.0}$.

Egg production was modelled by a non-linear function with five parameters [4]:

$$f(t) = \frac{100}{1 + a \cdot r^{\sqrt{t}}} - \left(b + c \cdot \sqrt{t} + d \cdot t\right)$$

As the first observed data points of the production cycle of a flock are quite specific for the overall data trend, the three datasets for which only partial data was available could not be used. From the datasets of the ten healthy flocks, seven were used for defining the initial prediction function and three were retained for validation purposes. With the more complex function class, more pseudo points were used to control the shape of the prediction functions. We positioned five pseudo points, at days $15, 50, 150, 250$ and 400. The first and second pseudo point, placed near the only inflection point of the

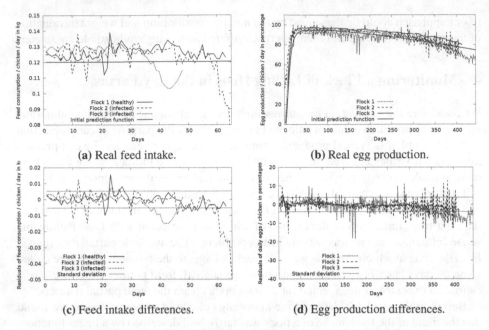

(a) Real feed intake.   (b) Real egg production.

(c) Feed intake differences.   (d) Egg production differences.

Fig. 5. Plots of the real data points of the considered trends (a) and (b), and differences between real data points and the prediction functions (c) and (d)

function, serve to ensure the expected sharp increase in egg production in the start-up phase of the production cycle; the remaining three pseudo points ensure the prediction functions to exhibit to a gradually strengthening decrease in production. The devaluation function for the pseudo points is taken to be $d(t) = \frac{4.0}{t+4.0}$. In the presence of five pseudo points providing control over the shape of the prediction functions, a fairly sharp weight decrease is appropriate. When fewer pseudo points had been used, the devaluation function should have described a more gradual weight decrease to ensure sufficient influence of the prior knowledge on the construction of the prediction functions.

The feed consumption and egg production of the three validation flocks are plotted in Fig. 5 (a) and (b) respectively. Fig. 5 (c) and (d) show the differences, defined as the real data points subtracted with the predicted points, and the average standard deviation expected for both trends. From Fig. 5 (a) and (b) it follows that the initial prediction function for feed intake does not yet model the three considered datasets well, while the initial prediction function for the egg production appears to be a fairly good fit to flocks 1 and 2 and is initially a bit too conservative for flock 3. For both feed consumption and egg production, a fairly quick adaptation of the initial prediction function results in small differences in Fig. 5 (c) and (d). However, the prediction function does not significantly change when confronted with large deviations of the flocks, as is confirmed by observing that similar deviations from the production graphs are visible in the difference graphs. Even when the feed consumption for the infected flocks drop significantly for several consecutive days, the prediction function hardly adapts to these observations, as desired. The peaks in the start up period of the egg production of flock 2 are not able to misguide the prediction function either.

To summarise, our method exhibits similar behaviour on this non-linear regression problem as on the linear one: the prediction function gradually adapts to newly observed data points, yet without following unexpected ones. Since the differences between the observed data points and the predicted ones better reflect the deviations from the actual trend of the flock at hand, the calculated residuals provide for a better detection of trend deviations. With our method, the CUSUM method and the Shewhart control chart, which are often used in veterinary science, are readily implemented for example and the resulting classification of each data point is compared with the classification provided by poultry experts. With a naive implementation, without parameter tuning, already a sensitivity between 0.8 and 1.0 and a specificity between 0.64 and 1.0 was achieved for the feed consumption trend.

5   Conclusions and Future Work

We presented a new method for real-time calculation of residuals, tailored to the monitoring of production processes involving natural variability. Initially controlled by prior knowledge of a general expected trend, our method predicts future data points, against which the true data points are compared. As further knowledge of the process instance becomes available, the expected trend is gradually adapted to the actually observed data trend. Experiments on both artificial and real-word data showed that our method results in residuals modelling chance fluctuations mostly, with a limited influence from natural variability. Our method can be practicably applied regardless of the function class of the expected data trend and of the regression type to be used: the only requirement is that both the function class and initial values for the function parameters are known. Our future work will aim at combining our method with known approaches to detecting trend deviations and at studying our method's performance for a range of problems in the applications field of animal production. Our investigations will also include the design of further guidelines for the use of our method and the definition of quality standards for deviation detection in view of natural variability.

Acknowledgement. We thank all reviewers for their useful comments and suggestions.

References

1. Bishop, C.M.: Pattern Recognition and Machine Learning. Springer, New York (2006)
2. Box, G.E.P., Jenkins, G.: Time Series Analysis, Forecasting and Control. Holden-Day (1990)
3. Koop, G.: Bayesian Econometrics. J. Wiley (2003)
4. Lokhorst, C.: Mathematical curves for the description of input and output variables of the daily production process in aviary housing systems for laying hens. Poultry Science, 838–848 (1996)
5. Mertens, K., et al.: An intelligent control chart for monitoring of autocorrelated egg production process data based on a synergistic control strategy. Computers and Electronics in Agriculture, 100–111 (2009)
6. Mertens, K., et al.: Data-based design of an intelligent control chart for the daily monitoring of the average egg weight. Computers and Electronics in Agriculture, 222–232 (2008)

7. Montgomery, D.C.: Statistical Quality Control. Wiley, Hoboken (2009)
8. Page, E.S.: Continuous inspection schemes. Biometrika (1954)
9. Neter, J., Kutner, M.H., Nachtsheim, C.J., Wasserman, W.: Applied Linear Statistical Models. Irwin, Chicago (1996)
10. Strathe, A.B., Danfæ, A.C.: A multilevel nonlinear mixed-effects approach to model growth in pigs. Journal of Animal Science, 638–649 (2010)
11. Val-Arreola, D., Kebreab, E., Dijkstra, J., France, J.: Study of the lactation curve in dairy cattle on farms in central Mexico. Journal of Dairy Science, 3789–3799 (2004)
12. Woudenberg, S.P.D., van der Gaag, L.C., Feelders, A., Elbers, A.R.W.: Real-time adaptive problem detection in poultry. In: Proceedings of the 21st European Conference on Artificial Intelligence (ECAI). IOS Press (to appear, 2014)

Author Index

Anderson, Blake 1
Arratia, Argimiro 13
Arzoky, Mahir 25
Azzopardi, George 345

Berthold, Michael R. 276
Boley, Mario 203
Borgelt, Christian 37
Boulicaut, Jean-François 84
Bringay, Sandra 239
Budka, Marcin 49

Cabaña, Alejandra 13
Cabaña, Enrique M. 13
Cain, James 25
Counsell, Steve 25, 61

da Costa Pereira, Célia 131
de Bono, Bernard 72
de Vries, Harm 345
Desmier, Élise 84
Dias, Pedro 96
Duivesteijn, Wouter 368
Duplisea, Daniel 298

Eastwood, Mark 49
Elbers, Armin R.W. 380

Feelders, Ad 380
Fitzgerald, Guy 61
Fournier-Viger, Philippe 108
Freire de Sousa, Jorge 227

Gabrys, Bogdan 49
Gama, João 227
Gomes, Luís 167
Grenon, Pierre 72
Grünwald, Peter 368
Gueniche, Ted 108

Hash, Curtis 1
Heard, Nicholas 321
Helmer, Sven 143
Helvensteijn, Michiel 72
Hielscher, Tommy 120

Hollmén, Jaakko 251
Höppner, Frank 286

Inthasone, Somsack 131

Jaber, Mohammad 143
Jiménez-Pérez, Pedro F. 155

Kadlec, Petr 49
Karimbi Mahesh, Kavitha 167
Kenny, Andrew 298
Kirkup, Don 309
Knobbe, Arno 345, 368
Koelewijn, André 345
Kok, Joost 72
Kokash, Natallia 72
Kotsifakos, Alexios 179
Kühn, Jens-Peter 120

Laclau, Charlotte 192
Lane, Terran 1
Lopes, José Gabriel P. 167

Magalhaes, Joao 96
Martin Salvador, Manuel 49
Mendes-Moreira, João 227
Moens, Sandy 203
Moniz, Nuno 215
Mora-López, Llanos 155
Moreira-Matias, Luis 227

Nadif, Mohamed 192
Neil, Joshua 321

Papapetrou, Panagiotis 143, 179
Pasquier, Nicolas 131
Peters, Jason 61
Picado-Muiño, David 37
Plantevit, Marc 84
Poncelet, Pascal 230

Quafafou, Mohamed 263

Rabatel, Julien 239
Read, Jesse 251

Robardet, Céline   84
Rochd, El Mehdi   263
Rodrigues, Fátima   215

Sampson, Oliver   276
Schwan, Stephanie   49
Schweier, Anke   286
Spiliopoulou, Myra   120
Swift, Stephen   25, 61

Tettamanzi, Andrea G.B.   131
Torgo, Luís   215
Trifonova, Neda   298
Tsakonas, Athanasios   49
Tseng, Vincent S.   108

Tucker, Allan   61, 298, 309
Turcotte, Melissa   321

Ukkonen, Antti   333

van der Gaag, Linda C.   380
Völzke, Henry   120

Weston, David J.   356
Witteveen, Jouke   368
Wood, Peter T.   143
Woudenberg, Steven P.D.   380

Zida, Souleymane   108
Žliobaitė, Indrė   49